SPRINGER PROTOCOLS HANDBOOKS

For further volumes:
http://www.springer.com/series/8623

Springer Protocols Handbooks collects a diverse range of step-by-step laboratory methods and protocols from across the life and biomedical sciences. Each protocol is provided in the Springer Protocol format: readily-reproducible in a step-by-step fashion. Each protocol opens with an introductory overview, a list of the materials and reagents needed to complete the experiment, and is followed by a detailed procedure supported by a helpful notes section offering tips and tricks of the trade as well as troubleshooting advice. With a focus on large comprehensive protocol collections and an international authorship, Springer Protocols Handbooks are a valuable addition to the laboratory.

Fundamental Approaches to Screen Abnormalities in Drosophila

Edited by

Monalisa Mishra

Neural Developmental Biology Lab, Department of Life Science, National Institute of Technology Rourkela, Rourkela, Odisha, India

Springer

Editor
Monalisa Mishra
Neural Developmental Biology Lab,
Department of Life Science
National Institute of Technology Rourkela
Rourkela, Odisha, India

ISSN 1949-2448 ISSN 1949-2456 (electronic)
Springer Protocols Handbooks
ISBN 978-1-4939-9755-8 ISBN 978-1-4939-9756-5 (eBook)
https://doi.org/10.1007/978-1-4939-9756-5

This Springer imprint is published by the registered company Springer Science+Business Media, LLC part of Springer Nature.
The registered company address is: 233 Spring Street, New York, NY 10013, U.S.A.

Contents

Contributors

JANMEJAYA BAG • *Neural Developmental Biology Lab, Department of Life Science, National Institute of Technology Rourkela, Rourkela, Odisha, India*
SAKSHI BANSAL • *RNA Biology Laboratory, Regional Centre for Biotechnology, Faridabad, India*
SRIRUPA BASU • *Neural Developmental Biology Lab, Department of Life Science, National Institute of Technology Rourkela, Rourkela, Odisha, India*
GEETANJALI CHAWLA • *RNA Biology Laboratory, Regional Centre for Biotechnology, Faridabad, India*
GYANASENI DHAR • *Neural Developmental Biology Lab, Department of Life Science, National Institute of Technology Rourkela, Rourkela, Odisha, India*
MADHAVI DUBEY • *Department of Life Sciences, Central University of Jharkhand, CTI Campus, Ranchi, Jharkhand, India*
HENA FIRDAUS • *Department of Life Sciences, Central University of Jharkhand, CTI Campus, Ranchi, Jharkhand, India*
PUJA KARMAKAR • *Neural Developmental Biology Lab, Department of Life Science, National Institute of Technology Rourkela, Rourkela, Odisha, India*
MONALISA MISHRA • *Neural Developmental Biology Lab, Department of Life Science, National Institute of Technology Rourkela, Rourkela, Odisha, India*
SUMIT MUKHERJEE • *Neural Developmental Biology Lab, Department of Life Science, National Institute of Technology Rourkela, Rourkela, Odisha, India*
KUMARI PRAGATI NANDA • *Department of Life Sciences, Central University of Jharkhand, CTI Campus, Ranchi, Jharkhand, India*
NIBEDITA NAYAK • *Neural Developmental Biology Lab, Department of Life Science, National Institute of Technology Rourkela, Rourkela, Odisha, India*
SANJEEV PAIKRA • *Neural Developmental Biology Lab, Department of Life Science, National Institute of Technology Rourkela, Rourkela, Odisha, India*
MANISH PANDEY • *RNA Biology Laboratory, Regional Centre for Biotechnology, Faridabad, India*
ABHINANDAN PATNAIK • *Neural Developmental Biology Lab, Department of Life Science, National Institute of Technology Rourkela, Rourkela, Odisha, India*
SUBHASHREE PRIYADARSINI • *Neural Developmental Biology Lab, Department of Life Science, National Institute of Technology Rourkela, Rourkela, Odisha, India*
RONNALY ROUT • *Neural Developmental Biology Lab, Department of Life Science, National Institute of Technology Rourkela, Rourkela, Odisha, India*
SWETAPADMA SAHU • *Neural Developmental Biology Lab, Department of Life Science, National Institute of Technology Rourkela, Rourkela, Odisha, India*

Chapter 1

Formulation of *Drosophila* Food for Various Feeding Studies

Subhashree Priyadarsini, Sumit Mukherjee, and Monalisa Mishra

Abstract

Drosophila melanogaster serves as an important model organism in the fields of genetics and developmental biology. Recent studies have indicated the use of *Drosophila* in a number of other studies such as metabolic, nanotoxicological and immunological studies. Rearing of fly in the laboratory condition is an important task to work with this organism. Different culture media are developed to maintain the growth and development of the flies where the components of the food play important roles. *Drosophila* used in various toxicological or metabolic studies often involve oral intake of various nanoparticles, pathogens, antibiotics and molecules to study metabolism, and this is done by mixing the compounds with fly food in proper concentration. This chapter describes the modification of fly food with various treatments such as the addition of nanoparticles, bacteria, antibiotics and AGE compounds. These protocols will help to use fly for various feeding experiments.

Key words Fly food, Nanoparticle-added food, AGE compound, Bacteria-contaminated food; antibiotics-treated food, High-fat food

1 Introduction

The fruit fly belongs to the genus *Drosophila*, which consists of a number of species widely used as a model organism in various fields of biology. Most of the *Drosophila* species have well-defined genetics, developmental stages and ecology. Besides variation among different species, the short generation time makes *Drosophila* as an ideal model for various studies [1]. Of the several species of *Drosophila*, *Drosophila melanogaster* is the most extensively used model [2]. Thus, it is essential to have a recipe so that the flies can be reared in laboratory condition.

Rearing of the fly is fairly simple in a natural environment. Thomas Hunt Morgan, the father of *Drosophila* biology, used banana as a culture medium to rear fly in the laboratory

Subhashree Priyadarsini and Sumit Mukherjee contributed equally with all other contributors.

Monalisa Mishra (ed.), *Fundamental Approaches to Screen Abnormalities in Drosophila*, Springer Protocols Handbooks, https://doi.org/10.1007/978-1-4939-9756-5_1,

[3]. Later, a number of modifications were made to develop a simple and economic media to rear fly [4]. The individual components present in the media have an impact on the growth and development of flies [4–6]. *Drosophila* can be raised easily in fermented food; however, the exact role of fermentation in *Drosophila* breeding is still not known. In laboratories, flies grow on solid media, where yeast is added to ferment the culture medium. Agar is used as a common ingredient to solidify the culture media. In the food media, sufficient amount of sugar is added for fermentation by yeast, as well as for the growth of the fly. Thus, fruits with high sugar content are often added in the food media [7]. Many culture media use sucrose as a sugar source. Another important component of the media is the corn meal, which provides fructose, a major carbohydrate source and also essential for cell proliferation. The other components are propionic acid, methyl paraben or nipagine. Propionic acid is used as an antifungal agent. Methyl paraben or nipagine is used as an antibacterial agent. Although this is the general preparation of food, depending on the aim of the study various modified food are used.

Drosophila is used in food industry, medicine and pharmacy for drug testing purpose. In all these studies, it is important to study the metabolism involved in this process [8–10]. Nanoparticles are widely used in various fields along with medicine [11], food [12, 13] and various dental materials [14]. Thus there is a chance of entering into the digestive system. Various studies used *Drosophila* as a model to study the deleterious effect of nanoparticles [14–18]. The method via which the nanoparticles are introduced into the food is described in this chapter. This protocol will be helpful to check the oral intake of nanoparticle by the fly. Besides nanotoxicity, *Drosophila* is used to study the effects of a number of important environmental toxicants including mercury, lead, arsenic and pesticides [19–21]. Antibiotics are natural compounds produced by bacteria that slow down the growth of other bacterial species in the surrounding. Although antibiotics undergo clinical trials before they are marketed, some of them may have adverse side effects that need to be tested [22]. Effect of antibiotics is also tested using the *Drosophila* model [23]. In this chapter, we are describing a protocol to treat *Drosophila* with gentamicin, kanamycin and neomycin mixed in food to check their potential side effects.

Drosophila is also used to check the pathogenicity and immunity against a number of foreign organisms [24, 25], such as bacteria, parasites virus, and fungal infections. The immune response seen in fly after the infection is evolutionarily conserved [26, 27]. *Drosophila* is infected with various pathogens by pricking or injection, thereby causing systemic infection [28–30]. However, many naturally occurring pathogenic bacteria enter into the gut via feeding on decomposed or rotten materials.

In such cases the immunocompetency of the gut plays a significant role [31–33]. Similarly, when a pathogen enters into the gut, it significantly affects the disease severity [34, 35] by activation of various immune pathways [36, 37]. In this chapter, we are describing the introduction of pathogenic bacteria with food. In this protocol we are describing two commonly occurring bacteria, *Pseudomonas aeruginosa* and *Escherichia. coli*. However this protocol is applicable to most of the bacteria except *Salmonella* species.

Drosophila is used as a model in various metabolic disorder studies such as diabetes, obesity-linked diseases such as cardiovascular disorders and ageing. It has been reported that high sugar and fat diet often induce conditions related to diabetes. Another metabolite that enhances the complicacy of diabetes is advanced glycation end products (AGE) compounds. AGE compounds are formed by a nonenzymatic reaction (called the Maillard reaction) between the free amino groups present within the proteins, lipids and nucleic acids by the reducing sugars such as glucose, sucrose and ribose [38]. This process occurs in the biological systems slowly during ageing and may cause several complications, such as diabetes, cardiovascular diseases, neurodegeneration and cancer [39, 40]. However, they can be also derived from exogenous sources, such as processed food [41]. Food that is processed at low temperature has a low concentration of AGE, whereas foods rich in lipid and protein content processed at high temperature have increased AGE content [42]. In animal models, chronic ingestion of oral AGEs causes various toxic effects including insulin resistance and diabetes in mice [43]. A similar effect of AGE compound on ageing is reported using *Drosophila* [44]. In the fly model, AGE compound can be added to the food media and various effects can be studied. In this chapter, we have given a brief description of the synthesis and feeding of the AGE compound to wild-type *Oregon-R* flies.

2 Materials

2.1 Standard Fly Food

1. Sucrose (Himedia-GRM3063).
2. Corn meal (From the local market).
3. Yeast extract (Subala yeast powder).
4. Type I agar (Himedia-GRM666).
5. Methyl paraben (Himedia-GRM1899): Prepare the solution by adding 0.25 g methyl paraben to 100 ml ethanol.
6. Propionic acid (Himedia-GRM3658).
7. Glass bottle (Schott Duran).
8. Glass vials (Riviera).

9. Cotton.
10. Autoclave bag.
11. Autoclave.
12. Weighing balance (Aczet).

2.2 Addition of Nanoparticles to the Standard Food

2.2.1 Nanoparticle Stock Solution

1. Fifteen millilitre tube (Tarson)
2. Synthesized Nanoparticle of interest
3. Probe sonicator
4. MilliQ water
5. Weighing balance (Aczet)

2.2.2 Fly Food

Prepare the standard food materials as mentioned in Sect. 2.1.

2.3 Addition of Antibiotic to the Standard Fly Food

1. Antibiotic stock solution (Gentamicin, Kanamycin, Neomycin)
2. Standard fly food
3. Control *Oregon-R* flies
4. Glass vials (Riviera)
5. Glass rod

2.4 Addition of Bacteria to the Standard Fly Food

2.4.1 Bacterial Culture

1. Luria-Bertani (LB) broth (Himedia—M1245)
2. Test tubes (Riviera)
3. Micropipette (Tarson)
4. Tips (Tarson)
5. MilliQ water
6. Cotton
7. Shaker incubator (Remi)
8. Stock culture of *Pseudomonas aeruginosa* and *Escherichia coli*

2.4.2 Fly Food

Prepare the standard food materials as mentioned in Sect. 2.1.

2.5 Addition of AGE Compound to the Fly Food

2.5.1 Fly Food

Prepare the standard food materials as mentioned in Sect. 2.1.

2.5.2 AGE Compound Synthesis

1. Bovine serum albumin (BSA) (Himedia—MB083).
2. Anhydrous glucose (Himedia—MB037).
3. 0.2M phosphate buffer (pH 7.4): Measure 21.8 g of disodium hydrogen phosphate (Na_2HPO_4, Mol. Wt. 177.99; Himedia—TC051) and 10.68 g of sodium dihydrogen phosphate (NaH_2PO_4, Mol. Wt. 137.99; Himedia—TC068), and

dissolve in 200 ml of milliQ water. Check the pH and add water to make the volume up to 1000 ml.
4. 15 ml tube (Tarson).
5. Parafilm M (Tarson).
6. Shaker incubator (Remi).
7. Glass bottle (Schott Duran).

2.5.3 AGE Purification

1. Dialysis membrane 110 (Himedia—LA395).
2. Dialysis clip (Himedia—LA404).
3. 0.5 M phosphate buffer (pH 7.4): Measure 54.50 grams of disodium hydrogen phosphate (Na_2HPO_4, Mol. Wt. 177.99; Himedia—TC051) and 26.74 grams of sodium dihydrogen phosphate (NaH_2PO_4, Mol Wt. 137.99; Himedia—TC068), and dissolve in 200 ml of milliQ water. Check the pH and make the volume up to 1000 ml.
4. Magnetic stirrer (Tarson).
5. Magnetic beads (Tarson).
6. Beaker (Borosil).

2.5.4 Dialysis Membrane Activation

1. Sodium carbonate ($NaHCO_3$) (Himedia--MB253)
2. Disodium EDTA (Himedia--MB011)
3. MilliQ water

3 Methods

3.1 Preparation of Standard Fly Food

1. To prepare the standard fly food (100 ml), weigh 4 g sucrose, 2.5 g yeast extract, 5 g corn meal and 0.8 g agar (type I) in a weighing balance.
2. Take 100 ml of milliQ water in a glass bottle and add all the components to it. Stir well with a glass rod.
3. Take ten clean glass vials and put a cotton plug at the mouth of each vial.
4. Keep the vials along with the food bottle in an autoclave bag, and autoclave for 45 min.
5. Take out the vials and the food bottle after 45 min from the autoclave and allow it to cool.
6. Take them inside a laminar air flow and add propionic acid and nipagine to the food, and mix it thoroughly (*see* **Note 4.1.1**).
7. Pour 10 ml of food in each vial from the food bottle and allow it to solidify (*see* **Note 4.1.2**).
8. Once the food is solidified, it can be used for the fly culture.

3.2 Addition of Nanoparticles to the Standard Food

3.2.1 Preparation of Nanoparticle Stock Solution

1. To prepare the stock solution (1 mg/ml), disperse the nanoparticle in milliQ water.
2. Sonicate the solution two times for 15 min each using a probe sonicator to disperse the nanoparticles properly (*see* **Note 4.2.1**).
3. The stock solution can be stored in room temperature and vortexed before the experiment.

3.2.2 Preparation of Fly Food and Addition of Nanoparticles

1. Prepare the standard fly food as per the protocol described in Sect. 3.1.
2. Allow the food to solidify.
3. Vortex the nanoparticle stock solution for 10 min.
4. Add the desired volume of the nanoparticle to 10 ml of fly food to obtain the desired concentrations.
5. Allow the food to dry at room temperature and transfer the flies to it (*see* **Note 4.2.2**).
6. The entire step for this is shown in Fig. 1.

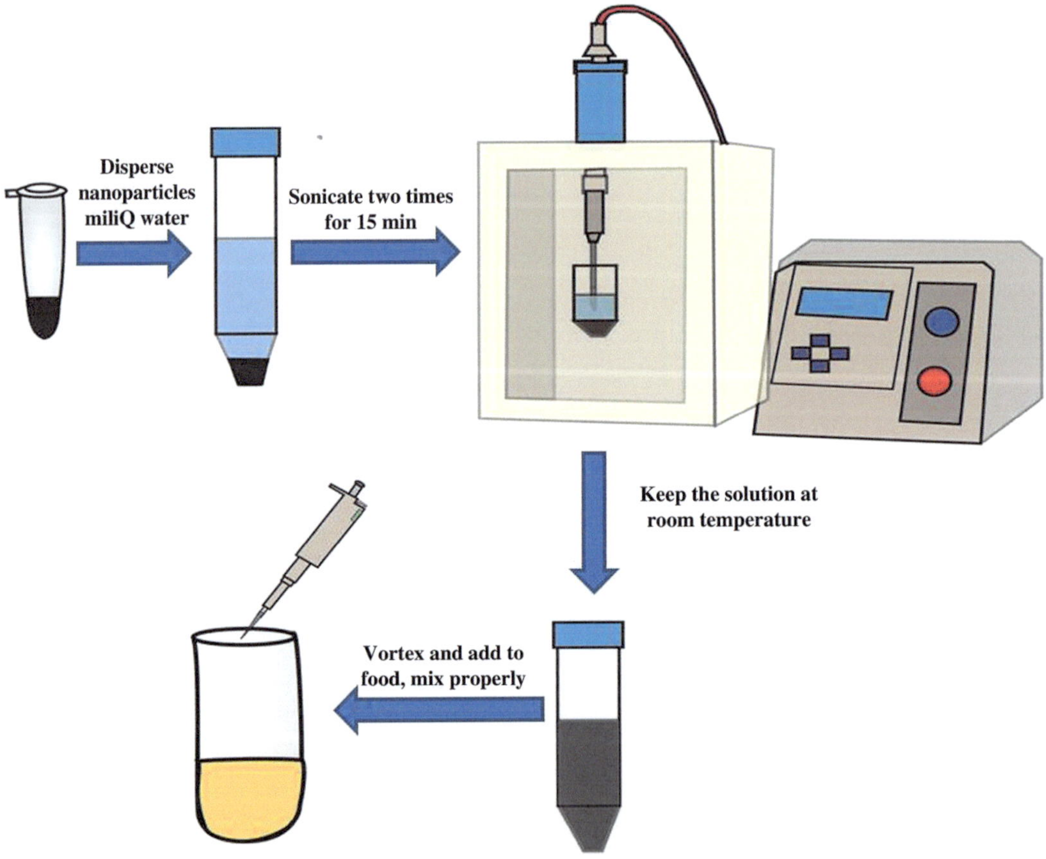

Fig. 1 Steps showing the addition of nanoparticles to standard fly food

3.3 Addition of Antibiotic to the Standard Fly Food

3.3.1 Preparation of the Antibiotic Stock Solution

1. Prepare an antibiotic stock solution of 50 mg/ml by weighing 50 mg of gentamicin/kanamycin/neomycin powder, and mix it with 1 ml of milliQ water.

3.3.2 Addition of Antibiotic to Fly Food

1. Prepare the standard fly food as per Sect. 3.1. Pour 10 ml of food in separate glass vials and allow them to cool.
2. Add the specific volume of antibiotic stock solution in the fly food to prepare different concentration (50, 100, 200, 500 μg/ml) of the treatment vials (Fig. 2) (*see* **Note 4.3.1**).
3. Mix well with a glass rod and allow the food to solidify at room temperature.

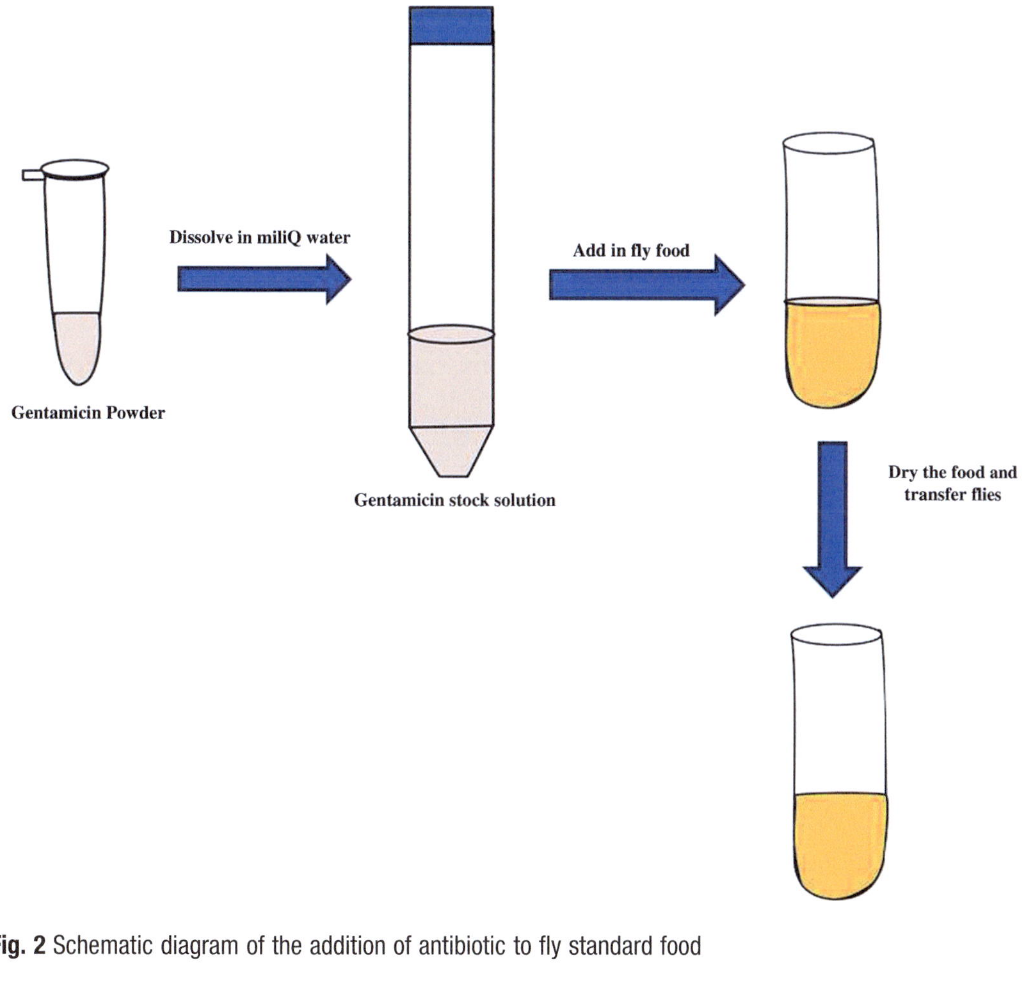

Fig. 2 Schematic diagram of the addition of antibiotic to fly standard food

4. After the solidification of food, transfer *Oregon-R* control flies (five females and three males) to each vial. The number of flies may vary according to culture size.
5. Keep the vials at 25 °C in 12 h light and dark cycle.
6. The other antibiotics can be added in the same way as mentioned here.

3.4 Addition of Bacteria to the Standard Fly Food

3.4.1 Preparation of Bacterial Culture

1. Inoculate 100 μl of *P. aeruginosa* or *E. coli* cultures in 10 ml of LB broth. Keep the broth at 37 °C and shake at 120 rpm for overnight. Grow the bacterial cultures up to an exponential phase, so that the bacteria are in rapidly replicating stage.
2. Subculture the overnight grown bacterial culture (100 μl) in a fresh LB broth (10 ml) (*see* **Note 4.4.1**).
3. Ensure the bacterial growth by determining the OD at 600 nm. Take the OD in every 30 mins till the OD reaches 0.6–0.8.

3.4.2 Infection of Flies

1. Starve the flies for 2–4 h before transferring them to the food vial to ensure oral infection.
2. Prepare the standard fly food as described in Sect. 3.1.
3. Once the food is prepared, take it inside laminar air flow.
4. Allow the food to cool down for some time, and then add propionic acid and nipagine to the food (*see* **Note 4.4.3**).
5. Pour the food to the vials and allow it to solidify.
6. Prepare the infection vials by adding different volumes of bacterial culture directly to the standard fly food, and allow it to be absorbed in the food media.
7. Once the food is dry, transfer the male and female flies to the food vial, and keep it at 25 °C for 12 h in light and dark conditions.
8. The entire protocol is shown in Fig. 3.

3.5 Addition of AGE Compound to the Fly Food

3.5.1 Preparation of AGE Compound

1. Prepare a solution of BSA (25 mg/ml) and glucose (0.5 M) in 50 ml of 0.2 M sodium phosphate buffer (pH 7.4).
2. Keep the solution in a glass bottle at 37 °C with the mouth sealed with Parafilm for 6 weeks without disturbing the solution (Fig. 4) (*see* **Note 4.5.1**).
3. Keep a control BSA solution of the same condition without glucose.
4. After incubation, take out the compound and perform dialysis to remove excess glucose.

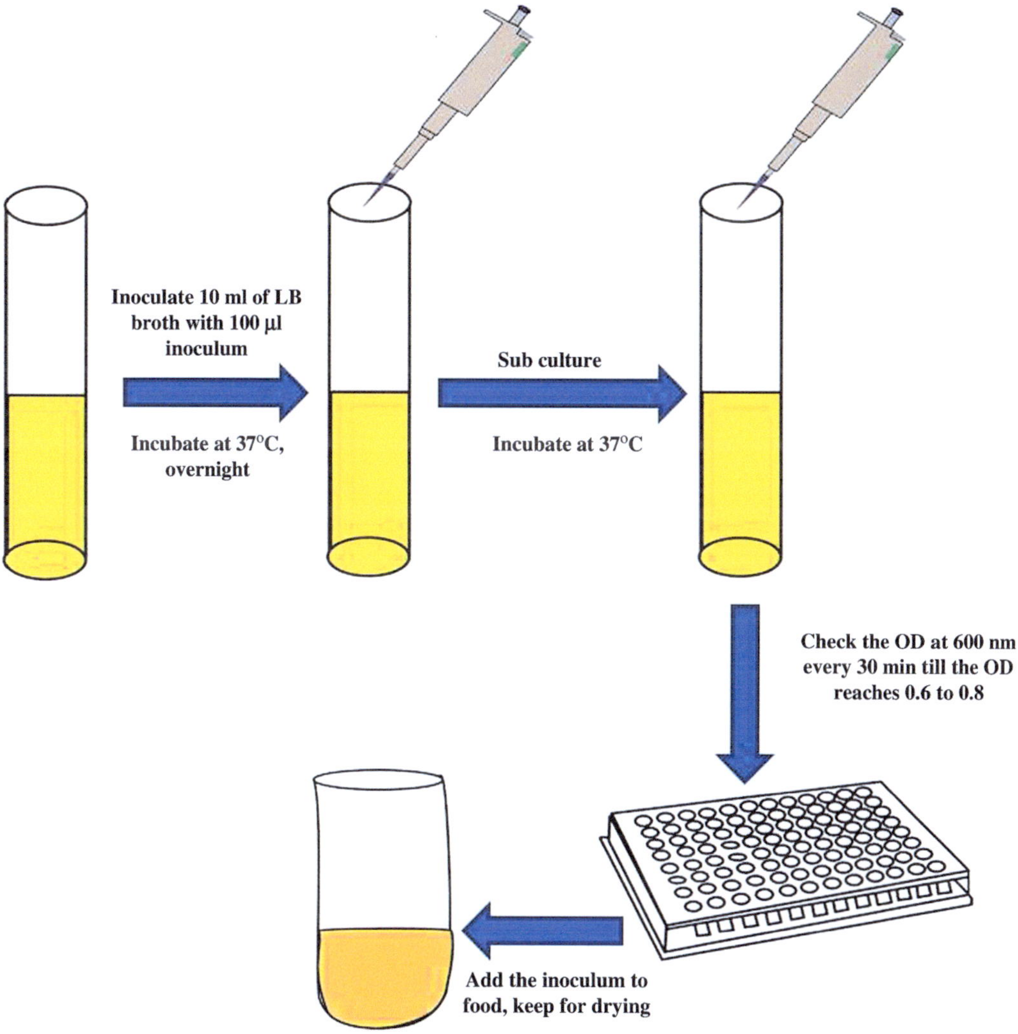

Fig. 3 Diagram showing preparation of bacterial culture and addition to the fly food

3.5.2 Purification of AGE Compound

Activation of Dialysis Membrane

1. For activation of the dialysis membrane, keep the dialysis tube under running water for 1–2 h.
2. Transfer the tubes in a solution of 100 mM sodium carbonate and 10 mM EDTA (pH 7.0) at 60 °C for 2 h with mild agitation.
3. Wash the dialysis tubes with milliQ water for several times. After activation of the tubes, do not let them dry. The same tube can be used two to three times for dialysis after proper washing.

Dialysis

1. Place the AGE compound in a dialysis tube with one end closed with a dialysis clip. Leave a little space at the top of the tube and put another clip.

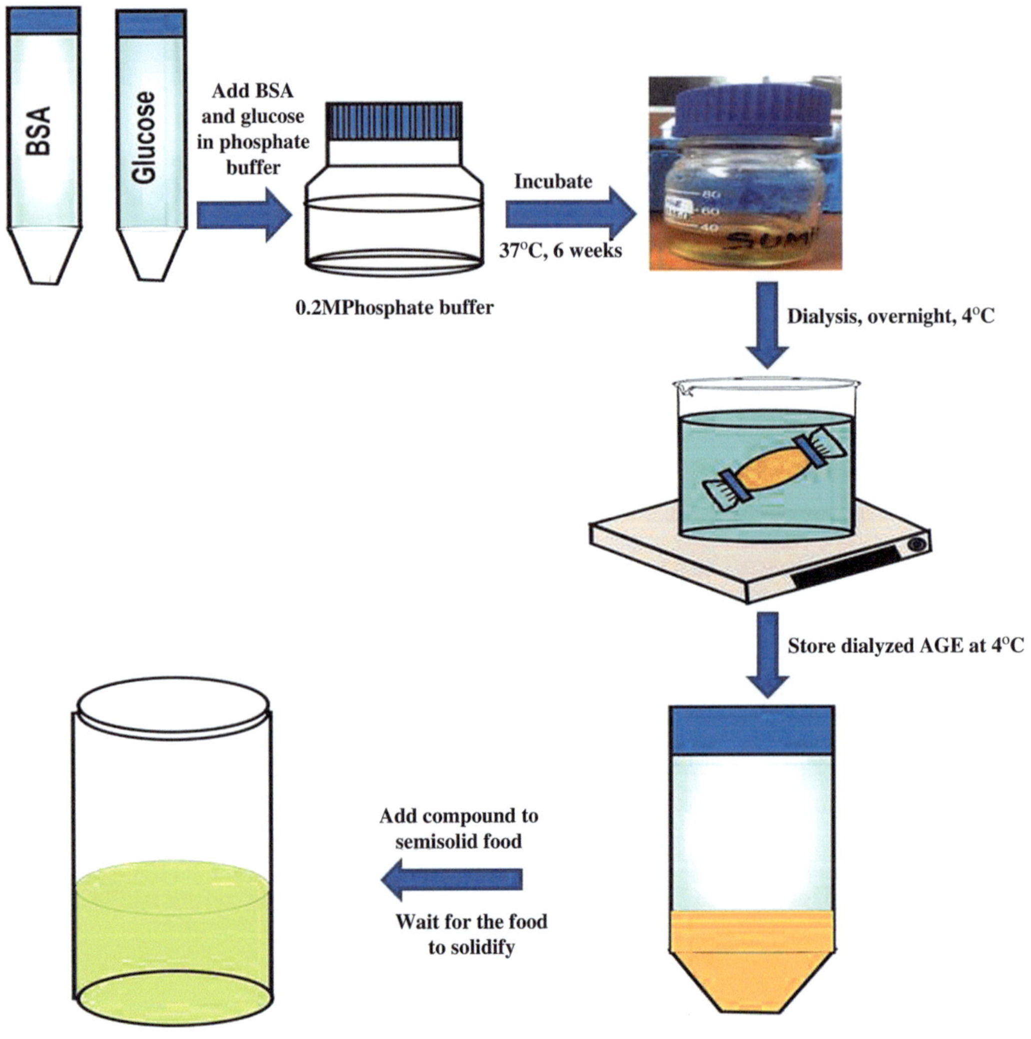

Fig. 4 Schematic diagram of AGE preparation, purification and addition to fly food

2. Place the tube in a beaker containing 500 ml of 0.5 M phosphate buffer. The volume of the buffer should be 100 times more than the solution in the dialysis bag.
3. Place the beaker on a magnetic stirrer for overnight at 4 °C (Fig. 4) (*see* **Note 4.5.2**).
4. Once dialysis is over, measure the protein concentration by the Bradford method.
5. The AGE compound after dialysis can be stored at 4 °C.

3.5.3 Preparation of Fly Food and Addition of AGE

1. Prepare the fly food as per the standard protocol mentioned in Sect. 2.1

2. After the preparation of food, add the AGE compound directly to the semisolid food in various amounts to achieve the desired concentration (Fig. 4).
3. Keep the food vials in room temperature so that the AGE compound will be absorbed within the food. This may take 2–3 days.
4. Once the food is dry, transfer the flies to the food vials.

4 Notes

4.1 Preparation of Standard Fly Food

1. Do not add nipagine and propionic acid to the hot food. First allow it to cool, but it should be added before solidification.
2. Do not put the cotton plug until the food is solidified; otherwise, there will be accumulation of water vapour inside the vials.

4.2 Addition of Nanoparticles to the Standard Food

1. Sonicate the nanoparticle stock solution well to disperse them properly.
2. Allow the food to dry completely before transferring the flies. Otherwise the flies will stick to the food.

4.3 Addition of Antibiotic to the Standard Fly Food

1. Do not add antibiotics to the hot food. It can seriously affect the efficacy of the antibiotic.

4.4 Addition of Bacteria to the Standard Fly Food

1. Always use fresh bacterial culture with an OD_{600} of 0.6–0.8, as infection requires a higher dose.
2. Generally we use propionic acid and nipagine as antibacterial agents to prevent the bacterial contamination of fly food. Some of the bacteria such as *Salmonella typhi* are susceptible to this chemical. So do not use these chemicals if you are adding *Salmonella* to the fly food.

4.5 Addition of AGE Compound to the Fly Food

1. Temperature is very essential for the synthesis of the AGE compounds. Make sure that the temperature is kept constant during the incubation period.
2. Do not perform dialysis for a longer time as excess water may cause lysis of the dialysis membrane.

Acknowledgements

SP and SM are thankful to MHRD for financial support. MM lab is supported by Grant No. BT/PR21857/NNT/28/1238/2017, EMR/2017/003054, Odisha DBT 3325/ST(BIO)-02/2017.

References

1. Markow TA, O'Grady P (2005) Drosophila: a guide to species identification and use. Elsevier, London
2. Ranganath H, Tanuja M (2000) Teaching and learning genetics with Drosophila 4. Pattern of inheritance of characters when there is interaction of genes or linkage of genes. RESONANCE
3. Pearl R (1926) A synthetic food medium for the cultivation of Drosophila: preliminary note. J Gen Physiol 9(4):513–519
4. Chattopadhyay D, James J, Roy D, Sen S, Chatterjee R, Thirumurugan K (2015) Effect of semolina-jaggery diet on survival and development of *Drosophila melanogaster*. Fly 9 (1):16–21
5. Guerra MJ, Mujica MV (2010) Physical and chemical properties of granulated cane sugar "panelas". Food Sci Technol 30(1):250–257
6. Avanesian A, Khodayari B, Felgner JS, Jafari M (2010) Lamotrigine extends lifespan but compromises health span in *Drosophila melanogaster*. Biogerontology 11(1):45
7. Stocker H, Gallant P (2007) Getting started: an overview on raising and using Drosophila. Humana Press, Springer
8. Konsolaki M (2013) Fruitful research: drug target discovery for neurodegenerative diseases in Drosophila. Expert Opin Drug Discovery 8 (12):1503–1513
9. Bell AJ, McBride SM, Dockendorff TC (2009) Flies as the ointment: Drosophila modeling to enhance drug discovery. Fly 3(1):39–49
10. Fernández-Hernández I, Scheenaard E, Pollarolo G, Gonzalez C (2016) The translational relevance of Drosophila in drug discovery. EMBO Rep 17(4):471–472
11. Douglas S, Davis S, Illum L (1987) Nanoparticles in drug delivery. Crit Rev Ther Drug Carrier Syst 3(3):233–261
12. Sanguansri P, Augustin MA (2006) Nanoscale materials development–a food industry perspective. Trends Food Sci Technol 17 (10):547–556
13. Krivorotova T, Cirkovas A, Maciulyte S, Staneviciene R, Budriene S, Serviene E, Sereikaite J (2016) Nisin-loaded pectin nanoparticles for food preservation. Food Hydrocoll 54:49–56
14. Priyadarsini S, Mukherjee S, Mishra M (2018) Nanoparticles used in dentistry: a review. J Oral Biol Craniofac Res 8(1):58–67
15. Sabat D, Patnaik A, Ekka B, Dash P, Mishra M (2016) Investigation of titania nanoparticles on behaviour and mechanosensory organ of *Drosophila melanogaster*. Physiol Behav 167:76–85
16. Mishra M, Sabat D, Ekka B, Sahu S, Unnikannan P, Dash P (2017) Oral intake of zirconia nanoparticle alters neuronal development and behaviour of *Drosophila melanogaster*. J Nanopart Res 19(8):282
17. Pappus SA, Ekka B, Sahu S, Sabat D, Dash P, Mishra M (2017) A toxicity assessment of hydroxyapatite nanoparticles on development and behaviour of *Drosophila melanogaster*. J Nanopart Res 19(4):136
18. Pappus SA, Mishra M (2018) A Drosophila model to decipher the toxicity of nanoparticles taken through oral routes. In: Cellular and molecular toxicology of nanoparticles. Springer, p 311–322
19. Rand MD, Montgomery SL, Prince L, Vorojeikina D (2014) Developmental toxicity assays using the Drosophila model. Curr Protoc Toxicol 59(1):1.12. 11–11.12. 20
20. Hirsch HV, Mercer J, Sambaziotis H, Huber M, Stark DT, Torno-Morley T, Hollocher K, Ghiradella H, Ruden DM (2003) Behavioral effects of chronic exposure to low levels of lead in *Drosophila melanogaster*. Neurotoxicology 24(3):435–442
21. Muñiz Ortiz JG, Opoka R, Kane D, Cartwright IL (2008) Investigating arsenic susceptibility from a genetic perspective in Drosophila reveals a key role for glutathione synthetase. Toxicol Sci 107(2):416–426
22. Jyoti S, Naz F, Siddique YH (2014) Toxic effects of gentamicin in the third instar larvae of transgenic *Drosophila melanogaster* (hsp70-lacZ) Bg 9. Toxicol Res 3(3):168–176
23. O'Shea KL, Singh ND (2015) Tetracycline-exposed *Drosophila melanogaster* males produce fewer offspring but a relative excess of sons. Ecol Evol 5(15):3130–3139
24. Hoffmann JA (2003) The immune response of Drosophila. Nature 426(6962):33
25. Buchon N, Silverman N, Cherry S (2014) Immunity in *Drosophila melanogaster*—from microbial recognition to whole-organism physiology. Nat Rev Immunol 14(12):796
26. Bergman P, Esfahani SS, Engström Y (2017) Drosophila as a model for human diseases—focus on innate immunity in barrier epithelia. In: Current topics in developmental biology, vol 121. Elsevier, pp 29–81
27. Siva-Jothy JA, Prakash A, Vasanthakrishnan RB, Monteith KM, Vale PF (2018) Oral bacterial infection and shedding in *Drosophila melanogaster*. J Visualized Exp 135

28. Apidianakis Y, Rahme LG (2009) *Drosophila melanogaster* as a model host for studying *Pseudomonas aeruginosa* infection. Nat Protoc 4 (9):1285
29. Khalil S, Jacobson E, Chambers MC, Lazzaro BP (2015) Systemic bacterial infection and immune defense phenotypes in *Drosophila melanogaster*. J Visualized Exp 99:e52613
30. Neyen C, Bretscher AJ, Binggeli O, Lemaitre B (2014) Methods to study Drosophila immunity. Methods 68(1):116–128
31. Nehme NT, Liégeois S, Kele B, Giammarinaro P, Pradel E, Hoffmann JA, Ewbank JJ, Ferrandon D (2007) A model of bacterial intestinal infections in *Drosophila melanogaster*. PLoS Pathog 3(11):e173
32. Sleiman MSB, Osman D, Massouras A, Hoffmann AA, Lemaitre B, Deplancke B (2015) Genetic, molecular and physiological basis of variation in Drosophila gut immunocompetence. Nat Commun 6:7829
33. Kuraishi T, Hori A, Kurata S (2013) Host-microbe interactions in the gut of *Drosophila melanogaster*. Front Physiol 4:375
34. Gupta V, Vasanthakrishnan RB, Siva-Jothy J, Monteith KM, Brown SP, Vale PF (2017) The route of infection determines Wolbachia antibacterial protection in Drosophila. Proc R Soc B 284(1856):20170809
35. Martins NE, Faria VG, Teixeira L, Magalhães S, Sucena É (2013) Host adaptation is contingent upon the infection route taken by pathogens. PLoS Pathog 9(9):e1003601
36. Ferreira ÁG, Naylor H, Esteves SS, Pais IS, Martins NE, Teixeira L (2014) The Toll-dorsal pathway is required for resistance to viral oral infection in Drosophila. PLoS Pathog 10(12): e1004507
37. Buchon N, Broderick NA, Poidevin M, Pradervand S, Lemaitre B (2009) Drosophila intestinal response to bacterial infection: activation of host defense and stem cell proliferation. Cell Host Microbe 5(2):200–211
38. Ulrich P, Cerami A (2001) Protein glycation, diabetes, and aging. Recent Prog Horm Res 56 (1):1–22
39. Takeuchi M, Yamagishi S-i (2009) Involvement of toxic AGEs (TAGE) in the pathogenesis of diabetic vascular complications and Alzheimer's disease. J Alzheimers Dis 16 (4):845–858
40. Nedić O, Rattan S, Grune T, Trougakos I (2013) Molecular effects of advanced glycation end products on cell signalling pathways, ageing and pathophysiology. Free Radic Res 47 (sup1):28–38
41. Uribarri J, Cai W, Sandu O, Peppa M, Goldberg T, Vlassara H (2005) Diet-derived advanced glycation end products are major contributors to the body's AGE pool and induce inflammation in healthy subjects. Ann N Y Acad Sci 1043(1):461–466
42. Assar SH, Moloney C, Lima M, Magee R, Ames JM (2009) Determination of N ε-(carboxymethyl) lysine in food systems by ultra performance liquid chromatography-mass spectrometry. Amino Acids 36 (2):317–326
43. Šebeková K, Hofmann T, Boor P, Šebeková K Jr, Og U, Erbersdobler HF, Baynes JW, Thorpe SR, Heidland A, Somoza V (2005) Renal effects of oral Maillard reaction product load in the form of bread crusts in healthy and subtotally nephrectomized rats. Ann N Y Acad Sci 1043(1):482–491
44. Tsakiri EN, Iliaki KK, Höhn A, Grimm S, Papassideri IS, Grune T, Trougakos IP (2013) Diet-derived advanced glycation end products or lipofuscin disrupts proteostasis and reduces life span in *Drosophila melanogaster*. Free Radic Biol Med 65:1155–1163

Chapter 2

Isolation of Gut, Imaginal Disc, Fat Body, Ovary and Testes in Various Developmental Stages of *Drosophila*

Nibedita Nayak, Gyanaseni Dhar, and Monalisa Mishra

Abstract

Drosophila melanogaster shares a remarkable resemblance in terms of its organ anatomy and function with that of mammals. Various organs of the fly are used as a model to study its pathologies and metabolic and therapeutic purposes. Gut, the largest organ of the fly body, shares similarity with mammalian intestine in terms of metabolic pathways, physiology and signalling. The fat body of the fly resembles the liver of mammals and helps in fat storage and nutrient sensing. Imaginal disc composed of 20–40 cells are used to check defect in early time point of development. The testes and ovary are used to study the reproductive defect. This chapter describes the dissection of the gut, fat body, ovary, testes, and imaginal discs from larvae and adult tissue.

Key words *Drosophila*, Gut, Fat body, Imaginal disc, Ovary, Testes

1 Introduction

Drosophila melanogaster is an excellent model organism to explore the mechanism underlying numerous pathologies related to metabolism, physiology and neurology. There is a striking similarity between fly and mammals in terms of pathways, molecular mechanism, anatomy and functioning of the organ. Thus, the various organs of the fly are used to understand the pathophysiology of human being. Below we are describing some of the tissues which are used as a model organ to understand the defects in human being.

1.1 Gut

The largest organ of *Drosophila* is the digestive tract or the gut. The gut is subdivided into the foregut, midgut and hindgut. The different types of cells vary within the digestive tract [1–4]. Although *Drosophila* undergoes various developmental stages (larva, pupa and adult), the gut division persists throughout the development [5]. However, the digestive tract of larva and adult differs anatomically and genetically [6, 7].

Monalisa Mishra (ed.), *Fundamental Approaches to Screen Abnormalities in Drosophila*, Springer Protocols Handbooks, https://doi.org/10.1007/978-1-4939-9756-5_2,

The fly gut retains various enzymes such as a-esterases, endo−/ exopeptidase, trypsin, mannosidases and lipases [8] for the digestion of carbohydrates, lipids and proteins [9–13]. Fly food also includes microorganisms. For the elimination of microorganisms, the gut preserves 15 lysozymes to digest the bacterial wall which is made up of peptidoglycan [14]. Alike mammals, the gut of *Drosophila* also reserves a plethora of microbes [15–18] which interfere with the amount and composition of food to be digested [19] and give protection against pathogens. The gut is encrusted with chitinous matrix (intestinal epithelium), impermeable cuticle (foregut and midgut) and peritrophic matrix (midgut) [20] as a defence mechanism. In *Drosophila*, the foregut resembles the oesophagus of mammals. The crop is similar to the mammalian stomach, and the anterior midgut is like mammalian small intestine. In the hindgut the electrolyte, nutrient and water absorption occur. The rectum is similar to the mammalian large intestine, and the undigested food is excreted from the anus [21, 22]. The fly gut shares anatomical similarity [4] in tissue and physiological function [23, 24] with mammals and is endothelial in origin [25, 26]. The intestinal stem cells are found in gut of both mammals and fly [27–31] harbour hormone-releasing enteroendocrine cells and nutrient-absorbing enterocytes [30, 32]. The secretory cells such as AMP (antimicrobial-producing), Paneth cells and mucus-producing goblet cells are present only in the mammalian gut but not in the fly gut. Mucus and AMPs are reported to be produced by fly gut [17, 33]. The cellular composition and regeneration of mammalian intestine and fly hindgut share similarity with each other [34]. Both fly and mammal share signalling pathways associated with intestinal regeneration, infection, tumour and ageing [32, 35–43]. Altogether *Drosophila* gut is used as a model to check ageing, infection, pathology, cancer, neurobiology and stem cell biology.

1.2 Fat Body

Drosophila fat body shares similarity with the mammalian adipose tissue in terms of the nutrition sensing, response to immune system and storage of energy [44]. It is found in both larvae and adult stage of the fly [45] is mesodermal in origin [46, 47]. The fat body senses nutritional condition [48] and helps in the development of brain and body size [49]. The metabolic process of the fat body resulted in many intermediate steps [50–52]. The fat body secretes certain proteins and peptides, for instance, insect-derived growth factor (IDFGs) and *Drosophila* insulin-like peptides (Dilps) that affect the growth of wing disc and insulin-producing cells in the brain simultaneously [50, 53]. The fat body helps in stress resistance and maturation of the ovary in female adult flies [54, 55]. During an energy deficit, fat get mobilized to target tissues through lipolysis [56–58]. *Drosophila* triglyceride lipase called as bummer (bmm), catalyse triglyceride to fatty acid, and bmm share functional similarity with the human lipase [59, 60]. Lipid droplets are present in the gut, Malpighian tubules and larval imaginal discs [57, 61]. Thus, *Drosophila* can be used as a model to investigate human lipometabolic disease and age-related diseases such as cancer, obesity and diabetes and discover novel therapeutics to combat the pathologies.

1.3 Imaginal Discs

Imaginal discs (shown in Fig. 1) are a bunch of 10 to 50 epithelial cells that segregate from the epidermis of the larval cavity which differentiates to the respective organs after metamorphosis [62, 63]. During transformation, from larva to a pupa, the larval tissues start degenerating to form the head, thorax, haltere, limbs, wings and genitalia, while the abdomen and the other adult structure arise from the pool of histoblast cells [64–66].

The larva has 19 imaginal discs, 9 bilateral discs form the epidermal structure, and the medial disc forms the genitalia [67]. The mouthpart of the fly arises from the labia and the disc, eye from the eye-antennal disc that lies in close proximity with the mouth hook and the optic lobe of the central nervous system [68–70]. The three pairs of leg discs at the larval stage come from the ventral ectoderm; the first pair is associated with the first thoracic segment T1, while the second and third pairs get associated with the second and third thoracic segment [71, 72]. The thoracic segment also yields the wing, leg and haltere imaginal discs [73, 74]. Eventually,

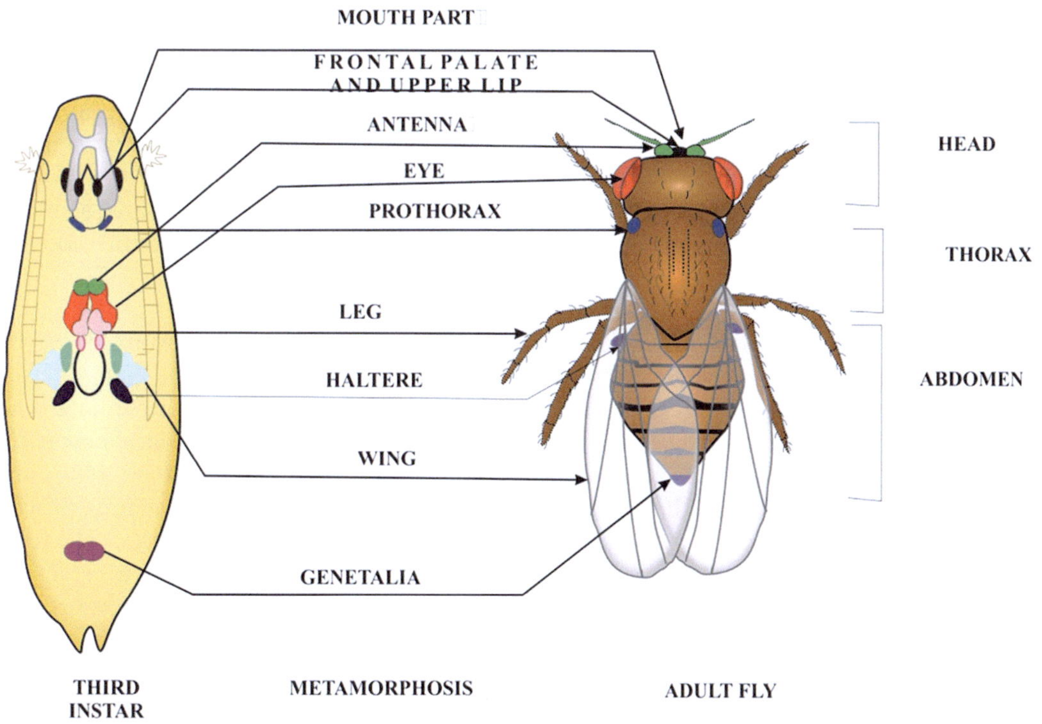

Fig. 1 Left-side image depicts the various imaginal discs of the third instar larva

the genitalia are developed from the medial disc extending from the eighth to the tenth abdominal segment [74, 75]. The imaginal discs are used for fate mapping [76–78]. The embryonic stage might have an effect on specific structures of the adult [79, 80]. Additionally, the disc primordia from an embryo can be cultured in vivo, transplanted to a larval host and upon metamorphosis developed to their determined structure [81].

1.4 Ovary

Drosophila oogenesis is a complex process which is regulated by various genes at different steps [82]. The *Drosophila* female has two ovaries with 16 to 21 ovarioles with a tubular structure. Each ovariole comprises 6–8 consecutively growing egg chambers [83]. In a developed ovary, each ovariole is composed of three regions such as the terminal filament, germarium and vitellarium. The *Drosophila* germarium comprises of four functional regions, 1, 2A, 2B and 3, and three types of stem cells, such as somatic stem cells, germ line stem cells and escort stem cells [84–86]. Stem cells provide an origin for ovarian germ line cells, whereas somatic cells develop the capsule of the germarium. In germarium, region 1 comprises of cystoblasts. The terminal filament cells, cap cells and germarium capsule cells develop a niche, confirming stimulation, development and maintenance of stem cells. The cap cells and terminal filament cells express the *Yb*, *wingless* (*wg*), *decapentaplegic* (*dpp*), *hedgehog* (*hh*) and *piwi* genes [82, 87, 88]. In *Drosophila*, the germ line is well-identified as a tissue lineage, and the development of germ line characterizes the development of a stem cell-derived tissue. Thus, the ovary provides the stem cell lineages and signalling pathways which control the regulation of maintenance of stem cell and fate of stem cell proliferation [83]. The follicle stem cell of the ovary of *Drosophila* has some similarity with the intestinal stem cell in mammals in various aspects. *Drosophila* ovary is a useful model to study the behaviour of stem cell [89].

1.5 Testis

Spermatogenesis includes the development of sperm from the primordial germ cell [90]. *Drosophila* male possesses a pair of testes. Each testis is a blind-ended, twisted tube around the seminal vesicle. The testis is formed of muscle and pigment cell, whereas the stem cells, spermatogenic cells and supporting somatic cells comprise the lumen of the testis [91]. The spermatogenesis process occurs in *Drosophila* in the cysts at the blind-ended testis. The testis possesses three types of cells: (1) hub cell (responsible for stem cell maintenance), (2) germ line stem cells and (3) somatic cyst progenitor cell. Insulin signalling plays a key role in spermatogenesis of *Drosophila*. The male germ line stem cells (GSCs) divide to generate the new germ line cysts for regeneration [92]. The epidermal growth factor receptor signalling pathway activation occurs in cyst cells that can inhibit the GSC proliferation [93, 94]. The stem cells have the capacity to renew and are capable of generating the cells which will able to move on a pathway to differentiate the sperm. To study the stem cell niche, *Drosophila* testes are used as the model system [95].

2 Materials and Chemicals Required

1. Female and male *Drosophila* (*Oregon R*).
2. Fly food material (yeast powder (SUBALA yeast, SYD-003/10), sucrose (FISHER SCIENTIFIC-qualigens-28,106), cornmeal (from local market), agar agar type1 (HiMedia-RM666), milliQ water).
3. Diethyl ether (HiMedia-AS047).
4. 4% paraformaldehyde (PFA)
5. 1X phosphate-buffered saline (PBS)
6. 1 N NaOH (HiMedia-MB095)
7. HCL (HiMedia-RM5955).
8. Eppendorf tube (1.5 ml).
9. Glass slide ((25mmX75 mm, ABDOS-G10101).
10. Cover slip (18 mm, BLUE STAR).
11. Forceps.
12. Needles.
13. Beaker (RIVERA).
14. Hot plate.
15. Stirring beads.
16. pH metre,
17. Stopwatch.
18. Pipette.
19. Dissection microscope/stereomicroscope.

3 Preparation of Chemicals

3.1 Preparation of 1X PBS

1. To prepare 100 ml of 1X PBS, add all the ingredients (Table 1) with 80 ml milliQ water, and allow it to dissolve.
2. Adjust the pH of the solution to 7.2 by adding HCl as per your requirement.
3. After maintaining the pH, adjust the volume to 100 ml by adding milliQ water. Autoclave the solutions prepared prior to use [96].

Table 1
Recipe for PBS preparation

	1X PBS (for 100 ml)	
	Chemical	**Amount (g)**
1	NaCl	0.8
2	KCl	0.02
3	Na_2HPO_4	0.14
4	KH_2PO_4	0.02

3.2 Preparation of 4% PFA

1. Heat 70 ml of 1X PBS in a glass beaker up to 60 °C on a hot plate with stirring beads.
2. Add 4 gram of paraformaldehyde to it, and stir continuously.
3. Add NaOH dropwise to make a completely clear solution.
4. Once this state is reached, cool and filter it.
5. Adjust the pH to 7.2 by adding HCl.
6. Now adjust the volume to 100 ml and make aliquots and finally store it in 4 °C.

4 Methodology

4.1 Dissection of Gut

1. Make standard fly food for experimental set-up of control and treated (*see* **Note 1**).
2. Transfer the male and female flies in the ratio of 2:3 or 5:7 by anaesthetizing flies with ether (*see* **Note 2**).
3. After 3–4 days, collect the third instar larva, and wash them in 1X PBS to remove food attached to their body (*see* **Note 3**).
4. Place the petri plate containing 1X PBS and larva under the stereomicroscope.
5. Hold both ends of the larva with the forceps (*see* **Note 4**).
6. Now slowly drag one end of the larva to pull out the gut (*see* **Note 5**).
7. Store the dissected gut in 4% PFA or place it on the slide for further needful processing (Fig. 2).
8. In the case of adult flies, all the steps are the same as that of larva with an additional step of beheading the fly before holding both ends of the fly with forceps (Fig. 3).

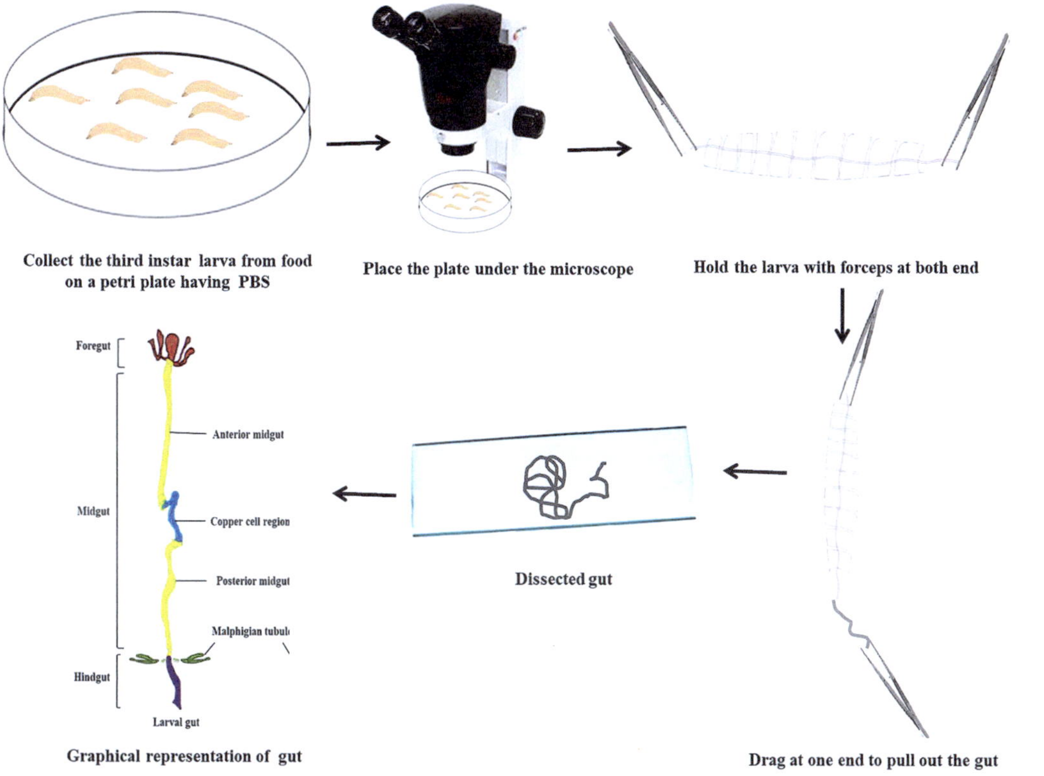

Fig. 2 Various steps involved in the dissection of the gut of the third instar larva

4.2 Dissection of Fat Body

1. Make standard fly food for experimental set-up of control and treated (*see* **Note 1**).
2. Transfer the male and female flies in the ratio of 2:3 or 5:7 by anaesthetizing flies with ether (*see* **Note 2**).
3. After 3–4 days, collect the third instar larva, and wash them in 1X PBS to remove food attached to their body (*see* **Note 3**).
4. Place the petri plate containing 1X PBS and larva under the stereomicroscope.
5. Hold both ends of the larva with the forceps (*see* **Note 4**).
6. Now slowly drag one end of the larva to pull out the gut and fat body that come out along with the gut (*see* **Note 5**).
7. Now slowly detach the fat body from the gut with the help of a small needle (*see* **Note 6**).
8. Store the dissected fat body in 4% PFA or place it on the slide for further needful processing (Fig. 4).
9. In the case of adult flies, all the steps are the same as that of larva with an additional step of beheading the fly before holding both ends of the fly with forceps (Fig. 5).

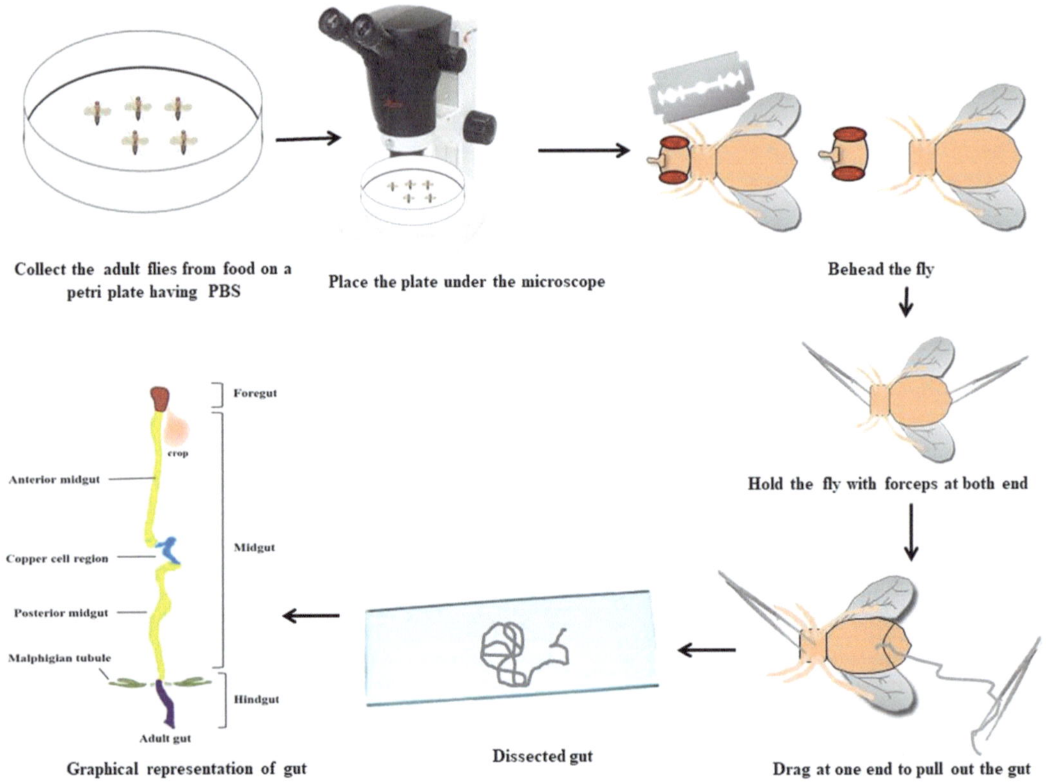

Fig. 3 Major steps involved in the isolation of the gut of an adult fly

4.3 Dissection of the Imaginal Disc

1. Take a vial with third instar larva. Collect the larva into another vial with the help of a paint brush.
2. Wash the larvae to get rid of food particles with distilled water and keep it on the ice bucket. This step helps to anaesthetize the larvae.
3. Put ice-cold 1X PBS on a slide and place the larva on it. Observe it under the microscope.
4. Using two fine forceps, hold the mouth hook at the anterior portion and the posterior portion of the larva and gently pull them apart; the larval body splits into one third from the anterior side of the mouth hook. This state is ideal for separating out all the imaginal discs as all of these are located at the anterior portion of the larval body (Fig. 6) (*see* **Note 7**).
5. Discard the unwanted tissues such as the gut and the posterior part of the larva with the help of needle and forceps. Look for the bunch of the imaginal discs attached to the central nervous system.
6. Separate them carefully and store it in 4% PFA for further experiments.

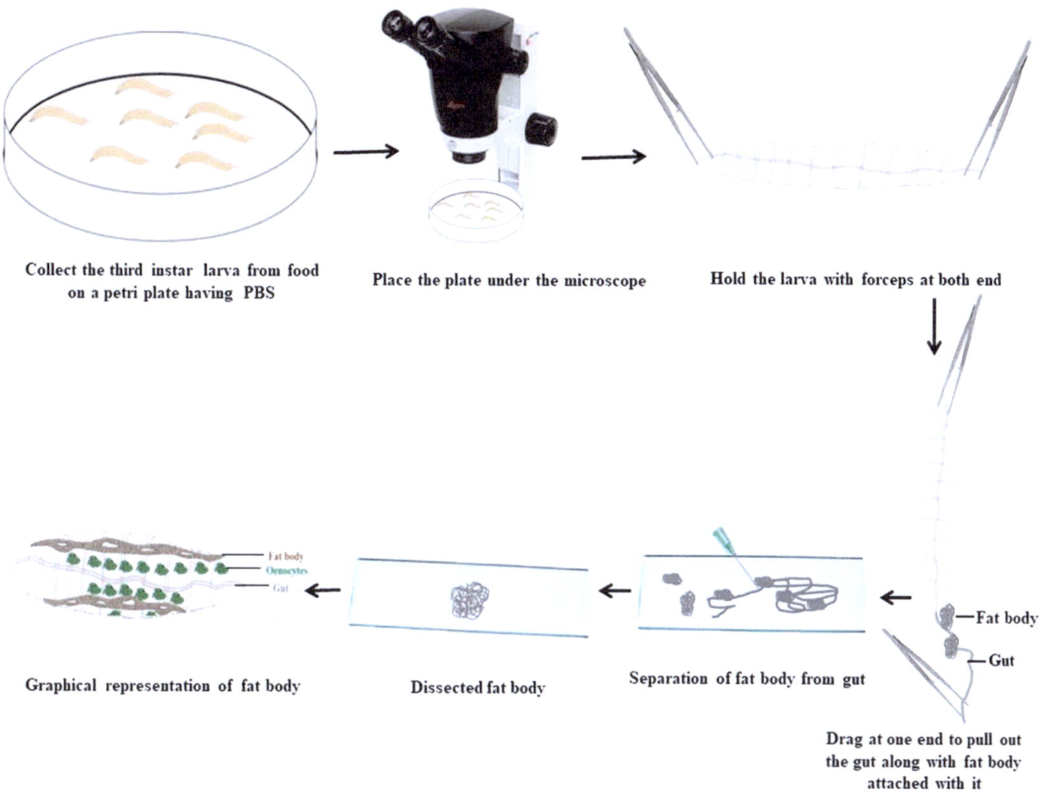

Fig. 4 Steps involved in fat body dissection of the third instar larva

4.4 Dissection of Ovary

1. Feed flies with yeast 1–2 days prior to dissecting them to fatten up the ovaries.
2. Anaesthetize flies using diethyl ether.
3. Using a pair of tweezers, select a female.
4. Dip the female fly into 1X PBS (Fig. 7).
5. Remove the lower thorax of the fly with a pair of tweezers.
6. Pull softly at the lower abdomen with another pair of tweezers until the internal organs in the abdomen are exposed.
7. Look for the pair of ovaries and detach it from other organs (e.g. the intestines) (*see* **Note 7**).
8. Keep ovaries in ice-cold 1X PBS while dissecting the next fly.

4.5 Dissection of Testis

1. Make a glass microscope slide with a black background.
2. Collect 1–3-day-old male flies.
3. Prepare ice-cold 1X phosphate-buffered saline (PBS) and dissecting needles.
4. Anaesthetize 50–100 flies using diethyl ether, and transfer them from the vial to the fly pad near the microscope slide.

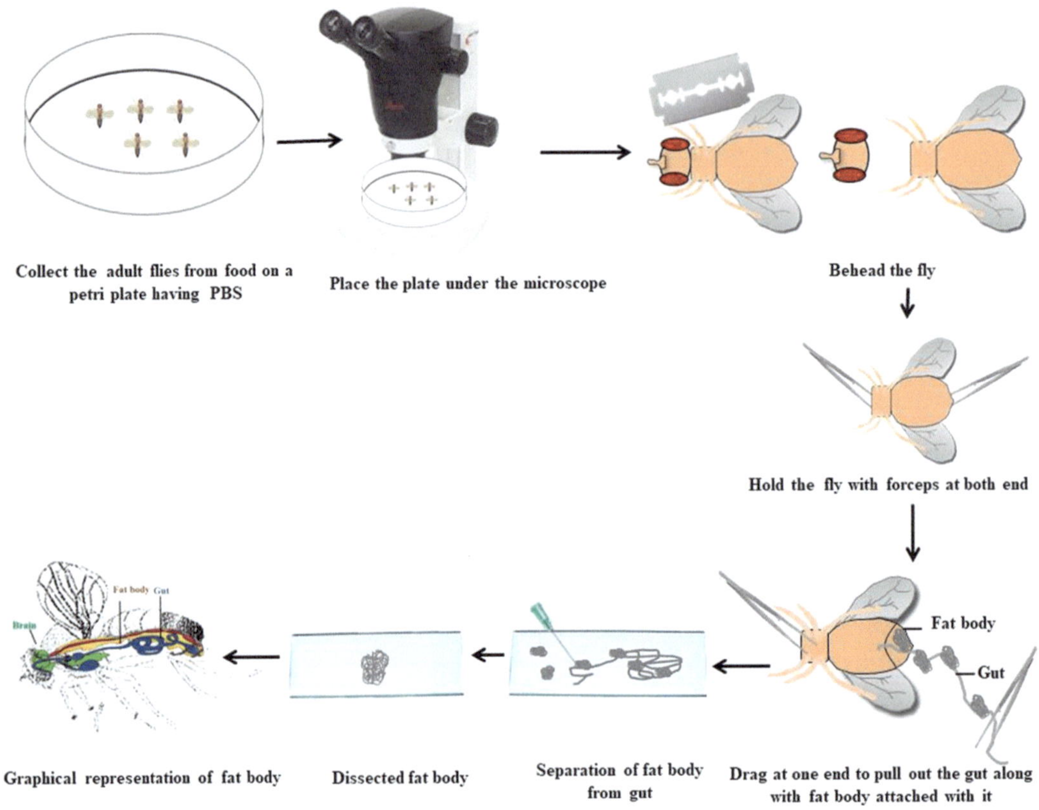

Fig. 5 Steps shown to dissect the fat body from an adult fly

5. Take approximately 50 μl of ice-cold PBS in a drop in the centre of the microscopic slide.
6. Gently take a male fly with the dissecting needle, dip the dissecting needle into PBS, gently touch the ventral side of the fly with the needle, and place it into the drop of PBS.
7. Place one needle on the middle of the fly, between the thorax and the abdomen, pushing down firmly. Place the other needle between the sixth (A6) and seventh (A7) abdominal segments, and pull the abdomen away from the thorax.
8. Take the testes apart from other organs, especially the gut (Fig. 8). Key differences are that testes always come in pairs; each testis has a spiral pattern and is whiter in colour (*see* **Note 7**).
9. Store the testes in ice-cold PBS.

5 Note

1. The fly should be maintained in optimal environmental conditions like temperature and light and humidity.

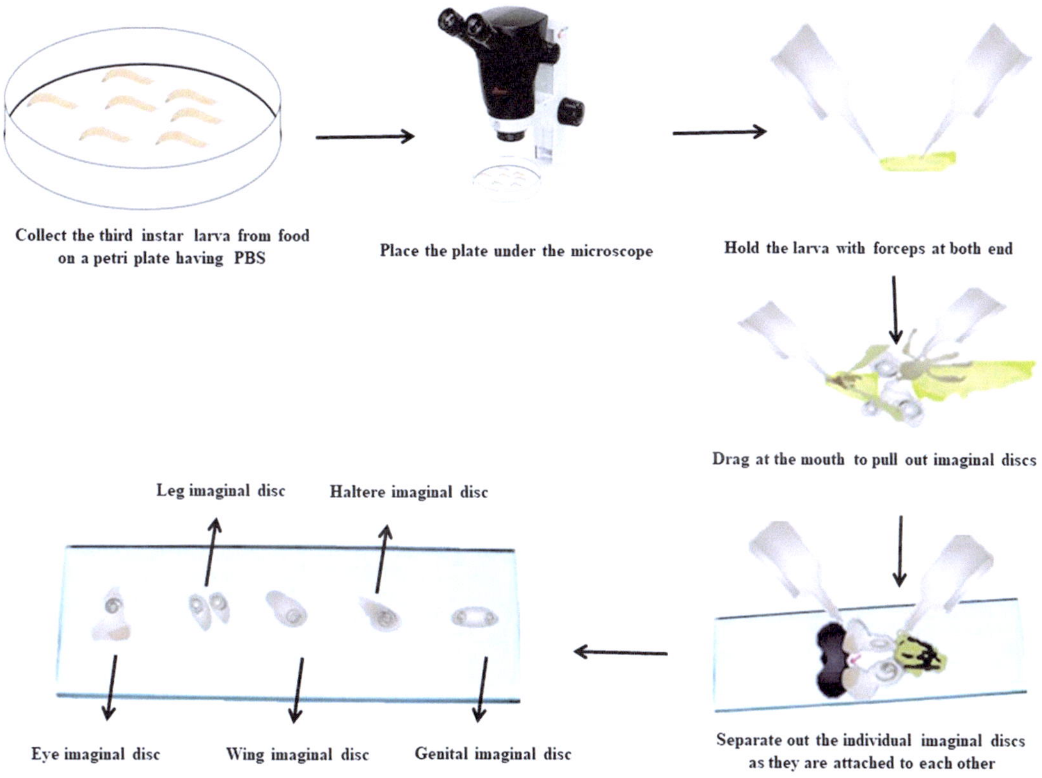

Fig. 6 Dissection of the eye, haltere, leg, wing, genital imaginal disc from the third instar larva

2. During anaesthetizing the fly, an ear bud has to be used. After dipping inside ether, the excess ether should be squeezed by pressing the bud at the tip of the ether tube. Excess dose of ether may lead to the death of fly, so once they are anaesthetized, we need to keep them for 20–30 min to be active for the next dose (if required).
3. Washing of larva and adult flies after removing them from food is essential because food may interfere with the assay.
4. The larva may be moving, making it difficult to hold in place, so to reduce larval movement, place the petri plate containing larva on ice pack for 5 min.
5. Carefully and slowly drag the gut to collect the full-length gut. Rough dragging may cause breakage of the gut.
6. Carefully and slowly detach the fat body to avoid damage to the tissue.
7. After dissection the imaginal disc, ovary or testis, should be handled carefully as it is transparent in colour.

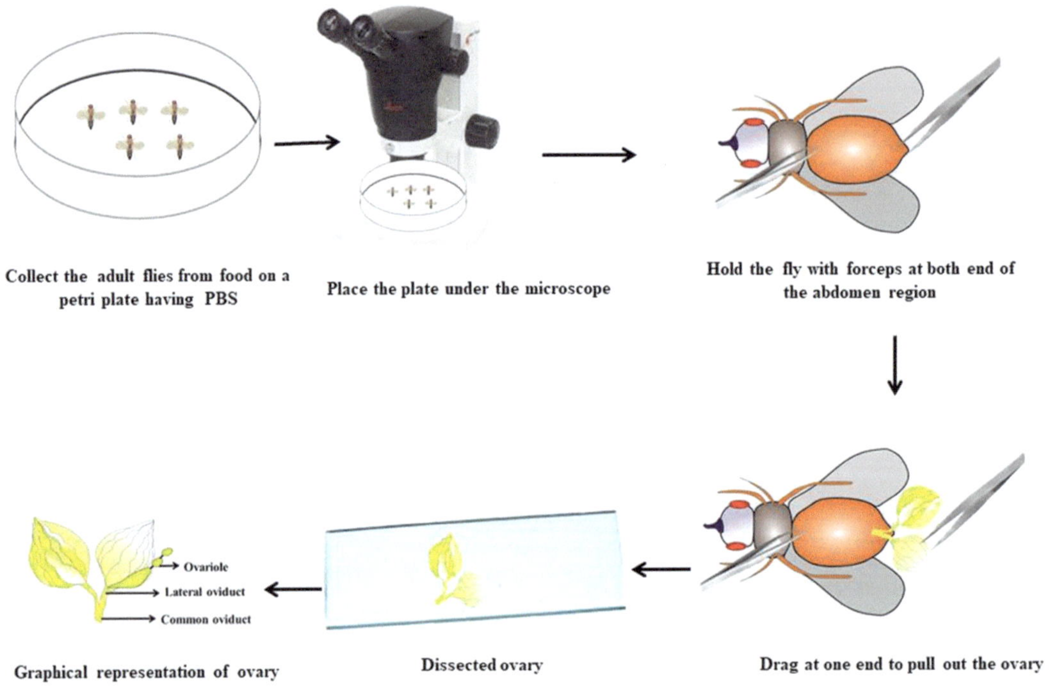

Fig. 7 Dissection of the ovary from an adult female

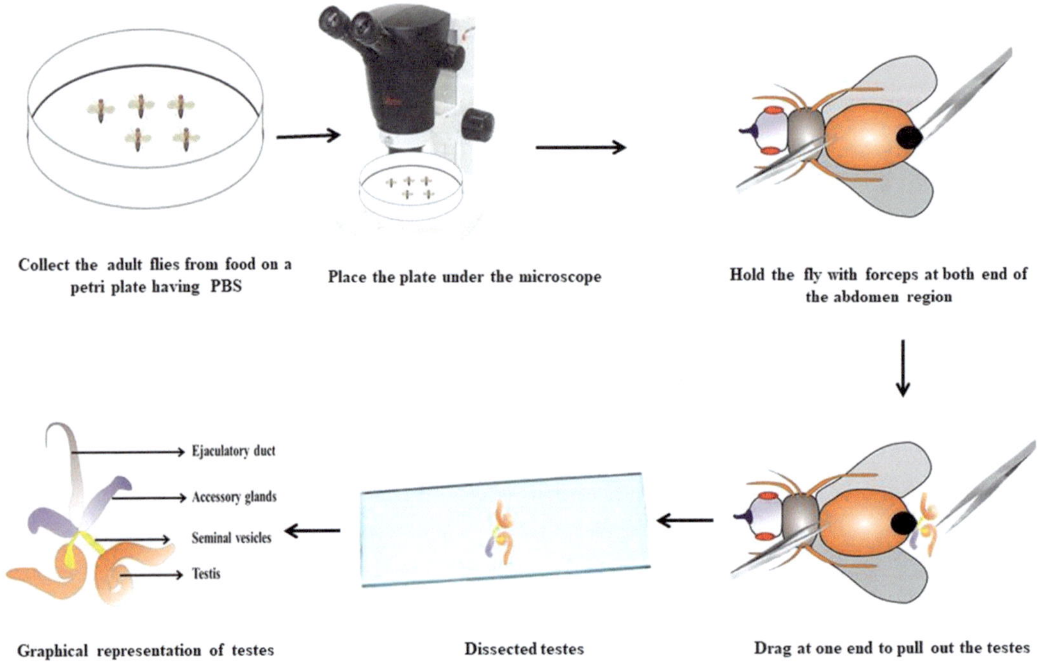

Fig. 8 Dissection of the testis from a male fly

Acknowledgements

NN and GD are thankful to DST/INSPIRE Fellowship for financial support. MM lab is supported by Grant No. BT/PR21857/NNT/28/1238/2017, EMR/2017/003054, Odisha DST 3325/ST(BIO)-02/2017.

References

1. Fristrom JW (1970) The developmental biology of Drosophila. Annu Rev Genet 4 (1):325–346
2. Gartner LP (1976) Fine structure of adult Drosophila midgut musculature. J Submicrosc Cytol
3. Gartner LP (1985) The fine structural morphology of the midgut of adult Drosophila: a morphometric analysis. Tissue Cell 17 (6):883–888
4. Shanbhag S, Tripathi S (2009) Epithelial ultrastructure and cellular mechanisms of acid and base transport in the Drosophila midgut. J Exp Biol 212(11):1731–1744
5. Takashima S, Younossi-Hartenstein A, Ortiz PA, Hartenstein V (2011) A novel tissue in an established model system: the Drosophila pupal midgut. Dev Genes Evol 221(2):69–81
6. Mathur D, Bost A, Driver I, Ohlstein B (2010) A transient niche regulates the specification of Drosophila intestinal stem cells. Science 327 (5962):210–213
7. Nakagoshi H (2005) Functional specification in the Drosophila endoderm. Develop Growth Differ 47(6):383–392
8. Buchon N, Osman D, David FP, Fang HY, Boquete J-P, Deplancke B, Lemaitre B (2013) Morphological and molecular characterization of adult midgut compartmentalization in Drosophila. Cell Rep 3(5):1725–1738
9. Carlson JR, Hogness DS (1985) The Jonah genes: a new multigene family in Drosophila melanogaster. Dev Biol 108(2):341–354
10. Ross J, Jiang H, Kanost MR, Wang Y (2003) Serine proteases and their homologs in the Drosophila melanogaster genome: an initial analysis of sequence conservation and phylogenetic relationships. Gene 304:117–131
11. Horne I, Haritos VS (2008) Multiple tandem gene duplications in a neutral lipase gene cluster in Drosophila. Gene 411(1–2):27–37
12. Horne I, Haritos VS, Oakeshott JG (2009) Comparative and functional genomics of lipases in holometabolous insects. Insect Biochem Mol Biol 39(8):547–567
13. Tamaki FK, Padilha MH, Pimentel AC, Ribeiro AF, Terra WR (2012) Properties and secretory mechanism of Musca domestica digestive chymotrypsin and its relation with Drosophila melanogaster homologs. Insect Biochem Mol Biol 42(7):482–490
14. Kylsten P, Kimbrell DA, Daffre S, Samakovlis C, Hultmark D (1992) The lysozyme locus in Drosophila melanogaster: different genes are expressed in midgut and salivary glands. Mol Gen Genet MGG 232 (3):335–343
15. Broderick NA, Lemaitre B (2012) Gut-associated microbes of Drosophila melanogaster. Gut Microbes 3(4):307–321
16. Ryu J-H, Kim S-H, Lee H-Y, Bai JY, Nam Y-D, Bae J-W, Lee DG, Shin SC, Ha E-M, Lee W-J (2008) Innate immune homeostasis by the homeobox gene caudal and commensal-gut mutualism in Drosophila. Science 319 (5864):777–782
17. Vodovar N, Vinals M, Liehl P, Basset A, Degrouard J, Spellman P, Boccard F, Lemaitre B (2005) Drosophila host defense after oral infection by an entomopathogenic Pseudomonas species. Proc Natl Acad Sci 102 (32):11414–11419
18. Apidianakis Y, Rahme LG (2009) Drosophila melanogaster as a model host for studying Pseudomonas aeruginosa infection. Nat Protoc 4(9):1285
19. Huang J-H, Douglas AE (2015) Consumption of dietary sugar by gut bacteria determines Drosophila lipid content. Biol Lett 11 (9):20150469
20. Hegedus D, Erlandson M, Gillott C, Toprak U (2009) New insights into peritrophic matrix synthesis, architecture, and function. Annu Rev Entomol 54:285–302
21. Edgecomb RS, Harth CE, Schneiderman AM (1994) Regulation of feeding behavior in adult Drosophila melanogaster varies with feeding regime and nutritional state. J Exp Biol 197 (1):215–235
22. Thompson D, Malagelada J (1981) Guts and their motions (gastrointestinal motility in

health and disease). J Clin Gastroenterol 3:81–87

23. Pitsouli C, Apidianakis Y, Perrimon N (2009) Homeostasis in infected epithelia: stem cells take the lead. Cell Host Microbe 6(4):301–307
24. Rubin DC (2007) Intestinal morphogenesis. Curr Opin Gastroenterol 23(2):111–114
25. Kedinger M, Simon-Assmann P, Haffen K (1987) Growth and differentiation of intestinal endodermal cells in a coculture system. Gut 28 (Suppl):237–241
26. Tepass U, Hartenstein V (1994) Epithelium formation in the Drosophila midgut depends on the interaction of endoderm and mesoderm. Development 120(3):579–590
27. Barker N, van de Wetering M, Clevers H (2008) The intestinal stem cell. Genes Dev 22 (14):1856–1864
28. Barker N, Van Es JH, Kuipers J, Kujala P, Van Den Born M, Cozijnsen M, Haegebarth A, Korving J, Begthel H, Peters PJ (2007) Identification of stem cells in small intestine and colon by marker gene Lgr5. Nature 449 (7165):1003
29. Micchelli CA, Perrimon N (2006) Evidence that stem cells reside in the adult Drosophila midgut epithelium. Nature 439(7075):475
30. Ohlstein B, Spradling A (2006) The adult Drosophila posterior midgut is maintained by pluripotent stem cells. Nature 439(7075):470
31. Sangiorgi E, Capecchi MR (2008) Bmi1 is expressed in vivo in intestinal stem cells. Nat Genet 40(7):915
32. Crosnier C, Stamataki D, Lewis J (2006) Organizing cell renewal in the intestine: stem cells, signals and combinatorial control. Nat Rev Genet 7(5):349
33. Buchon N, Broderick NA, Poidevin M, Pradervand S, Lemaitre B (2009) Drosophila intestinal response to bacterial infection: activation of host defense and stem cell proliferation. Cell Host Microbe 5(2):200–211
34. Takashima S, Mkrtchyan M, Younossi-Hartenstein A, Merriam JR, Hartenstein V (2008) The behaviour of Drosophila adult hindgut stem cells is controlled by Wnt and Hh signalling. Nature 454(7204):651
35. Lin G, Xu N, Xi R (2008) Paracrine wingless signalling controls self-renewal of Drosophila intestinal stem cells. Nature 455(7216):1119
36. Ohlstein B, Spradling A (2007) Multipotent Drosophila intestinal stem cells specify daughter cell fates by differential notch signaling. Science 315(5814):988–992
37. Choi NH, Kim JG, Yang DJ, Kim YS, Yoo MA (2008) Age-related changes in Drosophila midgut are associated with PVF2, a PDGF/VEGF-like growth factor. Aging Cell 7 (3):318–334
38. Uhlirova M, Jasper H, Bohmann D (2005) Non-cell-autonomous induction of tissue overgrowth by JNK/Ras cooperation in a Drosophila tumor model. Proc Natl Acad Sci 102 (37):13123–13128
39. Jiang H, Patel PH, Kohlmaier A, Grenley MO, McEwen DG, Edgar BA (2009) Cytokine/Jak/Stat signaling mediates regeneration and homeostasis in the Drosophila midgut. Cell 137(7):1343–1355
40. Lin G, Xu N, Xi R (2009) Paracrine unpaired signaling through the JAK/STAT pathway controls self-renewal and lineage differentiation of drosophila intestinal stem cells. J Mol Cell Biol 2(1):37–49
41. Buchon N, Broderick NA, Chakrabarti S, Lemaitre B (2009) Invasive and indigenous microbiota impact intestinal stem cell activity through multiple pathways in Drosophila. Genes Dev 23(19):2333–2344
42. Berkey CD, Blow N, Watnick PI (2009) Genetic analysis of Drosophila melanogaster susceptibility to intestinal Vibrio cholerae infection. Cell Microbiol 11(3):461–474
43. Iliopoulos D, Hirsch HA, Struhl K (2009) An epigenetic switch involving NF-κB, Lin28, Let-7 MicroRNA, and IL6 links inflammation to cell transformation. Cell 139(4):693–706
44. Zhang Y, Xi Y (2015) Fat body development and its function in energy storage and nutrient sensing in Drosophila melanogaster. J Tissue Sci Eng 6(1):1
45. Bharucha KN (2009) The epicurean fly: using Drosophila melanogaster to study metabolism. Pediatr Res 65(2):132
46. Hoshizaki DK (1994) Krüppel expression during postembryonic development of Drosophila. Dev Biol 163(1):133–140
47. Technau GM (1987) A single cell approach to problems of cell lineage and commitment during embryogenesis of Drosophila melanogaster. Development 100(1):1–12
48. Colombani J, Raisin S, Pantalacci S, Radimerski T, Montagne J, Léopold P (2003) A nutrient sensor mechanism controls Drosophila growth. Cell 114(6):739–749
49. Sousa-Nunes R, Yee LL, Gould AP (2011) Fat cells reactivate quiescent neuroblasts via TOR and glial insulin relays in Drosophila. Nature 471(7339):508
50. Géminard C, Rulifson EJ, Léopold P (2009) Remote control of insulin secretion by fat cells in Drosophila. Cell Metab 10(3):199–207

51. Arrese EL, Soulages JL (2010) Insect fat body: energy, metabolism, and regulation. Annu Rev Entomol 55:207–225
52. Baker KD, Thummel CS (2007) Diabetic larvae and obese flies—emerging studies of metabolism in Drosophila. Cell Metab 6 (4):257–266
53. Kawamura K, Shibata T, Saget O, Peel D, Bryant PJ (1999) A new family of growth factors produced by the fat body and active on Drosophila imaginal disc cells. Development 126 (2):211–219
54. Butterworth F, Emerson L, Rasch E (1988) Maturation and degeneration of the fat body in the Drosophila larva and pupa as revealed by morphometric analysis. Tissue Cell 20 (2):255–268
55. Aguila JR, Suszko J, Gibbs AG, Hoshizaki DK (2007) The role of larval fat cells in adult Drosophila melanogaster. J Exp Biol 210 (6):956–963
56. Athenstaedt K, Daum G (2006) The life cycle of neutral lipids: synthesis, storage and degradation. Cell Mol Life Sci CMLS 63 (12):1355–1369
57. Kühnlein RP (2011) The contribution of the Drosophila model to lipid droplet research. Prog Lipid Res 50(4):348–356
58. Brasaemle DL (2007) Thematic review series: adipocyte biology. The perilipin family of structural lipid droplet proteins: stabilization of lipid droplets and control of lipolysis. J Lipid Res 48 (12):2547–2559
59. Zimmermann R, Strauss JG, Haemmerle G, Schoiswohl G, Birner-Gruenberger R, Riederer M, Lass A, Neuberger G, Eisenhaber F, Hermetter A (2004) Fat mobilization in adipose tissue is promoted by adipose triglyceride lipase. Science 306 (5700):1383–1386
60. Zechner R, Strauss JG, Haemmerle G, Lass A, Zimmermann R (2005) Lipolysis: pathway under construction. Curr Opin Lipidol 16 (3):333–340
61. Palm W, Sampaio JL, Brankatschk M, Carvalho M, Mahmoud A, Shevchenko A, Eaton S (2012) Lipoproteins in Drosophila melanogaster—assembly, function, and influence on tissue lipid composition. PLoS Genet 8(7):e1002828
62. Beira JV, Paro R (2016) The legacy of Drosophila imaginal discs. Chromosoma 125 (4):573–592
63. Cohen S (1993) The development of Drosophila melanogaster, vol 2, ed Bate M, Arias AM. Cold Spring Harbor Press, New York
64. Ashe HL, Briscoe J (2006) The interpretation of morphogen gradients. Development 133 (3):385–394
65. Williams N (2010) Bugs in sight. Curr Biol 20 (10):R428–R429
66. Cohen B, Wimmer EA, Cohen SM (1991) Early development of leg and wing primordia in the Drosophila embryo. Mech Dev 33 (3):229–240
67. Robb JA (1969) Maintenance of imaginal discs of Drosophila melanogaster in chemically defined media. J Cell Biol 41(3):876–885
68. Aldaz S, Escudero LM, Freeman M (2010) Live imaging of Drosophila imaginal disc development. Proc Natl Acad Sci 107 (32):14217–14222
69. Repiso A, Bergantiños C, Corominas M, Serras F (2011) Tissue repair and regeneration in Drosophila imaginal discs. Develop Growth Differ 53(2):177–185
70. Karpen GH, Schubiger G (1981) Extensive regulatory capabilities of a Drosophila imaginal disk blastema. Nature 294(5843):744
71. Held LI Jr, Held LI Jr (2005) Imaginal discs: the genetic and cellular logic of pattern formation, vol 39. Cambridge University Press, Cambridge
72. García-Bellido A, Ripoll P, Morata G (1973) Developmental compartmentalisation of the wing disk of Drosophila. Nat New Biol 245 (147):251
73. Abbott LC, Karpen GH, Schubiger G (1981) Compartmental restrictions and blastema formation during pattern regulation in Drosophila imaginal leg discs. Dev Biol 87(1):64–75
74. Adams MD, Celniker SE, Holt RA, Evans CA, Gocayne JD, Amanatides PG, Scherer SE, Li PW, Hoskins RA, Galle RF (2000) The genome sequence of Drosophila melanogaster. Science 287(5461):2185–2195
75. Alexandre C, Baena-Lopez A, Vincent J-P (2014) Patterning and growth control by membrane-tethered wingless. Nature 505 (7482):180
76. Bergantiños C, Vilana X, Corominas M, Serras F (2010) Imaginal discs: renaissance of a model for regenerative biology. BioEssays 32 (3):207–217
77. Hadorn E (1968) Developmental capacity of embryonal blastema in Drosophila following cultivation in an adult host. Rev Suisse Zool 75:557–569
78. Slack J (2003) Regeneration research today. Dev Dyn 226(2):162–166
79. Bryant PJ (1975) Pattern formation in the imaginal wing disc of Drosophila melanogaster: fate map, regeneration and duplication. J Exp Zool 193(1):49–77

80. Schubiger G (1971) Regeneration, duplication and transdetermination in fragments of the leg disc of Drosophila melanogaster. Dev Biol 26(2):277–295
81. Usui K, Simpson P (2000) Cellular basis of the dynamic behavior of the imaginal thoracic discs during Drosophila metamorphosis. Dev Biol 225(1):13–25
82. Ogienko A, Fedorova S, Baricheva E (2007) Basic aspects of ovarian development in Drosophila melanogaster. Russ J Genet 43 (10):1120–1134
83. Eliazer S, Buszczak M (2011) Finding a niche: studies from the Drosophila ovary. Stem Cell Res Ther 2(6):45
84. Spradling A (1993) Developmental genetics of oogenesis. The development of Drosophila melanogaster
85. Decotto E, Spradling AC (2005) The Drosophila ovarian and testis stem cell niches: similar somatic stem cells and signals. Dev Cell 9 (4):501–510
86. Lin H, Spradling AC (1993) Germline stem cell division and egg chamber development in transplanted Drosophila germaria. Dev Biol 159(1):140–152
87. Cox DN, Chao A, Lin H (2000) Piwi encodes a nucleoplasmic factor whose activity modulates the number and division rate of germline stem cells. Development 127(3):503–514
88. King FJ, Szakmary A, Cox DN, Lin H (2001) Yb modulates the divisions of both germline and somatic stem cells through piwi-and hh-mediated mechanisms in the Drosophila ovary. Mol Cell 7(3):497–508
89. Kirilly D, Xie T (2007) The Drosophila ovary: an active stem cell community. Cell Res 17 (1):15
90. Mena Jr V (2012) Characterization of the stem cell niche in Drosophila testes
91. Papagiannouli F (2014) Male stem cell niche and spermatogenesis in the Drosophila testis—a tale of germline-soma communication. In: Adult stem cell niches. IntechOpen
92. Ueishi S, Shimizu H, Inoue YH (2009) Male germline stem cell division and spermatocyte growth require insulin signaling in Drosophila. Cell Struct Funct:0904160039–0904160039
93. Kiger AA, White-Cooper H, Fuller MT (2000) Somatic support cells restrict germline stem cell self-renewal and promote differentiation. Nature 407(6805):750
94. Schulz C, Wood CG, Jones DL, Tazuke SI, Fuller MT (2002) Signaling from germ cells mediated by the rhomboid homolog stet organizes encapsulation by somatic support cells. Development 129(19):4523–4534
95. de Cuevas M, Matunis EL (2011) The stem cell niche: lessons from the Drosophila testis. Development 138(14):2861–2869
96. Wilk R, Murthy SU, Yan H, Krause HM (2010) In situ hybridization: fruit fly embryos and tissues. Current Protocols Essential Laboratory Techniques 4(1):9.3. 1-9.3. 24

Chapter 3

Hemolymph Analysis of *Drosophila melanogaster* by Giemsa Staining

Janmejaya Bag and Monalisa Mishra

Abstract

Drosophila hemolymph is analogous to vertebrate blood. It also provides an innate immune system which possesses three different types of cells. The cells stain with Giemsa solution which is composed of Azure B, eosin and methylene blue. This stain is usually used to stain the blood samples to identify the various blood cell types. Mostly it stains the G-banding of chromosomes, where it strongly binds with the phosphate group of DNA. In human blood cells, erythrocytes stain pink, monocyte cytoplasm stains pale blue, platelets stain light pale pink, leukocytes and nuclear chromatin stain magenta and lymphocyte cytoplasm stains sky blue. In *Drosophila* hemolymph, three types of cells are found. These are plasmatocytes, crystallocytes and lamellocytes. Where plasmatocytes stain with light pink stain, crystal cells stain dark blue, and lamellocytes stain light cytoplasm with dark nuclei. The current study explains the analysis of haemocyte of *Drosophila*.

Key words Giemsa staining, Hemolymph, Crystallocyte, Lymphocyte, Plasmatocyte

1 Introduction

Drosophila melanogaster is a widely used model organism in the field of neuroscience, genetics and developmental biology. Seventy-five percent of the human disease genes share homology with *Drosophila* offering it as a model to study various human diseases. Besides gene various amino acids (glutamine, taurine and glutamate) which serve as neurotransmitters are also conserved in *Drosophila*. More importantly, neurotransmitters associated neurodegenerative disorder of vertebrates are present in *Drosophila* [1, 2]. Several blood disorders are studied by using a fly as a model [3]. The disease-causing genes responsible for acute myeloid leukaemia are expressed in flies.

Drosophila possesses hemolymph, the analogous of vertebrate's blood. Like blood it comes in direct contact with the tissue during circulation. However, it is a colourless transparent fluid with many haemocytes. These haemocytes play an important role in the

Monalisa Mishra (ed.), *Fundamental Approaches to Screen Abnormalities in Drosophila*, Springer Protocols Handbooks, https://doi.org/10.1007/978-1-4939-9756-5_3, © Springer Science+Business Media, LLC, part of Springer Nature 2020

defence mechanism of the *Drosophila*; thus its number rises or falls depending on the physiological requirement of the body [4–6]. Hemolymph contains haemocyanin as the pigment responsible for carrying the oxygen. It is a protein which contains copper and becomes blue when it comes in contact with the oxygen. Hemolymph contains water, amino acid, protein, carbohydrate and lipid like human blood. It also contains many ions like calcium, chlorine, sodium and potassium for various physiological activities [7].

1.1 Blood as an Immune System

Haematopoiesis process is conserved from *Drosophila* to vertebrate [8]. Blood helps to (1) maintain the innate immune system and (2) distribute haemocytes to various organs during different developmental processes [9]. Haemocytes help in the elimination of pathogens from the body [10]. However, *Drosophila* does not possess an adaptive immune system like vertebrates. It has only an innate immune system [11] which includes humoral- and cell-mediated immune system.

Drosophila developmental cycle includes embryonic stage, larval stage, pupal stage and later adult stage. The types of blood cells vary in different developmental stages. In the embryonic stage, the circulatory body fluids contain only crystal cells and plasmatocytes, and this is the first part of haematopoiesis [12]. In the larval stage, the lymph glands comprise the haematopoietic progenitors that will produce three types of blood cells later [13]. Plasmatocytes develop first, which are the homologous structure of monocytes [14]. Ninety percent of the cells present in the hemolymph are plasmatocytes. Next, crystal cells develop, which is responsible for melanization and homologous structure of platelets of a human being [15]. Lastly, lamellocytes developed to engulf large foreign particles [16] and share functional homology with eosinophils [11]. Lamellocytes are thin and flat cells, found in the larval stage and absent in the adult stage. Adult *Drosophila* possesses secretory, crystal cells, lamellocytes and plasmatocytes. Secretory cells are the specialized cells, which never secreted to the hemolymph; else it sends a signal for the cell differentiation. Posterior signalling cetre (PSCs) mediate the haematopoietic activity of the fly. In adults, lymphoid gland possesses a pair of primary and several pairs of small secondary lobes that have some pro-haemocytes [17]. The anterior-most primary lobes have three zones: (1) cortical zone (CZ), (2) medullary zone (MZ) and (3) posterior signalling centre (PSC) [13]. PSCs are involved in Notch, JAK-STAT and Hedgehog (Hh) signalling.

Like human blood, various parameters can be detected from the hemolymph to know the physiological status of the animal. Various amino acids (indicator of monoamine) percentage can be detected from the hemolymph to know the neurochemistry of the brain. Nitrogen metabolite waste like urea, uric acids and ammonia can be determined from the blood. Glucose, trehalose, protein and lipid present in the hemolymph indicate the metabolism of the animal. Thus, hemolymph is an indispensable tissue to analyse the physiological status of the animal.

Various antioxidant levels can be measured from the hemolymph. ROS which itself act as a signalling molecule can be measured from the hemolymph. Metabolites for the glucose, trehalose and protein can be quantified from haemocytes. Consequently, any type of defect in the haemocyte can be detected from the hemolymph [4]. Type of haemocytes present in the hemolymph can be quantified using the Giemsa stain. Giemsa is a solution mixture of Azure, eosin and methylene blue. Giemsa binds to the phosphate group of DNA. Human blood cells stained as pink for erythrocytes, sky blue for lymphocytes, pale blue for monocytes and the cytoplasm, magenta colour for leucocyte and the platelets stained as pale pink [18–20]. Similarly, *Drosophila* larvae's hemolymph shows various shape, size and colour with the Giemsa stain [21]. Crystallocytes stained as pale pink and large round in size, lamellocytes stained fade blue and flat in shape and plasmatocytes are stained as light pink and appear small and round in shape [22, 23]. By counting the number and type of hemocytes within the body, the physiological state of the body can be known. Therefore, *Drosophila* hemolymph screening is the best way to determine the cellular responses and the genes which are regulated for the cellular and humoral responses of the body.

2 Materials

2.1 Hemolymph Collection

1. Petri plate (90 mm diameter × 15 mm width)
2. 1× PBS
3. 0.5 and 1.5 mL Eppendorf tube
4. MilliQ water
5. Fine needle/pins (Soccerene Premium Quality pin size 1.40 mm long × 0.35 mm diameter)
6. Seventy percent ethanol
7. Ice-box
8. Centrifuge (ThermoFisher SCIENTIFIC, Heraeus Megafuge 16R)
9. −20 °C freezer

2.2 GIEMSA Stain

1. Sterilized pricking needle (Soccerene Premium Quality pin size 1.40 mm long × 0.35 mm diameter)
2. 1× PBS
3. *Drosophila* third instar larvae.
4. Grease-free slide (ABDOS, G10101 25 × 75 mm)

5. Giemsa staining solution (Nova Biotech, S-121)
6. Compound microscope

3 Methods

3.1 Giemsa Staining of isolated hemolymph

1. Take out 20–25 third instar larvae from the rearing media, and rinse with 1× PBS.
2. Dry them with tissue paper and transfer them to a 0.5 mL centrifuge tube.
3. Make a small cut at the bottom of the centrifuge tube with a fine razor blade (*see* **Note 1**).
4. Put the 0.5 mL centrifuge tube into a 1.5 mL centrifuge tube and transfer it into the ice-box.
5. Prick the larvae with the fine pins carefully. Use 70% of ethanol to clean the pins (*see* **Notes 2 and 3**).
6. After that centrifuge in 10,000 rpm for 5 min at 4 °C.
7. Collect the hemolymph and store it in −20 °C for further use (*see* **Note 6**) (Fig. 1).
8. To prevent melanization perform the hemolymph collection on ice (*see* **Note 7**).
9. Take 10 μL of hemolymph on a clean glass slide and add 10 μL of 1× cold PBS to it.
10. Make a smear with another glass slide by keeping at an angle of 45° with the first slide.

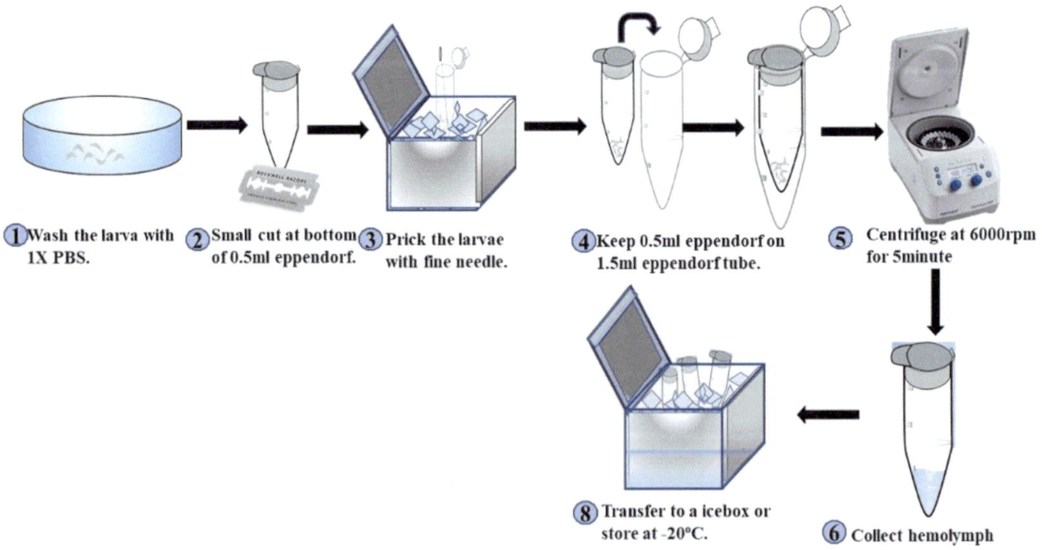

Fig. 1 Collection of hemolymph from third instar larvae

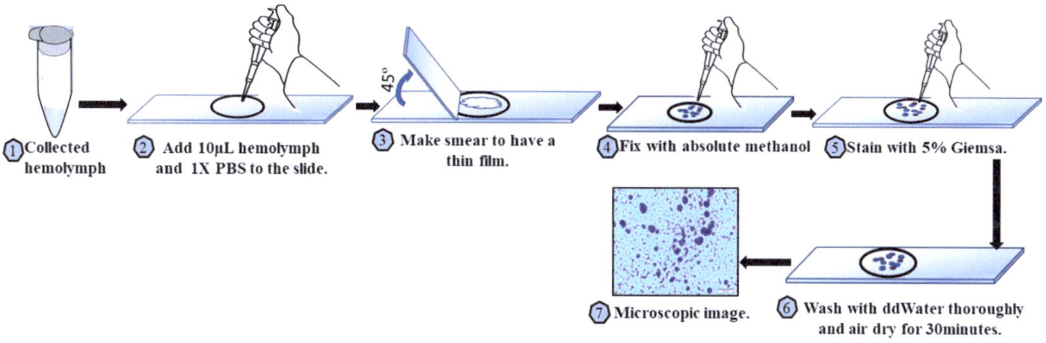

Fig. 2 Giemsa staining procedures from collected hemolymph

11. Air-dry for 10 min and fix it with absolute methanol for another 5 min.
12. For the preparation of 5% Giemsa, dilute the Giemsa's staining solution either with milliQ water (0.5:10) or with 1× PBS.
13. Add 5% Giemsa for 2 min.
14. Wash the slide with distilled water and air-dry for 10 min (*see* **Notes 4 and 5**).
15. Examine under compound microscope at 400× magnification (Fig. 2).

3.2 Hemolymph Giemsa Stain by bleeding of larvae

1. Preparation of 1× PBS: From the stock solution of 10× PBS, dilute it ten times with milliQ water to make it 1× PBS. For 100 ml of 1× PBS, take 10 ml of 10× PBS and dilute it with 90 mL of milliQ water.
2. Collect five third instar larvae from the rearing media.
3. Wash the larvae with 1× PBS thoroughly.
4. Mark a circle at the centre of a clean slide and put three to four larvae on that.
5. Take a fine pin to prick the larvae and add 10 μL of cold PBS to the hemolymph on the slide (*see* **Notes 1–3**).
6. Make smear as mentioned above in Sect. 3.1.
7. Air-dry the slide for 15 min and then fix it with methanol for 10 min.
8. After that stain with 5% Giemsa and air-dry for 2 min.
9. Then wash the stained slides in distilled water and air-dry for 15 min.
10. Examine under the compound microscope at 400× magnification (Fig. 3).

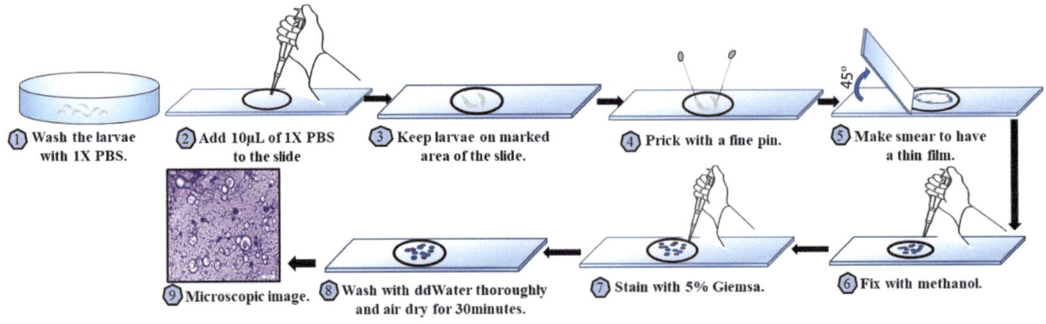

Fig. 3 Giemsa staining of hemolymph of third instar larva

Table 1
% count for various hemocytes

Sample	% plasmatocytes	% crystallocytes	% lamellocytes
S1	36	25.54	37.57
S2	32.67	38.08	29.23
S3	38.34	23.54	36.93
S4	32	34.56	33
S5	35.45	34.65	29
S6	32.5	37.94	29.56

3.3 Cell Counting

1. Count six slides from each treatment group under a bright field microscope.
2. If the microscope is attached with a camera, take 20–30 images from each slide, at the same magnification.
3. Randomly take 100 cells from each slide.
4. Identify and count the different types of cells present in those 100 cells (Table 1).
5. Repeat Steps 2–4 for all the six slides (S1–S6).
6. Calculate the standard mean value of each type of cell for all the six slides, and plot the graph (Fig. 4).

Total no. of cells = (S1 + S2 … + S6) until you reach 600.

Calculate the % of cells present in the total cell population as shown in Table 1.

You can put the values in the graph as well.

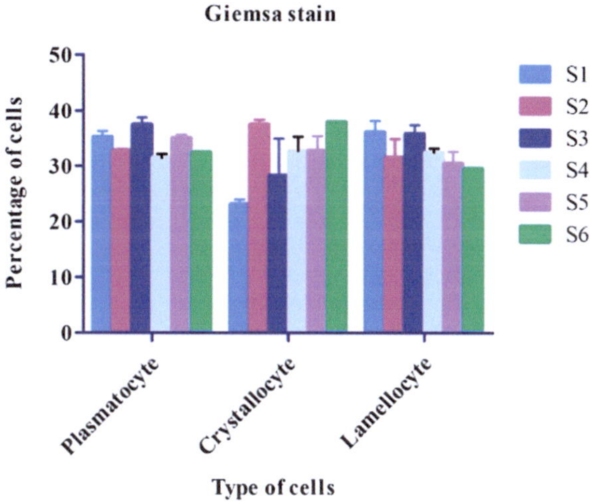

Fig. 4 Sample graph showing the percentage of different blood cells in control and treated flies

4 Notes

1. During hemolymph collection, avoid the mixing of dust particles; the glass slide should be wiped with methanol before the experiment.
2. Use the fine and clean needle/razor for pricking the larvae.
3. While pricking, do not break the other parts of the body.
4. Washing and drying of the sample must be under observation; avoid overwashing and overdrying.
5. Always use the gloves.
6. Hemolymph collection should be performed in the ice box.
7. If your sample undergoes melanization, use CO_2 to prevent it.
8. If there is an issue of aggregation, then dilute the hemolymph with anticoagulant buffer (41 mM citric acid, 98 mM NaOH, 186 mM NaCl, 1.7 mM EDTA, pH 4.5).

Acknowledgements

JB is thankful to BT/PR21857/NNT/28/1238/2017 for financial support. MM lab is supported by Grant No. BT/PR21857/NNT/28/1238/2017, EMR/2017/003054, Odisha DBT 3325/ST(BIO)-02/2017.

References

1. Shank R, Aprison M (1981) Present status and significance of the glutamine cycle in neural tissues. Life Sci 28(8):837–842
2. Piyankarage SC, Augustin H, Grosjean Y, Featherstone DE, Shippy SA (2008) Hemolymph amino acid analysis of individual Drosophila larvae. Anal Chem 80(4):1201–1207
3. Sorrentino RP, Melk JP, Govind S (2004) Genetic analysis of contributions of dorsal group and JAK-Stat92E pathway genes to larval hemocyte concentration and the egg encapsulation response in Drosophila. Genetics 166 (3):1343–1356
4. Ratcliffe N (1979) Role of hemocytes in defense against biological agents. Insect hemocytes: development, forms, functions, and techniques, pp 331–414
5. Hultmark D (2003) Drosophila immunity: paths and patterns. Curr Opin Immunol 15 (1):12–19
6. Hoffmann JA (2003) The immune response of Drosophila. Nature 426(6962):33
7. Wyatt GR (1961) The biochemistry of insect hemolymph. Annu Rev Entomol 6(1):75–102
8. Choma MA, Suter MJ, Vakoc BJ, Bouma BE, Tearney GJ (2011) Physiological homology between Drosophila melanogaster and vertebrate cardiovascular systems. Dis Model Mech 4(3):411–420
9. Franc NC, Dimarcq J-L, Lagueux M, Hoffmann J, Ezekowitz RAB (1996) Croquemort, a novel Drosophila hemocyte/macrophage receptor that recognizes apoptotic cells. Immunity 4(5):431–443
10. Hetru C, Troxler L, Hoffmann JA (2003) *Drosophila melanogaster* antimicrobial defense. J Infect Dis 187(Supplement_2):S327–S334
11. Sackton TB, Lazzaro BP, Schlenke TA, Evans JD, Hultmark D, Clark AG (2007) Dynamic evolution of the innate immune system in Drosophila. Nat Genet 39(12):1461
12. Makhijani K, Brückner K (2012) Of blood cells and the nervous system: hematopoiesis in the Drosophila larva. Fly 6(4):254–260
13. Jung S-H, Evans CJ, Uemura C, Banerjee U (2005) The Drosophila lymph gland as a developmental model of hematopoiesis. Development 132(11):2521–2533
14. Pearson AM, Baksa K, Rämet M, Protas M, McKee M, Brown D, Ezekowitz RAB (2003) Identification of cytoskeletal regulatory proteins required for efficient phagocytosis in Drosophila. Microbes Infect 5(10):815–824
15. Holz A, Bossinger B, Strasser T, Janning W, Klapper R (2003) The two origins of hemocytes in Drosophila. Development 130 (20):4955–4962
16. Lanot R, Zachary D, Holder F, Meister M (2001) Postembryonic hematopoiesis in Drosophila. Dev Biol 230(2):243–257
17. Márkus R, Laurinyecz B, Kurucz É, Honti V, Bajusz I, Sipos B, Somogyi K, Kronhamn J, Hultmark D, Andó I (2009) Sessile hemocytes as a hematopoietic compartment in *Drosophila melanogaster*. Proc Natl Acad Sci 106 (12):4805–4809
18. Perry P, Wolff S (1974) New Giemsa method for the differential staining of sister chromatids. Nature 251(5471):156
19. Fischer AH, Jacobson KA, Rose J, Zeller R (2008) Hematoxylin and eosin staining of tissue and cell sections. Cold Spring Harb Protoc 2008(5):pdb. prot4986
20. van Furth R, Raeburn JA, van Zwet TL (1979) Characteristics of human mononuclear phagocytes. Blood 54(2):485–500
21. Silva J, Boleli I, Simões Z (2002) Hemocyte types and total and differential counts in unparasitized and parasitized *Anastrepha obliqua* (Diptera, Tephritidae) larvae. Braz J Biol 62 (4A):689–699
22. Meister M, Lagueux M (2003) Drosophila blood cells. Cell Microbiol 5(9):573–580
23. Meister M (2004) Blood cells of Drosophila: cell lineages and role in host defence. Curr Opin Immunol 16(1):10–15

Chapter 4

Histological Analysis of the *Drosophila* Head with Special Reference to the Eye and Brain

Puja Karmakar and Monalisa Mishra

Abstract

Drosophila brain and eye are a well-known model to analyse the effect of different genes and environmental factors. Several genes associated with neurodegenerative diseases are conserved between *Drosophila* and human being. Thus, structural and functional abnormality due to loss of function of these genes can be studied in the brain and eye of *Drosophila*. The structural alteration includes a change in shape, size, number and arrangement of the photoreceptor cells within the eye. All these gross structural defects of the fly eye and brain can be easily visualized under a light microscope with proper fixation, sectioning and staining. The current chapter describes the dissection, fixation, sectioning, staining and mounting of the *Drosophila* head to detect the structural defect of eye and brain under a light microscope.

Key words Compound eye, Histology, Paraffin section, Haematoxylin, Eosin staining

1 Introduction

Drosophila possesses compound eye on either side of the head. The eye develops from the eye-antenna imaginal disc of the third instar larvae.The compound eye possesses approximately 800 functional units called ommatidia. The ommatidia are arranged across the compound eye to give an oval-shaped structure [1]. In between the ommatidia, mechanosensory bristles are present. Each ommatidium consists of four crystalline cone cells, eight photoreceptor cells, one pair of primary (1°) pigment cells, six secondary (2°) pigment cells and three tertiary (3°) pigment cells [2]. The photoreceptors (R1–R8) are arranged to form a trapezoid, and they are arranged in two different arrays.The photoreceptors (R1–R6) form the peripheral layer and the photoreceptor R7 and R8 forms the central layer. The apical part of the photoreceptor contains finger-like projections known as microvillus. There are 80,000 microvilli in each ommatidium. Each microvillus houses light-sensing pigments known as opsin.

Monalisa Mishra (ed.), *Fundamental Approaches to Screen Abnormalities in Drosophila*, Springer Protocols Handbooks, https://doi.org/10.1007/978-1-4939-9756-5_4, © Springer Science+Business Media, LLC, part of Springer Nature 2020

The eye development begins from the late third instar larval eye imaginal disc. During development, the photoreceptor R8 is first recruited followed by recruitment of other photoreceptors. After the recruitment of eight photoreceptors, four cone cells (CC) are recruited. The cone cells are named as the anterior/posterior (a/p) pair and the equatorial/polar (eq/pl) pair. Both the pair of cone cells form the quartet. The cone cells secret the cornea. After puparium formation (APF), both the primary pigment cells (1°PCs) are recruited to enwrap the CC quartet. This step results in the formation of PR/CC/1°PC core. Next step is the formation of a hexagonal ring of interommatidial cells (IOCS). The IOCS is comprised of 2°PCs, 3°PCs and bristle cells. The cells which remain unrecruited are eliminated by programmed cell death [3]. Ras and Notch signalling are responsible for patterning of pigment cells which form the organized lattice in the fly eye [4]. Any abnormality in the recruitment of cells during the developmental process results in a malformed eye. The neurons of the ommatidia pass through the basal lamina and go to the brain.

The fly neurons possess 50 neuropils; each has a cell body. Out of 50 neuropils, 4 contributes to the formation of the optic lobe. The optic lobe comprised of the outer lamina, middle medulla and the inner lobula complex. The lobula is also divided into the anterior and posterior region and is known as lobula and the lobula plate, respectively [5]. The lamina of each ommatidium is associated with motion vision [6]. The medulla receives direct connections from colour photoreceptors and also from lamina neurons [7].

Besides all these structures, cellular junctions also has a role in maintaining the structure of the eye. The cornea, cone cell and rhabdom are connected both apically and basally by means of septate junctions [8]. Thus, the role of the septate junction in regulating the structure of the eye needs further investigation. Many factors such as gene and the environment may affect the cone cell formation resulting in a defective eye and brain in *Drosophila*. We are describing some of the factors below.

1.1 Factors Affecting the Eye and Brain of Drosophila

Drosophila eye is a complex organ, and various intrinsic as well as extrinsic factors can alter the development. The key intrinsic pathways which regulate eye development of *Drosophila* are Ras and Notch. These two pathways are conserved from invertebrate to vertebrate. However, the direct target of Ras and Notch signalling is Prospero and dPax2. These two transcription factors act on cone cells, which is essential for the secretion of the cornea, the lens of the eye [9, 10]. Cone cells also thought to regulate the morphogenesis of photoreceptor cells. Apical contact switching between both the CC pairs reorients the apical surface of the photoreceptor by 90° [11]. This step guides the photoreceptors to expand perpendicularly to the apical-basal axis of the retina [11]. There are many mutants where the rhabdomere expand only half of the retina

[12–14]. Besides the length, in some mutants the rhabdomere also changes its shape and produces spaghetti-shaped rhabdom [12–14].

Different environmental and genetic factors can alter the structure of the *Drosophila* brain. Alteration of these factors resulted many neurodegenerative diseases. Fly brain and eye development are affected by a number of environmental factors including physicochemical stress and social interactions [15]. Heat stress affects the volume of the mushroom bodies resulting in a reduced mushroom body [15]. The reduced size of the mushroom body affects and reduces the sensation of odour [15]. High larval crowding further reduces the volume of the optic lobes, central complex and antennal lobes [15]. Altogether these results suggest that with the response to stressors, the structure of the brain alters. By sectioning of brain tissue, we can see the alteration of such types of changes.

Many genes associated with the neurodegenerative disease are conserved in *Drosophila*. Parkinson's disease is due to the reduction of dopaminergic neurons in the substantia niagra [16]. The sole gene responsible for Parkinson's disease is *α-synuclein* which is present in *Drosophila*. Thus, *Drosophila* is used to understand the mechanism of neurodegeneration in Parkinson's patients [17]. Alzheimer's disease, another neurodegenerative disease, is caused by the accumulation of extracellular β-amyloid and intracellular Tau-containing neurofibrillary tangles [18]. Alzheimer's disease is also modelled in *Drosophila* [19]. With the sectioning of the brain, the morphological abnormalities can be checked [20]. Histology is a less expensive and easy technique to find out the structural abnormalities in the brain and eye of *Drosophila* using a light microscope. The current chapter describes the sectioning of the *Drosophila* head to analyse it under the light microscope.

2 Materials

2.1 Dissection and Fixation of the Dissected Head

1. Ethyl ether (SRL-64665(052943)
2. Cotton swabs
3. Zero-sized paint brush
4. Petri plate (85 × 13 mm)
5. 1 × Phosphate buffer saline (PBS)
6. Stereomicroscope
7. Forceps
8. Razor blade
9. 5 ml Storage vials (Tarsons-523070)
10. 4% PFA
11. 1 × PBST

12. MilliQ water/distilled water
13. Ethanol (Merck-K47072383539)
14. Xylene (Merck-IA51640368)

2.2 Embedding and Trimming

1. Forcep
2. Beaker
3. Small porcelain cup (CoorsTek)
4. Scalpel (razor blade)
5. Handmade paper boats/moulds
6. Spirit lamp
7. Paraffin wax (52–60 °C melting point)
8. Stereomicroscope
9. Oven

2.3 Sectioning and Affixation

1. Scalpel
2. A rotary microtome (York Scientific Industries)
3. Adhesive (egg albumin) or gelatin
4. Acetone
5. Cleaned slides of 25 × 75 mm (Abdos)
6. MilliQ water
7. Hot plate (Tarsons)
8. Triangle-shaped cut Whatman 40 filter paper

2.4 Staining

1. Haematoxylin stain (Solarbio Life Science)
2. Eosin stain (HIMEDIA-GRM115-10G)
3. Xylene (Merck-IA51640368)
4. Alcohol/ Ethanol (Merck-K47072383539)
5. Coupling jar (Abdos-P90101)

2.5 Mounting

1. DPX (SIGMA- 06522-500 μl)
2. 18 mm cover slip (Blue star)

3 Methods

3.1 Dissection and Fixation of the Dissected Head

1. Anaesthetize 25 number of flies by using diethyl ether through the cotton swabs.
2. Wash the flies with 1 × PBS in a petriplate for removal of unnecessary food particles from their body with the help of a zero-sized brush.

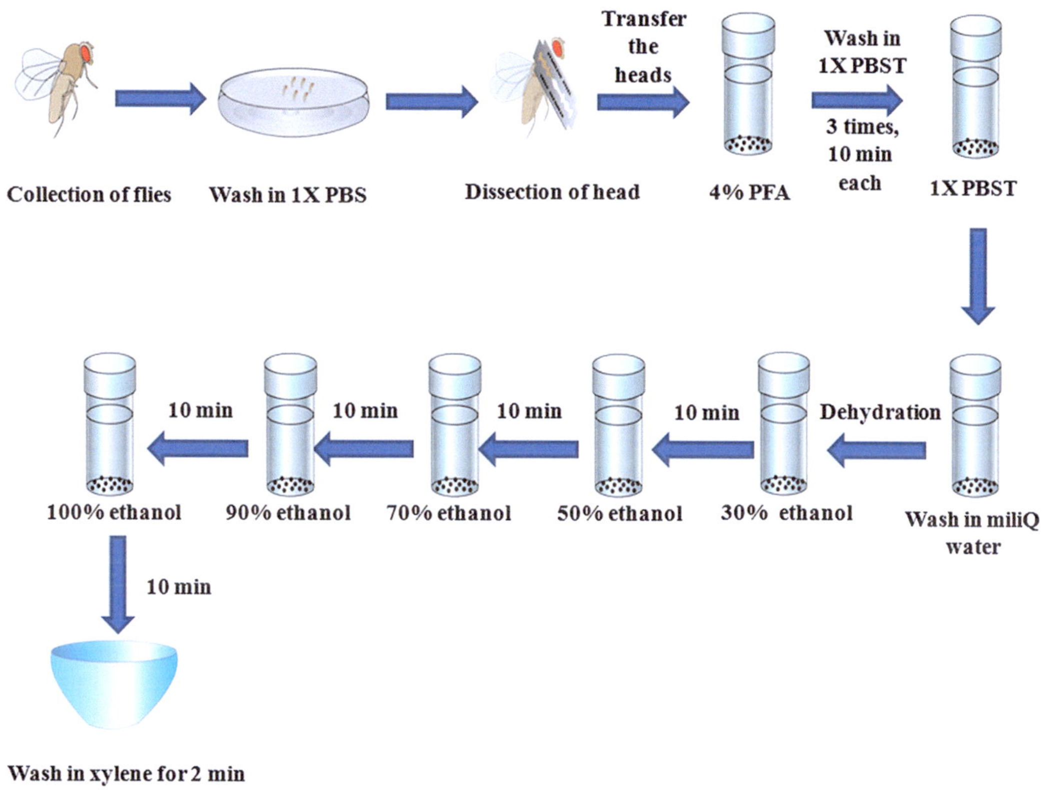

Fig. 1 Dissection and fixation of the head tissue of *Drosophila melanogaster*

3. Put the flies under a stereomicroscope. Hold the thorax of the fly with the help of a forceps and dissect the head with the help of a razor blade.
4. Transfer the required number of dissected heads into a 5 ml glass vial with 4% PFA (*see* **Note 4.4.1.2**).
5. Wash the heads in 2–4 ml of 1X PBST for three times, 10 min each.
6. Discard the PBST and rinse the heads with 2–4 ml of milliQ water two times, 10 min each wash.
7. Dehydrate the sample with a series of ethanol (30%, 50%, 70%, 90% and 100%) by keeping for 10 min in each concentration.
8. Then dip the heads in xylene for 1–2 min for clearing the tissues (Fig. 1).

3.2 Embedding and Trimming

1. Keep the paraffin wax on incubator at 52–60 °C prior to starting of the dehydration procedure (*see* **Note 4.4.2.1).**
2. Melted paraffin is now ready to use.

3. Prepare two sets of xylene and paraffin mixture in a ratio of 3:1 and 2:2 in porcelain cups, and keep it in the oven for approximately 30 min whose temperature must be maintained at 52–60 °C (*see* **Note 4.4.2.2**).
4. Now put the dehydrated heads in 3:1 solution for 30 min followed by 2:2 for 2 h.
5. Transfer the heads to the pure melted wax. Repeat this step three times in every 1 h interval.
6. Prepare the paper boats and fill it with melted paraffin very carefully (*see* **Note 4.4.2.4**).
7. Before solidification of the upper layer, transfer heads carefully to the centre of the upper layer of a paper boat.
8. Change the orientation of samples with the help of a warm needle (*see* **Note 4.4.2.5**).
9. Mount the heads in a way that the eyes are facing upward.
10. Align the heads six in each row. Mount the heads in the form of a rectangle.
11. Keep the heads for half an hour for proper solidification of the samples.
12. Store the block at 4 °C till next use.
13. Trim the solid paraffin block with a razor blade to form a rectangle or square. Trim closer to the sample so that there will be less paraffin surrounding it (Fig. 2).

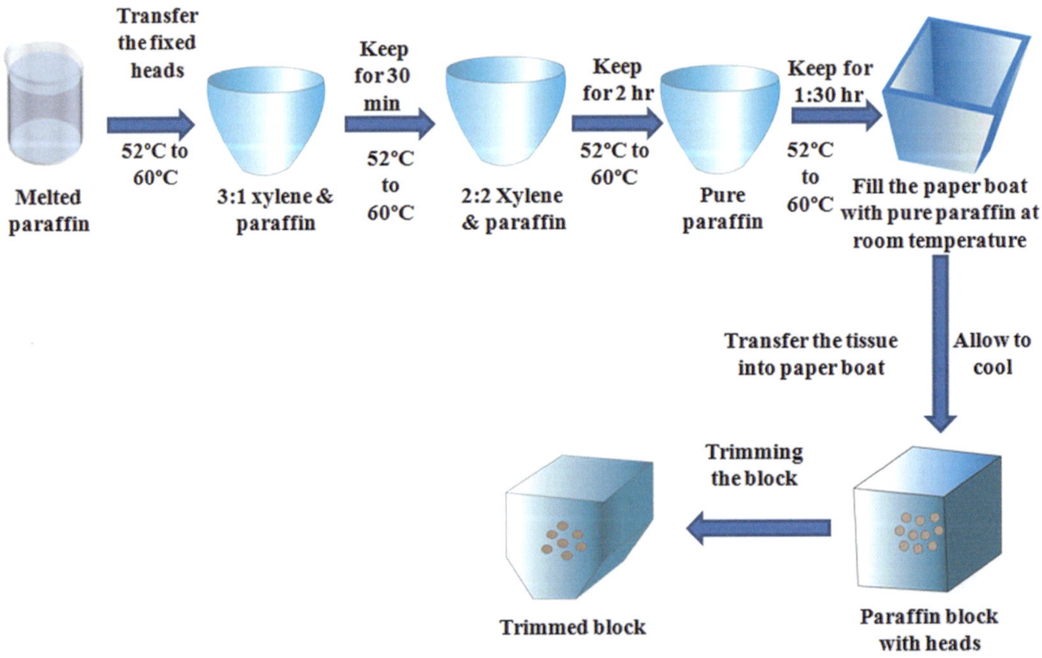

Fig. 2 Embedding of the head tissue into the paraffin and trimming of the paraffin block

3.3 Sectioning and Affixation

1. Fix the trimmed paraffin block containing tissue to the centre of a block holder.
2. Fit the block holder into the object holder of the microtome (*see* **Note 4.4.3.1).**
3. Fit a sharp razor in the knife carriage in an angle of 45° by adjusting the screws.
4. Move the knife including its carriage towards the tissue blocks to touch the block during the downward movement.
5. Cut 15–20-μm-thick sections until the tissue section appears like a ribbon (*see* **Note 4.4.3.5).**
6. With the best fixation, 5–6 μm sections can also be obtained.
7. Move the drive wheel clockwise to get a ribbon.
8. Carefully remove the long continuous ribbon with the help of a forcep.
9. Before elongation of the ribbon, clean the slides properly by using acetone and mark properly before placing the ribbon.
10. Take a small drop of an adhesive like egg albumin on the top of the slide, and rub on the surface with the clean fingertip.
11. Heat the slide over a flame to burn the albumin. (Steps 10 and 11 can be replaced by using gelatin-coated slides or commercially available coated slides.)
12. Put a few drops of floating medium (milliQ water) on the surface of the slide.
13. Take out a small piece of tissue ribbon carefully with the help of a forecep and place it over the floating medium of the slide.
14. Transfer the slide to a warm hot plate which is enough to stretch the paraffin (*see* **Note 4.4.3.6**).
15. Arrange the sections properly and remove the floating medium carefully.
16. Dry the slides in the air by placing it in an inclined position or by using triangle shaped Whatman −40 filter paper. Keep tissues for air-drying overnight for proper attachment of tissue sections to the slide (Fig. 3).

3.4 Staining

Stain the slides obtained from previous section with haematoxylin-eosin. Haematoxylin stains the nucleolus of the cells and makes the cell appear blue. Eosin turns the cytoplasmic components into pink, red or orange. These two stains give a complete structure of the tissue. For the staining, the following steps will be performed:

1. Deparaffinize the tissue section by dipping the slides in xylene-containing coupling jar for 1 min.
2. Rehydrate the slides through graded alcohols (absolute, 90%, 70%, 50%, 30%, respectively), 10 min each.

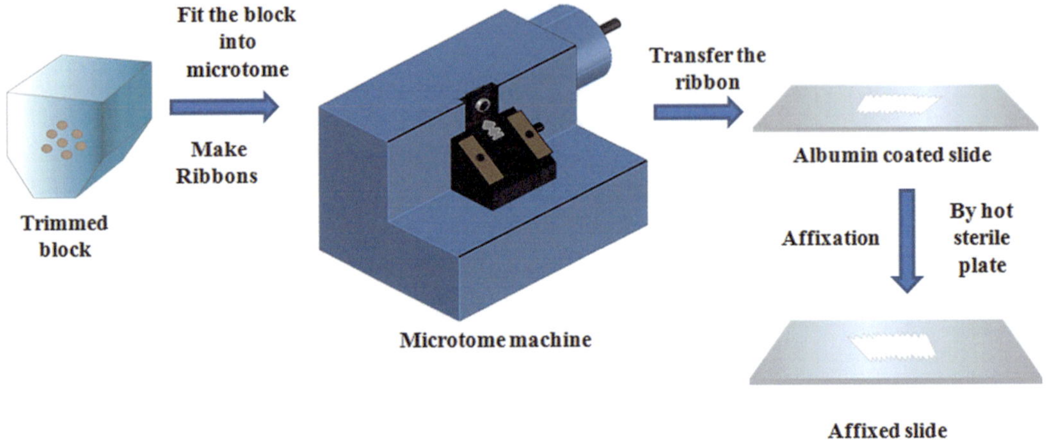

Fig. 3 Sectioning of the paraffin block by using microtome machine and then affixing the ribbon section into the albumin-coated slide

3. Stain the slides in haematoxylin for 5 min.
4. Then wash the slides in water for 2 min twice. If the stain is excess, wash with running water for a few minutes (*see* **Note 4.4.4**).
5. Transfer the slides to upgraded alcohol concentrations (30%, 50%, 70% and 90%) for 10 min each.
6. Counterstain the slides with eosin for 1 min. Wash the excess eosin through 90% alcohol.
7. Dehydrate completely in absolute alcohol for 5–15 min and clear in xylene for few seconds (Fig. 4).

3.5 Mounting

1. Immediately after removing from xylene, mount the stained slides with the mounting media DPX.
2. Put a drop of DPX on the slides to eliminate the chances of the sample being dried. Hold the coverslip at 45° angle. Gently place it over the slide and allow the medium to spread slowly (*see* **Note 4.4.5**).
3. Observe the slide under the light microscope (Fig. 5).

4 Notes

4.1 Dissection and Fixation of the Fly Head

1. Do not touch and damage the intact eye which will later be sectioned.
2. Carefully transfer the fly heads from one storage vials to another, and keep the tissue in vials as the time mentioned in the protocol.

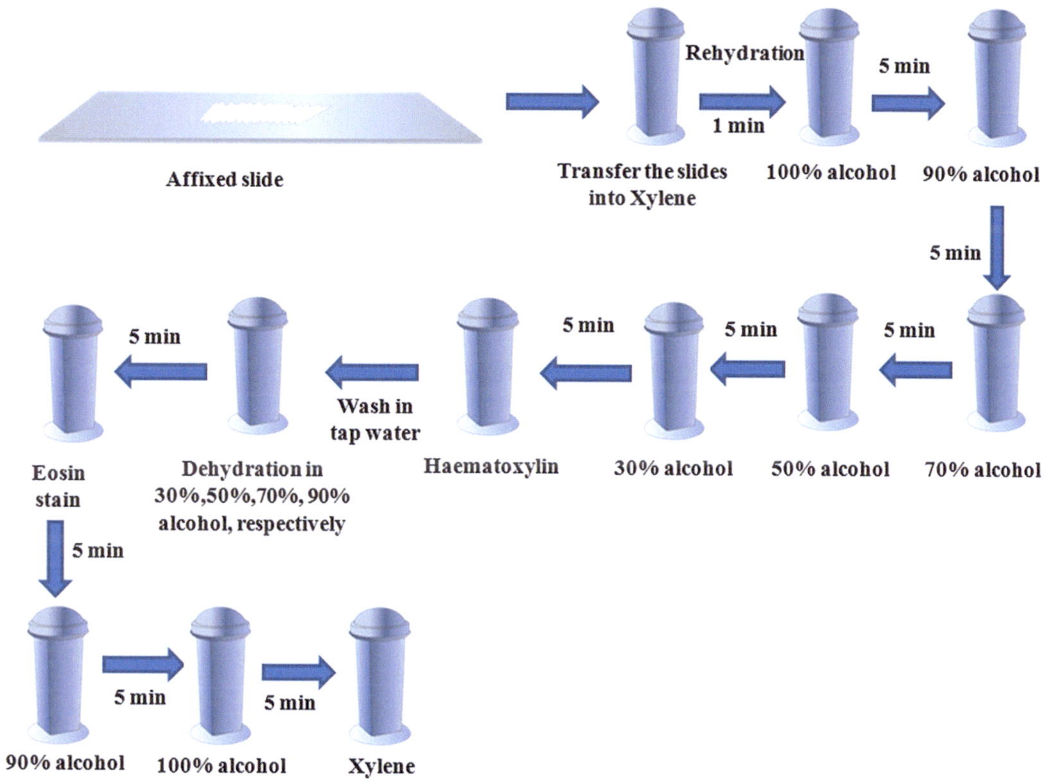

Fig. 4 Staining of the tissue section

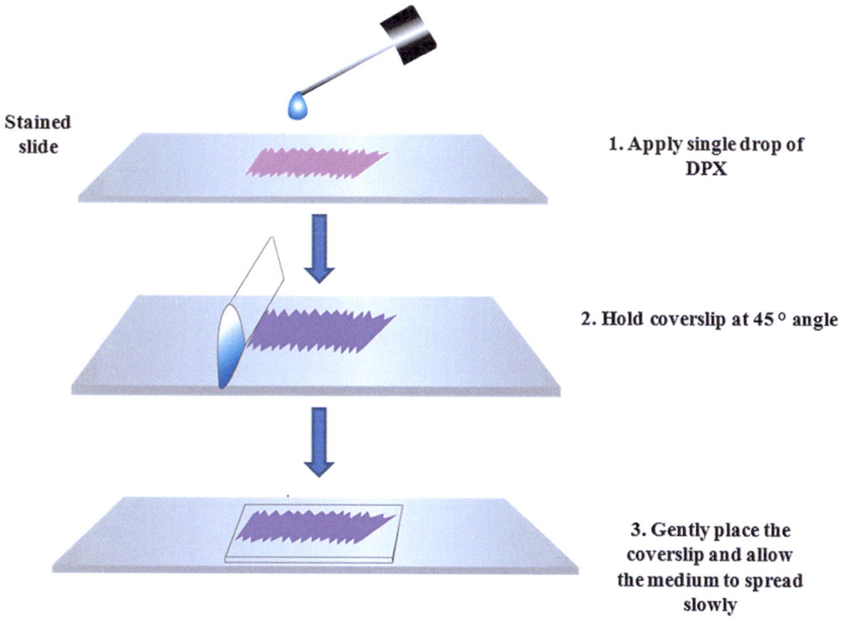

Fig. 5 Mounting of the stained slides

4.2 Embedding and Trimming

1. Do not over heat your sample while keeping it in wax in the oven. This will result in the powder while sectioning.
2. Do not keep the sample in xylene for very long time; this will make the sample black.
3. Any altered steps in the dehydration will not allow ribbon formation.
4. Fill the paper boat very carefully with paraffin to avoid the formation of air bubbles.
5. Apply warm scalpel to prevent the cooling of the upper surface.
6. Align all heads appropriately and then remove the needle very carefully from the paraffin.
7. If there will be any misalignment of heads, remove the head with the dissecting needle, and then position the head as desired.
8. Do the steps quickly because paraffin wax has the property to solidify in room temperature.
9. Trim the block properly since it helps to form the proper ribbon.

4.3 Sectioning and Affixation

1. All screws of microtome should be checked and tightened properly before moving the drive wheel.
2. Razor angle must be correct. Paraffin must not be very hard otherwise proper ribbon formation will not occur.
3. Razor must be cleaned and sharp to avoid curved ribbon formation.
4. Edge of the block must be uniform and parallel for getting a proper ribbon.
5. Any kind of grit must not be present in a block; otherwise, splitted ribbons may form.
6. At the time of affixation, carefully maintain the temperature of the hot plate.

4.4 Staining

Do not keep the slides in the staining solution for a long time. This will result in a strong stain and the proper structure will not be visible.

4.5 Mounting

Cover the slide very carefully to avoid bubble formation, and then kept it in room temperature for a few minutes to dry.

Acknowledgements

PK is thankful toEMR/2017/003054 for financial support. MM lab is supported by Grant No. BT/PR21857/NNT/28/1238/2017, EMR/2017/003054, Odisha DBT 3325/ST(BIO)-02/2017.

References

1. Cagan R (2009) Principles of Drosophila eye differentiation. Curr Top Dev Biol 89:115–135
2. Kumar JP (2012) Building an ommatidium one cell at a time. Dev Dyn 241(1):136–149
3. Charlton-Perkins M, Cook T (2014) Prospero and Pax2 regulate the adhesion and patterning of the Drosophila eye. In: Control of Drosophila eye specification, patterning and function by the transcription factors prospero and Pax2, p 103
4. Doroquez DB, Rebay I (2006) Signal integration during development: mechanisms of EGFR and notch pathway function and cross-talk. Crit Rev Biochem Mol Biol 41(6):339–385
5. Morante J, Desplan C (2004) Building a projection map for photoreceptor neurons in the Drosophila optic lobes. In: Seminars in cell & developmental biology. Elsevier
6. Meinertzhagen I, Sorra K (2001) Synaptic organization in the fly's optic lamina: few cells, many synapses and divergent microcircuits. In: Progress in brain research. Elsevier, pp 53–69
7. Charlton-Perkins M, Brown NL, Cook TA (2011) The lens in focus: a comparison of lens development in Drosophila and vertebrates. Mol Gen Genomics 286(3–4):189
8. Mishra M, Knust E (2012) Analysis of the Drosophila compound eye with light and electron microscopy. In Retinal degeneration. Springer, pp 161–182
9. Charlton-Perkins MA et al (2017) Multifunctional glial support by semper cells in the Drosophila retina. PLoS Genet 13(5):e1006782
10. Morrison CA et al (2018) Glass promotes the differentiation of neuronal and non-neuronal cell types in the Drosophila eye. PLoS Genet 14(1):e1007173
11. Longley RL Jr, Ready DF (1995) Integrins and the development of three-dimensional structure in the Drosophila compound eye. Dev Biol 171(2):415–433
12. Pham H, Yu H, Laski FA (2008) Cofilin/ADF is required for retinal elongation and morphogenesis of the Drosophila rhabdomere. Dev Biol 318(1):82–91
13. Bulgakova NA, Knust E (2009) The crumbs complex: from epithelial-cell polarity to retinal degeneration. J Cell Sci 122(15):2587–2596
14. Jemc J, Rebay I (2006) Characterization of the split ends-like gene spenito reveals functional antagonism between SPOC family members during Drosophila eye development. Genetics 173(1):279–286
15. Wang X et al (2018) Environmental effects on Drosophila brain development and learning. J Exp Biol 221(1):jeb169375
16. Cookson MR, Bandmann O (2010) Parkinson's disease: insights from pathways. Hum Mol Genet 19(R1):R21–R27
17. Mizuno H et al (2011) α-Synuclein transgenic Drosophila as a model of Parkinson's disease and related synucleinopathies. Parkinson's Dis 2011
18. Chakraborty R et al (2011) Characterization of a Drosophila Alzheimer's disease model: pharmacological rescue of cognitive defects. PLoS One 6(6):e20799
19. Tan FHP, Azzam G (2017) *Drosophila melanogaster*: deciphering Alzheimer's disease. Malays J Med Sci 24(2):6
20. Sunderhaus ER, Kretzschmar D (2016) Mass histology to quantify neurodegeneration in Drosophila. J Vis Exp (118):e54809

Chapter 5

Methodology to Detect the Abnormality of *Drosophila* Gut by Various Staining Techniques

Subhashree Priyadarsini, Sumit Mukherjee, and Monalisa Mishra

Abstract

The gastrointestinal tract or gut is a vital organ of the body that not only mediates digestion but also senses to modulate nutrient storage as well as food intake. Gut also serves as the first line of defence against a variety of pathogens. Thus, the gut abnormality should be investigated carefully to understand the harmful effects of any potential toxicant exposed to the body via the oral route. The gut of *Drosophila melanogaster*, a well-known model organism, possesses striking similarities with that of humans, thereby modeled to check the toxic effect of nanoparticles or pathogens taken through oral route. Oral intake of nanoparticles in *Drosophila* resulted in the formation of micronuclei, alters arrangement of gut cells, induces cell death and generates ROS. In order to study the gut abnormalities in *Drosophila*, various staining procedures are used. The present chapter briefly describes a detailed staining methodology to detect gut abnormality and localization of trace elements like iron and copper within the gut.

Key words Prussian blue, Copper cell, Trypan blue, DCFH-DA, DAPI, Hoechst

1 Introduction

The origin of the gastrointestinal tract within the body is one of the major developments in an animal that allows the transition from an intracellular to an extracellular mode of digestion [1, 2]. The gut is one of the largest organs present within the body cavity. It has central roles in digestion and absorption of the nutrients. Besides this, the gut serves as the first line of defence against a wide variety of pathogens [1, 2]. The gut is the main source to transmit neuronal and endocrine signals [3, 4]. These signals control nutrient storage or food intake by regulating the activity of other organs [5]. Thus, besides digestion, the gut regulates multiple biological processes. Strikingly, the human gut shares many structural and functional similarities with the fruit fly, *Drosophila melanogaster* [6]. Furthermore, various genes and transcription factors involved in the proper development of *Drosophila* gut are conserved from *Drosophila* to a human being [7–9]. This makes the *Drosophila* gut

Monalisa Mishra (ed.), *Fundamental Approaches to Screen Abnormalities in Drosophila*, Springer Protocols Handbooks, https://doi.org/10.1007/978-1-4939-9756-5_5, © Springer Science+Business Media, LLC, part of Springer Nature 2020

as a good model to study the interactions between stress, repair and immune responses of a variety of pathogens and toxic substances.

Although there is a number of methods to study gut abnormalities, various staining methods have been proved to be useful to determine the abnormality of the gut. The abnormalities may include an alteration of some of the important metals, stress protein, damaged cell, the arrangement of nuclei, and level of fat deposition within the gut [10, 11]. Common staining techniques used to detect the gut abnormality include trypan blue staining, Prussian blue staining, copper cell staining, DAPI, Hoechst and DCHFDA staining [10–15]. These techniques are easy to do within the laboratory, inexpensive and useful for the early detection of gut damage. Using these techniques the stained gut can be visualized either by (1) bright-field microscope or (2) fluorescent microscope depending on the types of stain used for the staining process.

1.1 Bright-Field Staining

1.1.1 Prussian Blue Staining

Prussian blue (PB) stain is commonly used for the detection of iron granules within the tissues. It is an extremely sensitive test to detect the level of iron within the tissue [16]. The staining reaction detects the presence of ferric ions (Fe^{3+}) within the tissue. Excess iron within the cell is toxic for the organism [17]. Under the acidic conditions, Prussian blue reacts, and iron is released in the reaction in the form of ferrocyanide ions which comes from the potassium ferrocyanide solution. This step resulted in the formation of hexaferrocyanide $Fe(CN)^{6+}$, an insoluble product that yields a bright blue precipitate and gets deposited as blue granule within the tissue. PB mainly stains the midgut of *Drosophila* where the iron cells are present. Thus, any abnormal accumulation of iron in the midgut cells can be detected.

1.1.2 Copper Cell Staining

Copper cells are found in the gut throughout the development. The differentiation starts in the embryonic stage by the action of homeotic gene *labial* [18]. These cells persist in the midgut of larvae as well as in the adult fly. Around 100 cells are found in the middle midgut region. For the first time, these cells were identified by Strasburger in 1932 [19]. Later, the properties of these cells was established, and they were named as cuprophilic cells [20, 21]. The apical part of these cells possesses microvilli [21, 22]. The cytoplasm of these cells is filled with mitochondria. Foods present near the copper cells are highly acidic $pH < 2.3$ [19, 23]. In the absence of copper cell, the acid secretion stops in the stomach. Thus, copper cells share functional similarity with parietal cells of mammals. Copper cells can be distinguished from other cells by staining with phalloidin. Phalloidin binds to the apical part of the actin present in the copper cell [24]. The other method is by feeding the copper sulphate to the larvae. The copper cells appear orange under a fluorescent microscope [20, 25]. Copper binds to the metallothionein protein family and thus produces orange colouration [25]. The orange luminescence of the copper cell can be detected under UV light when the larvae are fed with copper sulphate. Consequently, if there is any defect in the copper cell, there will be reduced intensity in the orange colour when the larvae will feed with copper sulphate.

1.1.3 Trypan Blue Staining

Trypan blue assay or dye exclusion test is a simple and useful assay to detect the dead cells among a cell population. This dye is used to stain dead cells based on their inability to exclude the dye out of the cell. As a result, the dye remains inside the cell making them appear blue in colour. Live cells are not stained blue as the cell membrane does not allow the dye to pass through the membrane and thus excludes the dye. So in a mixed population of live and dead cells, the dead cells can be distinguished from the live cells. This technique is often used to stain the gut of *Drosophila* to detect any tissue damage [14, 26, 27].

1.2 Fluorescence Staining

1.2.1 DCFH-DA Staining

In the *Drosophila* model, the most widely used fluorescent probe to determine ROS at tissue level is dichlorodihydrofluorescein diacetate (DCFH-DA). This dye detects the cellular oxidative stress by measuring the intracellular redox level directly. The DCFH-DA/DCFH behaves differently with respect to various cells since the level of a permeability barrier is different for different tissues. Thus, to detect the activity of DCFH-DA/DCFH in different cells is a prerequisite before using the dye in different tissues. The principle of working is based on the oxidation of DCFH-DA. The probe is permeable to the cell, and once it entered into the cell, it is converted to nonfluorescent $DCFH^-$ (intermediated free radical), when hydrolyzed by the esterase. The $DCFH^-$ easily diffuses inside where it is oxidized by non-specific ROS to be converted into green fluorescent DCF probe. The fluorescence intensity of the DCF is directly proportional to the oxidative stress level in the cell [28]. In this chapter, we have described a method to detect ROS level in the gut by DCFH-DA staining.

1.2.2 Hoechst Staining

Hoechst is positively charged and can be easily accessible to live cells. It has a high affinity for AT-rich regions of DNA in fixed cells. After binding, the fluorescence increases up to 30-fold. It has less toxic effects than DAPI, another dye that also binds to DNA and stains the nuclei. The excitation and emission wavelengths are 360 and 460 nm, respectively. It emits blue colour like that of DAPI and can be excited under UV filter in a fluorescence microscope [29]. It is used to check any damage occurred within the gut due to the feeding of any substances which is harmful to the body.

1.2.3 DAPI Staining

4′, 6-diamidino-2–phenylindole (DAPI) is a commonly used fluorophore which detects both live and dead cells [30]. After binding the fluorescence increases up to 20 times [31]. DAPI bears chemical formula, $C_{16}H_{15}N_5{\bullet}2HCl$ and its molecular weight is 350.25 g/mol. The excitation and emission wavelengths are

340 and 488 nm, respectively [32]. The fluorescence blue colour can be detected under UV light. It binds to double-stranded DNA strongly, and the excitation reaches to 360 nm, while emission shifts to 460 nm [32]. It binds to mostly the minor groove region of DNA double-strand with higher specificity to AT-rich regions [33]. DAPI fades faster than other dyes [34]. DAPI staining is useful to identify the layering of the gut and if there is any fragmentation of nuclei within the gut.

The current protocol uses various dyes to detect the structural, elemental as well as the stress level of the gut.

2 Materials

2.1 Larval Dissection

1. Third instar larva.
2. Phosphate-buffered saline (PBS).
3. Fine forceps (SCHIFIC, stainless steel tweezer).
4. Fine needles (Soccerene Premium Quality Pin 1.40mm long X 0.35mm diameter).
5. Stereomicroscope.
6. Petri plate.
7. 0 sized paint brush
8. 1.5 ml tube.

2.2 Tissue Sample Fixation and Staining

1. 4% paraformaldehyde (PFA) (HiMedia-MB059)
2. DAPI (HiMedia-MB097).
3. Hoechst.
4. DCFH-DA.
5. Trypan blue.
6. Prussian blue.
7. Copper sulphate ($CuSO_4$).

2.3 Slide Preparation and Microscopy

1. PBS-Tween 20 (PBST).
2. Glass slide (25 mm x 75 mm; Abdos- G10101).
3. Coverslip (18 mm; Blue star).
4. Tissue paper.
5. 20% glycerol (HiMedia-MB060)
6. Bright-field microscope.
7. Fluorescence microscope.

2.4 Other Chemicals

1. Sodium chloride (HiMedia-GRM853).
2. Potassium chloride (HiMedia-MB043).

3. Disodium hydrogen phosphate (HiMedia-MB024).
4. Potassium dihydrogen phosphate (HiMedia-TC011).
5. Potassium ferrocyanide.
6. Glycerol (HiMedia-MB060).
7. Tween 20 (Sigma-P9416).
8. Hydrochloric acid (HiMedia-AS003).
9. Sodium hydroxide (HiMedia-MB095).
10. Paraformaldehyde.
11. Whatman filter paper.

3 Methods

3.1 Preparation of PBS (1x) and PBST (1x)

1. To prepare 100 ml of 1x PBS, add 0.8 g NaCl, 0.02 g KCl, 0.14 g Na_2HPO_4, 0.02 KH_2PO_4 in 80 ml milliQ water.
2. Mix thoroughly and adjust the pH to 7.2.
3. Make the volume up to 100 ml.
4. Add 1% Tween 20 to 1x PBS (100ml).

3.2 Preparation of 4% PFA

1. Heat 70 ml of 1x PBS in a glass beaker up to 60 °C on a hot plate with stirring beads.
2. Add 4gm of paraformaldehyde to it and stir continuously.
3. Add NaOH dropwise to make a completely clear solution.
4. Keep the solution for cooling and filter it.
5. Adjust the pH to 7.2 by adding HCL dropwise.
6. Make the volume up to 100 ml with 1X PBS and store at 4 °C.

3.3 Preparation of 20% Glycerol

1. Add 2 ml of glycerol to 8 ml of milliQ water for 10 ml of 20% glycerol solution.
2. Mix the solution slowly by pipetting and store at 4 °C in an amber coloured tube.

3.4 Preparation of Working Solution of DAPI

1. Prepare a stock solution of 1 mg/ml DAPI by diluting in autoclaved milliQ water.
2. Store the solution in 1.5 ml amber coloured tube in 4 °C.
3. Make the working solution of 1 μg/ml, by taking 1 μl of stock solution, and dilute it with 1 ml of 1x PBS. Keep the solution in a 1.5 ml amber tube at 4 °C.

3.5 Preparation of Trypan Blue Solution (0.2%)

1. Add 2 mg of trypan blue powder in 100 ml of 1X PBS.
2. Mix well and filter the solution to remove precipitates (*see* **Note 5.4.2.1**).

3.6 Preparation of DCFH-DA

1. Prepare a stock solution of 1 mg/ml DCFH-DA by diluting the powder in 1 ml milliQ water.
2. Prepare the working solution of 1 μg/ml, by diluting 1 μl of the stock solution with 1 ml of 1X PBS. Store it in 4 °C in 1.5 ml amber tube.

3.7 Preparation of Prussian Blue Solution

1. Prepare a 2% solution of potassium ferrocyanide and 2% HCl.
2. Add both the solutions in equal volume and mix well.

3.8 Preparation of Copper Sulphate Solution

1. Add 10 mg of copper sulphate to 10 ml of milliQ water to prepare 1 mg/ml stock solution.
2. For the working solution, dilute the stock solution to 0.5 mg/ml.

3.9 Dissection of the Larva

1. Take six third instar larvae from a food vial, and wash them in 1x PBS in a petri plate.
2. In a slide take a drop of 4% PFA at the centre and place a larva in it.
3. Dissect the larvae with the help of fine forceps.
4. Take out the gut, and transfer it in 4% PFA solution in a 1.5 ml tube for fixation.

3.10 Fixation and Staining

1. Fix the gut with 4% PFA by keeping it for 30 min in room temperature. The gut can be kept in PFA for overnight at 4 °C (Fig. 1).
2. After 30 min, remove the PFA from the tube, and wash the guts three times in PBS for 5 min each.
3. Remove the PBS solution, and wash the samples with PBST two times for 15 min each. PBST helps the stain to permealize the gut cells (Fig. 1).
4. Remove PBST and add 10 μl of the staining solution (*see* **Note 5.4.3.1**).
5. For DAPI and Hoechst keep the samples for 5 min in the dark. For DCFH-DA staining, keep the samples for 30 min in the dark (*see* **Note 5.4.3.2**).
6. Remove stain, and wash the samples with 1x PBS two times for 5 min.

3.11 Slide Preparation for Microscopy

1. Take a glass slide and clean it using acetone (*see* **Note 5.4.1.1**).
2. Put the stained sample along with 1x PBS at the centre.
3. Remove the PBS solution carefully by wiping with a tissue paper (*see* **Note 5.4.1.3**).
4. Put one drop of (~ 20 μl) of 20% glycerol to the samples at the centre.

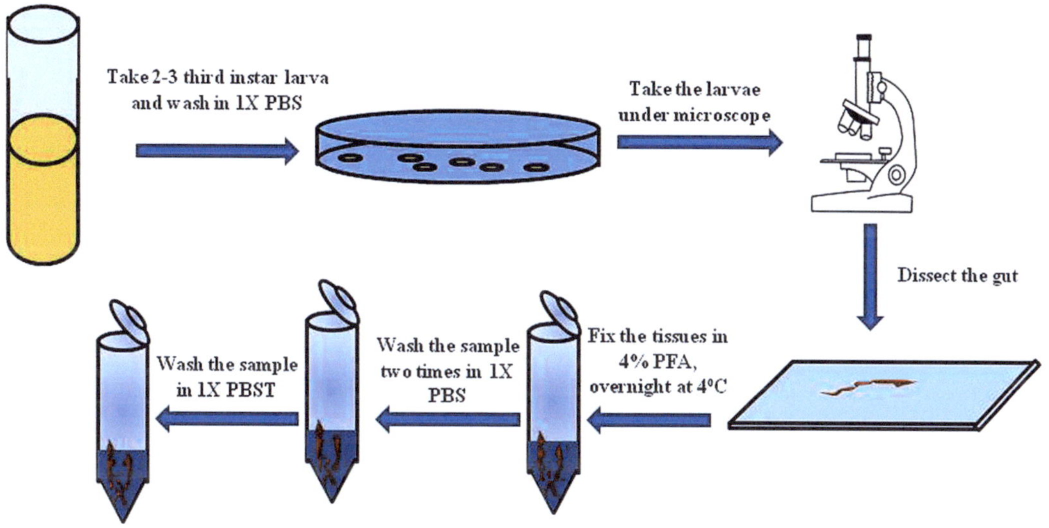

Fig. 1 Fixation of the gut tissues in paraformaldehyde and processing with PBST for staining

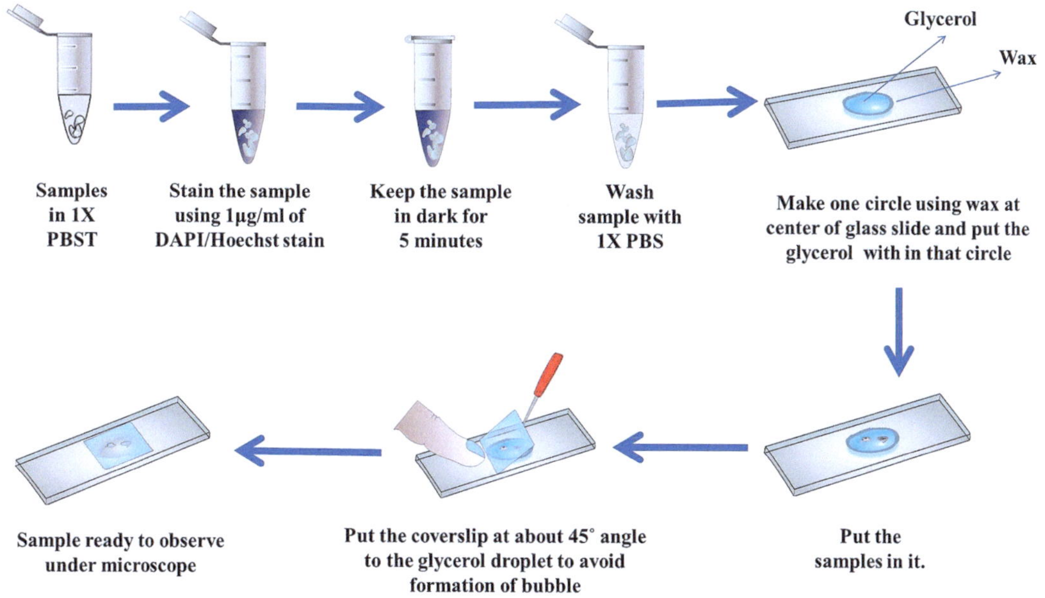

Fig. 2 Preparation of slides for observation of samples under the microscope

5. Take a clean coverslip, and put it above the sample by placing it at 45^0 using a needle (Fig. 2).
6. Wipe out the excess glycerol by tissue paper.
7. Observe the slides under a fluorescence microscope (*see* **Note 5.4.3.3**).
8. The images of DAPI-stained gut (Fig. 3a), Hoechst (Fig. 3b) and DCFH-DA staining are shown (Fig. 4a, b).

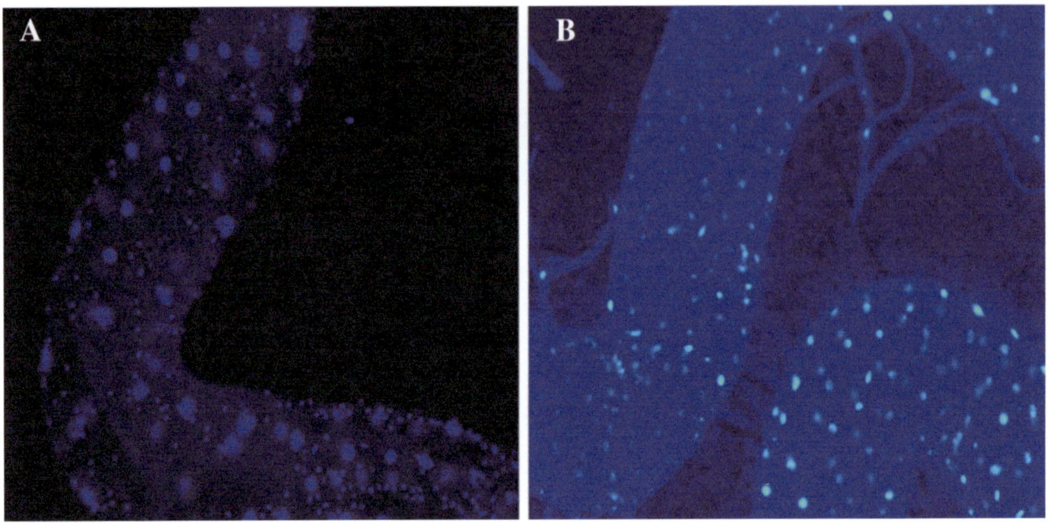

Fig. 3 Images showing fluorescence staining of the larval gut. (**a**) DAPI staining of the larval gut, (**b**) Hoechst staining

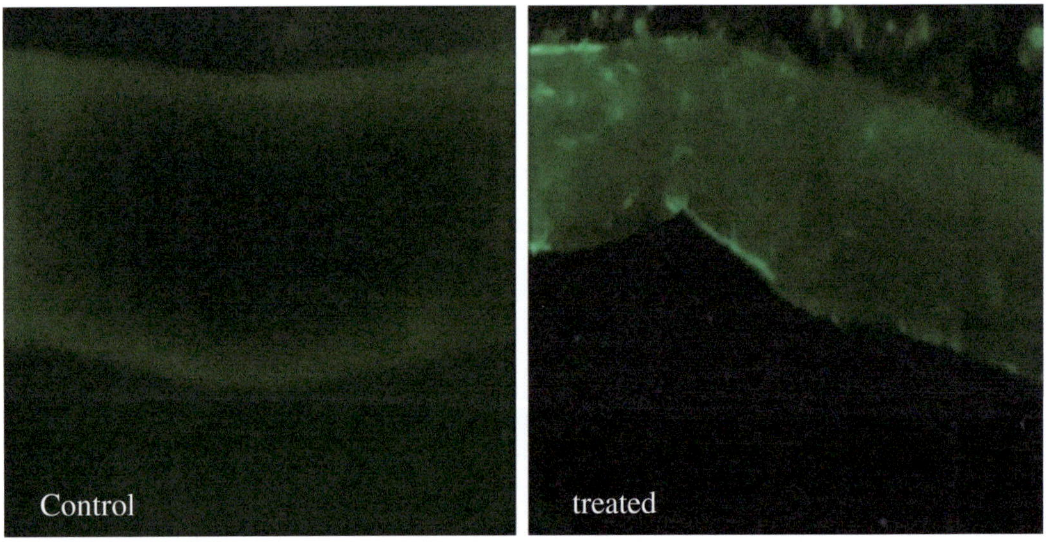

Fig. 4 DCFH-DA staining of the larval gut. Note the strong intensity of the treated gut

3.12 Copper Cell Staining

1. Take six third instar larvae and wash them in 1X PBS.
2. Incubate larvae in 500 mg/L CuSO4 solution for 30 min.
3. Then dissect the gut of these larvae by using forceps under a microscope.
4. Mount it on the slide with 20% glycerol.
5. Take images under bright-field microscope (Fig. 5).
6. The dark orange staining of the copper cells can be seen under the microscope (Fig. 6).
7. To get a bright orange colour of the Copper cells, image it under UV light.

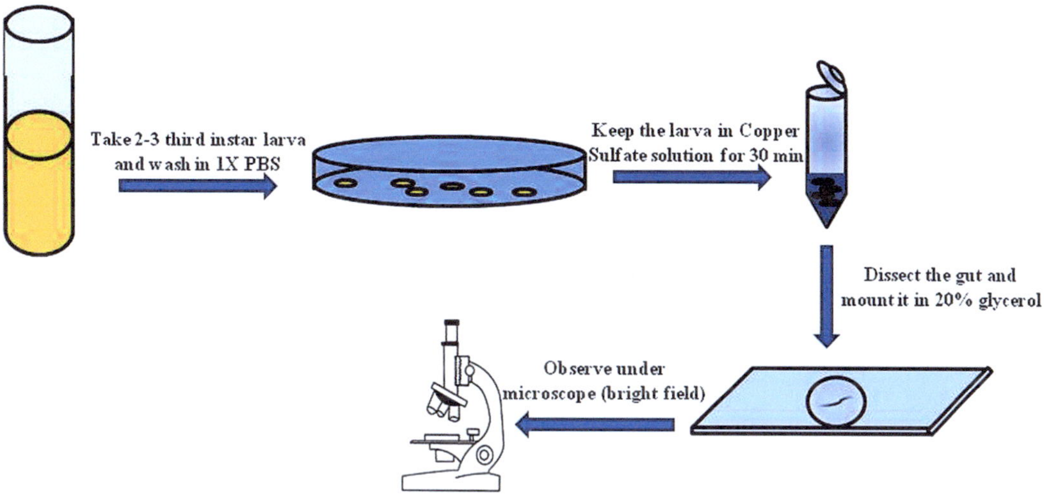

Fig. 5 Schematic diagram to show steps involved in copper cell staining

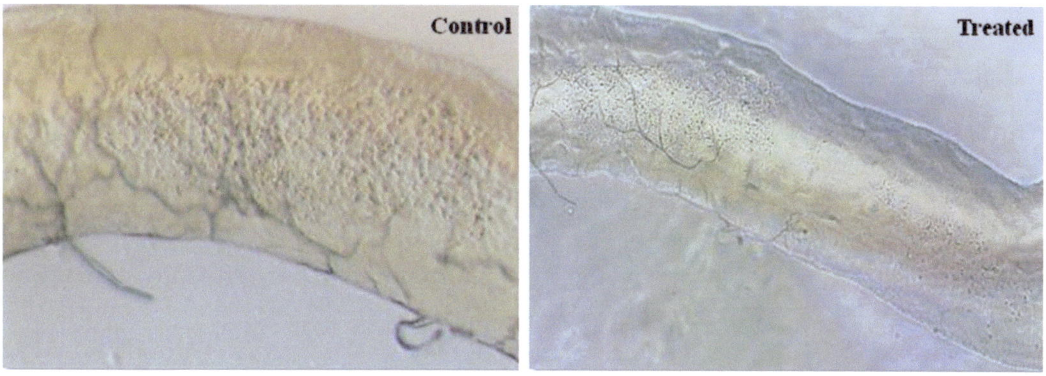

Fig. 6 Image showing the larval gut after copper cell staining in control and treated under a light microscope. Note the more orange colour intensity of copper cells in the control gut

3.13 Prussian Blue Staining

1. Take six third instar larvae in a petri plate, and wash them in 1X PBS (Fig. 7).
2. Take a drop of cold PBS (1X) at the centre of a slide, and put a larva on it.
3. Dissect the larva under a stereomicroscope to remove the gut out of it (Fig. 7).
4. Keep the gut in a 1.5 ml tube in 4% PFA for 30 min to fix the tissues.
5. After 30 min, remove the PFA solution, and wash the gut with PBS three times for 5 min each.
6. Remove the PBS, and wash the gut two times in PBST for 15 min each.

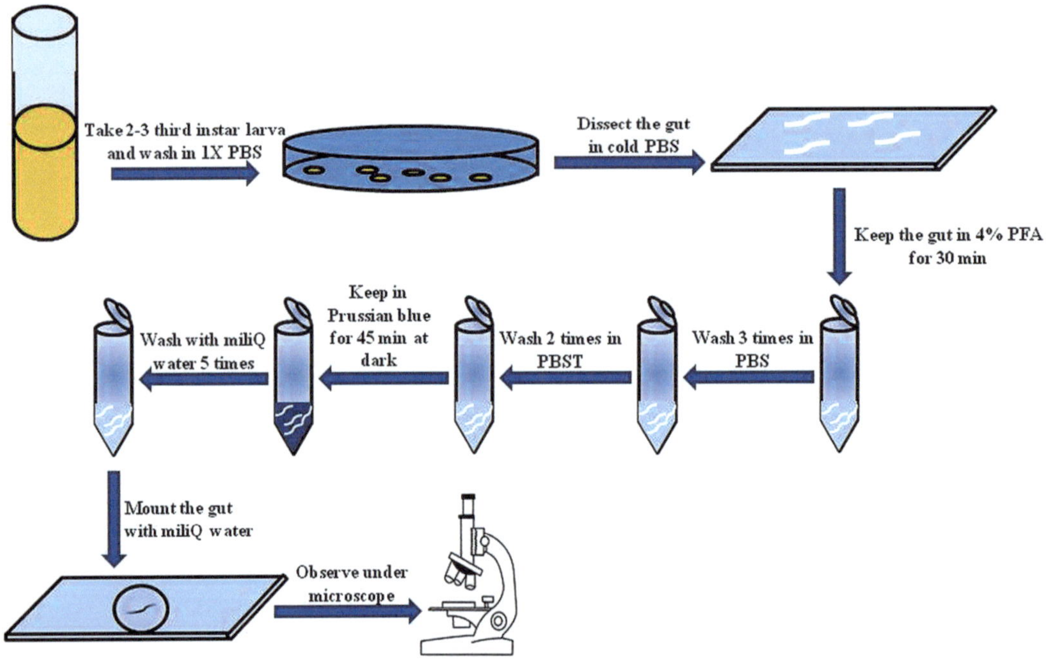

Fig. 7 Image showing the steps involved in Prussian blue staining

7. Remove the PBST solution, and take 200 μl of Prussian blue solution.
8. Keep the gut for staining at dark for 45 min at room temperature (Fig. 7) [35].
9. After the incubation, remove the Prussian blue solution, and wash the gut five times with milliQ water.
10. Mount the gut at the centre of a clear slide with milliQ water (*see* **Note 5.4.2.3**).
11. Take a coverslip, and put it above the sample by placing it at a 45^0 angle with the help of a needle (Fig. 7).
12. Wipe out the excess water by tissue paper.
13. Observe the gut under a bright-field microscope to see iron deposition in the cells (Fig. 8).

3.14 Trypan Blue Staining

1. Take five third instar larvae from the food vial, and wash them with 1X PBS to remove the food particles (Fig. 9).
2. Transfer the larvae to a 1.5 ml tube containing 1 ml of 0.2% trypan blue solution (Fig. 9).
3. Keep the tube in a shaker for 30 min at room temperature (*see* **Note 5.4.2.2**).
4. Remove the stain solution, and wash the larvae thrice (10 min each) with PBS on a rocker (Fig. 9).

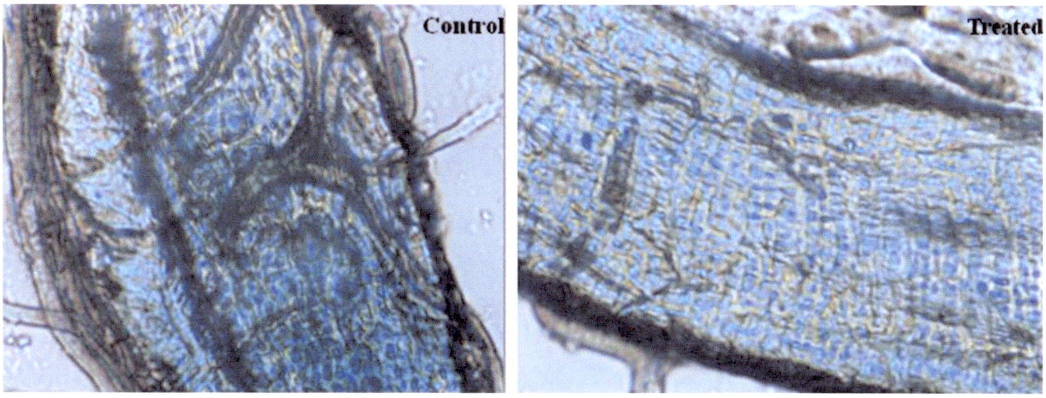

Fig. 8 Prussian blue staining of the control and treated larva

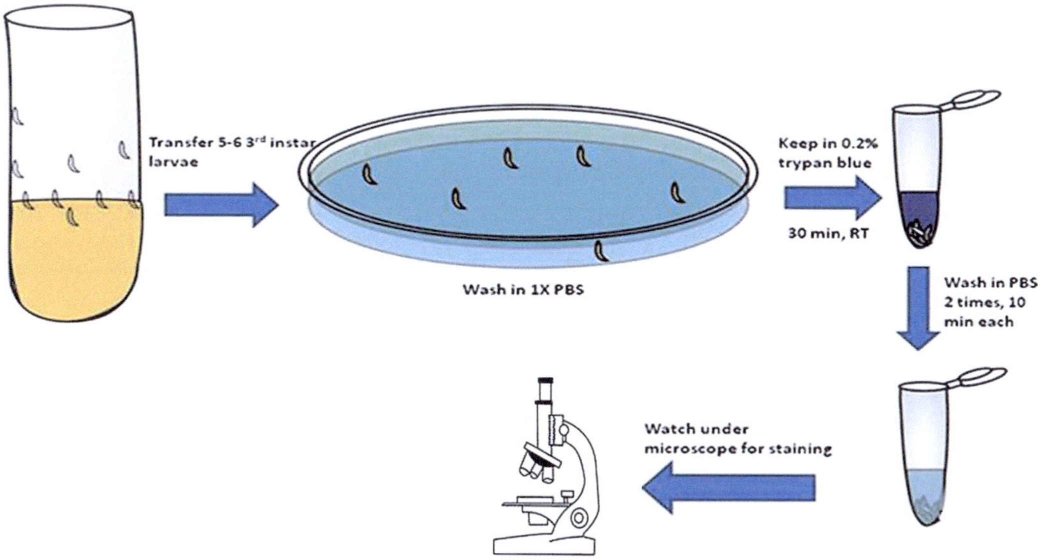

Fig. 9 Trypan blue assay to see the damage in the gut cells: Figure showing the steps involved in trypan blue staining

5. Keep them in PBS for 10 min to remove the excess dye.
6. Take the larvae under a stereomicroscope, and image them. (Fig. 9).
7. Transfer the image to the computer, and search for the blue stain in the gut.
8. The cell mortality or damage can be scored based on staining, given that '0' for no stain, '+' for light stain, '++' for blue stain in whole tissue and '+++' for dark blue staining of the whole tissue (Fig. 10).

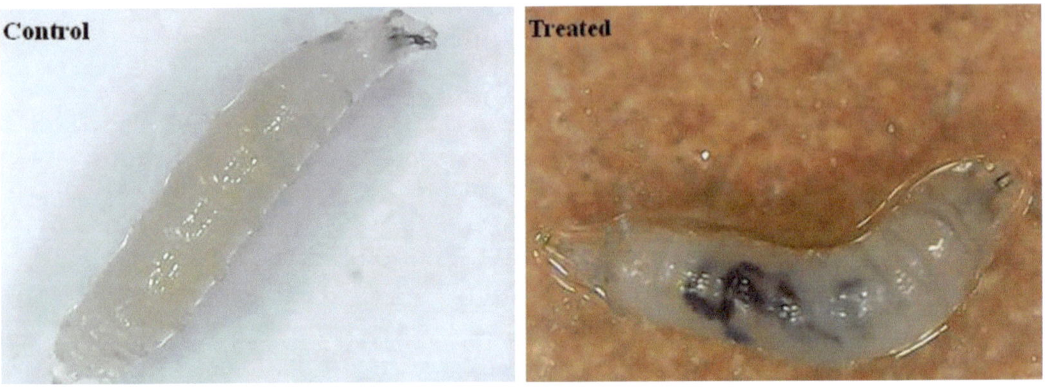

Fig. 10 Images showing larval gut damage by trypan blue staining

4 Notes

4.1 Slide Preparation

1. Take a clear glass slide free of any dust particles.
2. Place the samples at the centre of the slide by carefully looking at the microscope.
3. After mounting the samples, wipe out the excess mounting media with tissue paper.

4.2 Bright-Field Staining

1. Trypan blue acts as a potential carcinogen, so always wear gloves while working with it.
2. Do not keep the larva for a longer time in trypan blue solution as it may give a false result.
3. In case of Prussian blue staining, mount the samples in water in place of glycerol.

4.3 Fluorescent Staining

1. The concentration of the stain should be maintained appropriately as higher concentration causes overstaining, which is difficult to be washed off.
2. As the fluorescent dyes are light-sensitive, the dye should be always kept and used in the dark.
3. In the fluorescence microscope, the exposure time of the light should be fixed to maintain uniformity, and eliminate the chance of error.

Acknowledgements

SP and SM are thankful to MHRD for financial support. We are thankful to Nibedita Nayak for DCHFDA gut image and Swetapadma Sahu for DAPI and Hoechst image used in this study.

MM lab is supported by Grant No. BT/PR21857/NNT/28/1238/2017, EMR/2017/003054 and Odisha DBT 3325/ST (BIO)-02/2017.

References

1. Lemaitre B, Miguel-Aliaga I (2013) The digestive tract of Drosophila melanogaster. Annu Rev Genet 47:377–404
2. Miguel-Aliaga I, Jasper H, Lemaitre B (2018) Anatomy and physiology of the digestive tract of Drosophila melanogaster. Genetics 210 (2):357–396
3. Stengel A, Taché Y (2018) Gut-brain neuroendocrine signaling under conditions of stress—focus on food intake-regulatory mediators. Front Endocrinol 9
4. Takashima S, Adams KL, Ortiz PA, Ying CT, Moridzadeh R, Younossi-Hartenstein A, Hartenstein V (2011) Development of the Drosophila entero-endocrine lineage and its specification by the notch signaling pathway. Dev Biol 353(2):161–172
5. Liu Q, Jin LH (2017) Organ-to-organ communication: a Drosophila gastrointestinal tract perspective. Front Cell Dev Biol 5:29
6. Apidianakis Y, Rahme LG (2011) Drosophila melanogaster as a model for human intestinal infection and pathology. Dis Model Mech 4 (1):21–30
7. Liu Q, Jin LH (2017) Tissue-resident stem cell activity: a view from the adult Drosophila gastrointestinal tract. Cell Commun Signal 15 (1):33
8. Li H, Jasper H (2016) Gastrointestinal stem cells in health and disease: from flies to humans. Dis Model Mech 9(5):487–499
9. Allocca M, Zola S, Bellosta P (2018) The fruit fly, Drosophila melanogaster: modeling of human diseases (Part II). In: Drosophila melanogaster-model for recent advances in genetics and therapeutics. IntechOpen
10. Pappus SA, Mishra M (2018) A Drosophila model to decipher the toxicity of nanoparticles taken through oral routes. In: Cellular and molecular toxicology of nanoparticles. Springer, pp 311–322
11. Micchelli CA (2014) Whole-mount immunostaining of the adult Drosophila gastrointestinal tract. Methods 68(1):273–279
12. Mishra M, Sabat D, Ekka B, Sahu S, Unnikannan P, Dash P (2017) Oral intake of zirconia nanoparticle alters neuronal development and behaviour of Drosophila melanogaster. J Nanopart Res 19(8):282
13. Pappus SA, Ekka B, Sahu S, Sabat D, Dash P, Mishra M (2017) A toxicity assessment of hydroxyapatite nanoparticles on development and behaviour of Drosophila melanogaster. J Nanopart Res 19(4):136
14. Sabat D, Patnaik A, Ekka B, Dash P, Mishra M (2016) Investigation of titania nanoparticles on behaviour and mechanosensory organ of Drosophila melanogaster. Physiol Behav 167:76–85
15. Carmona ER, Escobar B, Vales G, Marcos R (2015) Genotoxic testing of titanium dioxide anatase nanoparticles using the wing-spot test and the comet assay in Drosophila. Mut Res Genet Toxicol Environ Mutagen 778:12–21
16. Pearse AG (1953) Histochemistry theoretical and applied. Little, Brown and Company, Boston
17. Tang X, Zhou B (2013) Iron homeostasis in insects: insights from Drosophila studies. IUBMB Life 65(10):863–872
18. Hoppler S, Bienz M (1994) Specification of a single cell type by a Drosophila homeotic gene. Cell 76(4):689–702
19. Strasburger M (1932) Bau, Funktion und Variabilität des Darmtraktus von Drosophila melanogaster Meigen. Akad. Verlagsges, Leipzig
20. Poulson D, Bowen V, Hilse R, Rubinson A (1952) The copper metabolism of Drosophila. Proc Natl Acad Sci U S A 38(10):912
21. Filshie B, Poulson D, Waterhouse D (1971) Ultrastructure of the copper-accumulating region of the Drosophila larval midgut. Tissue Cell 3(1):77–102
22. Dubreuil RR, Grushko T, Baumann O (2001) Differential effects of a labial mutation on the development, structure, and function of stomach acid-secreting cells in Drosophila melanogaster larvae and adults. Cell Tissue Res 306 (1):167–178
23. Dubreuil RR (2004) Copper cells and stomach acid secretion in the Drosophila midgut. Int J Biochem Cell Biol 36(5):742–752
24. Lee JK, Coyne RS, Dubreuil RR, Goldstein L, Branton D (1993) Cell shape and interaction defects in alpha-spectrin mutants of Drosophila melanogaster. J Cell Biol 123(6):1797–1809

25. McNulty M, Puljung M, Jefford G, Dubreuil RR (2001) Evidence that a copper-metallothionein complex is responsible for fluorescence in acid-secreting cells of the Drosophila stomach. Cell Tissue Res 304(3):383–389
26. Krebs RA, Feder ME (1997) Tissue-specific variation in Hsp70 expression and thermal damage in Drosophila melanogaster larvae. J Exp Biol 200(14):2007–2015
27. Siddique YH (2012) Protective role of curcumin against the toxic effects of cyclophosphamide in the third instar larvae of transgenic Drosophila melanogaster (hsp70-lacZ) Bg9. Alternative Medicine Studies 2(1):2
28. Wang H, Joseph JA (1999) Quantifying cellular oxidative stress by dichlorofluorescein assay using microplate reader1. Free Radic Biol Med 27(5–6):612–616
29. Cai K, Yang J, Guan M, Ji W, Li Y, Rens W (2005) Single UV excitation of Hoechst 33342 and propidium iodide for viability assessment of rhesus monkey spermatozoa using flow cytometry. Arch Androl 51(5):371–383
30. Dmitrieva NI, Burg MB (2008) Analysis of DNA breaks, DNA damage response, and apoptosis produced by high NaCl. Am J Physiol Renal Physiol 295(6):F1678–F1688
31. Lakowicz JR, Gryczynski I, Malak H, Schrader M, Engelhardt P, Kano H, Hell SW (1997) Time-resolved fluorescence spectroscopy and imaging of DNA labeled with DAPI and Hoechst 33342 using three-photon excitation. Biophys J 72(2. Pt 1):567
32. Gryczynski I, Malak H, Lakowicz JR (1996) Multiphoton excitation of the DNA stains DAPI and Hoechst. Bioimaging 4(3):138–148
33. Tanious FA, Veal JM, Buczak H, Ratmeyer LS, Wilson WD (1992) DAPI (4′, 6-diamidino-2-phenylindole) binds differently to DNA and RNA: minor-groove binding at AT sites and intercalation at AU sites. Biochemistry 31 (12):3103–3112
34. Gallardo-Escarate C, Álvarez-Borrego J, Von Brand E, Dupré E, Del Rio-Portilla MA (2007) Relationship between DAPI-fluorescence fading and nuclear DNA content: an alternative method to DNA quantification? Biol Res 40 (1):29–40
35. Mehta A, Deshpande A, Bettedi L, Missirlis F (2009) Ferritin accumulation under iron scarcity in Drosophila iron cells. Biochimie 91 (10):1331–1334

Chapter 6

Cryodissection and Tissue Preparation of *Drosophila* Thorax for Indirect Flight Muscle Imaging

Madhavi Dubey, Kumari Pragati Nanda, and Hena Firdaus

Abstract

Drosophila melanogaster has elaborate musculature comparable to the vertebrate specialized muscles, i.e. skeletal, cardiac and smooth. One of the largest fibrillar muscles found in the adult fly thorax is the indirect flight muscles (IFM). IFM is routinely used by the biologists to study muscle development, muscle attachment to the epidermis, cell biology, neuromuscular interaction, regulation of muscle contraction and signalling. This chapter gives an overview of the structure and development of IFM along with the procedure of dissection and imaging of IFM. This protocol can be used for morphological and phenotypic characterization of *Drosophila* muscle which will greatly help in our understanding of muscle biology pertaining to the myopathies and their aetiology.

Key words *Drosophila*, Indirect flight muscles (IFM), Pupal dissection, Fly thorax transverse section, Polarized light imaging, Confocal microscopy

1 Introduction

1.1 IFM as a Model Tissue

The indirect flight muscles (IFM) form the bulk of the thoracic musculature in *Drosophila* and consist of two sets of muscles, dorsal longitudinal muscles (DLM) and dorsoventral muscles (DVM). The nomenclature is based on the orientation in the adult fly thorax (Fig. 1). They are the adult set of musculature formed during the second wave of myogenesis which happens during the pupal metamorphosis to form adult flies. These two sets of muscles show opposing contraction leading to thoracic deformation which indirectly powers flight by enabling wing movement. The IFM fibres are well-arranged structures which are reflected in the defined numbers of fibres (six for DLMs and three bundles of two or three fibres each in DVM) with neuronal connectivity, fixed orientation and attachment to tendon cells, thereby establishing a specific pattern [1]. They are fibrillar muscles, and ultrastructure assembly is similar to skeletal muscles of vertebrates [2]. Functionally, they are asynchronous muscles, and their regulation involves parameters similar to that of

Monalisa Mishra (ed.), *Fundamental Approaches to Screen Abnormalities in Drosophila*, Springer Protocols Handbooks,
https://doi.org/10.1007/978-1-4939-9756-5_6,

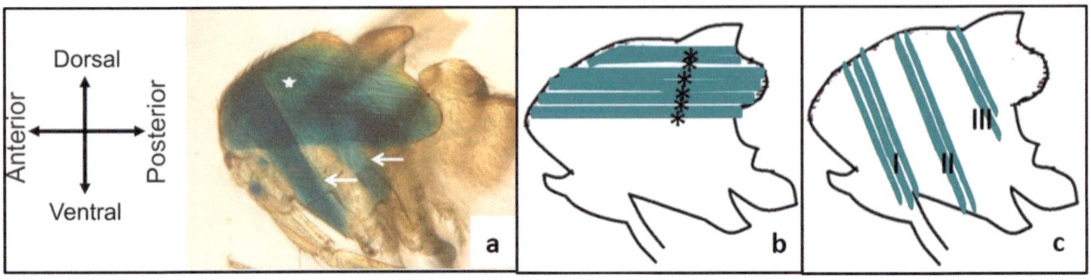

Fig. 1 IFM organization in *Drosophila* hemithorax. (**a**) A sagittal view of hemithorax showing IFM expressing reporter lac-Z gene stained with X-gal. One of the DLM fibres running along the anterior-posterior axis of the thorax is marked with asterisk and arrows point to DVM. (**b**, **c**) Schematic showing organization and number of DLM and DVM fibres

cardiac muscles [3]. The spatio-temporal developmental profile of IFM has been minutely studied [4]. The high-throughput imaging coupled with tissue-specific expression of fluorophores has greatly helped in real-time *in vivo* imaging [5]. Null mutations of structural protein isoform specific to IFM yield viable fly and have significantly assisted in studying stoichiometric roles of these proteins in muscle development [6–9]. Functional and developmental mutant scoring for IFM defects could be achieved by simple flight test behaviour, thus enabling the study of a large number of genetic loci involved in muscle development and function. Altogether, ease of muscle localization, patterning, documented cellular events, ease of mutant scoring and availability of vast information in the annotated database (FlyBase) make IFM an ideal tissue to study muscle biology.

1.2 IFM Development and Myofibrillogenesis

Cell autonomous and extrinsic cues synergistically sculpt the IFM tissue development allowing it to acquire distinctive features enabling high wing beating frequency of about 200 Hz during flight. IFM, like other skeletal muscles, are syncytia of many fused adult myoblasts. These adult myoblasts are delineated from the mesodermal lineage during embryonic muscle development and remain linked primarily with wing imaginal disc during larval life. They proliferate avoiding fusion until pupal metamorphosis. Afterwards, they leave the disc and migrate along the motor neuron to reach their site of fusion. DLMs develop on partially histolysed larval oblique muscle (LOM) template by the fusion of new adult myoblast, whereas DVM development involves *de novo* fusion and emergence of fibre [4, 10]. At 10 h after puparium formation (APF), a pool of myoblast surrounding LOM is ready to fuse. By 20 h APF, longitudinal splitting of the three LOM fibres gives rise to six templates of DLM. Myoblast fusion continues post splitting to give rise to much elongated fibre. Filopodia are developed at muscle ends, and reciprocal PS integrin expression at muscle and tendon cell leads to the firm attachment of a muscle.

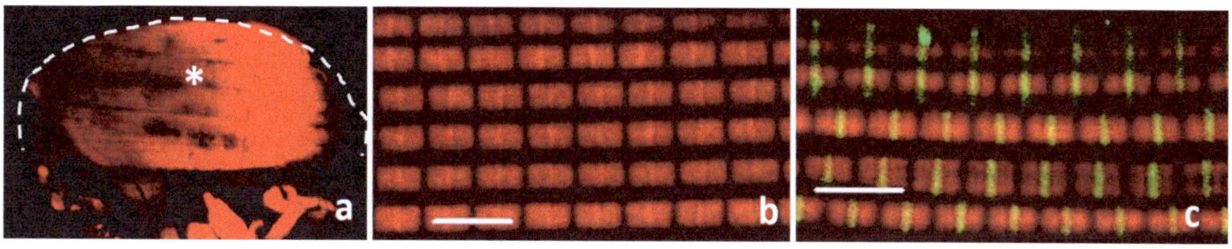

Fig. 2 Confocal images of *phalloidin-TRITC* stained DLM fibre. (**a**) Hemithorax whole mount showing DLM. (**b**) Ultrastructure of myofibrils displaying sarcomere arrangement. (**c**) Z-disc localizing protein (green) in *phalloidin-TRITC* stained myofibre (red). Scale bar denotes 5 μm

Expression of structural proteins marks the end of myoblast fusion, and myofibrils constrict to one-third of their original size, thereby probably stretching their attachment checking the stability of muscle-tendon attachment site. By 42 h APF, rearranged sarcomeres are visible in developing myofibril. On the other hand, DVM fibres are seeded by dumfounded (Duf) expressing myoblast acting as the founder myoblast [11]. Other myoblasts fuse with this founder myoblast giving rise to the nascent fibres. DLM as well as DVM fibre morphogenesis goes simultaneously, thus attaining growth and fibre mass increase until the adult fly ecloses out of the pupal case.

Myofibrillar assembly is initiated by the actin and its associated protein [2] at the periphery attached to the plasma membrane. These stress fibrelike structures are similarly conserved in vertebrate [12]. Structural proteins organized in sarcomeres like actin, myosin, troponin T, troponin I, troponin C, tropomyosin, α-actinin, etc. are conserved in vertebrates, and our invertebrates model *Drosophila* and perform similar functions [6, 7]. Sarcomere length increases as fibre mature and adult fly sarcomere length are roughly around 3.4 μm (Fig. 2). However unlike vertebrates, in *Drosophila*, sarcomere number remains constant, and there is no addition of new sarcomere as development proceeds. Myofibre girth is increased by the addition of new myofilaments at the periphery, and adult myofibre is roughly 35 thick filaments wide [2]. Maintenance of stoichiometric ratio of thick and thin filament-associated proteins during IFM development dictates coordinated regulation of expression of these proteins [8, 9].

1.3 Drosophila *IFM Genetics*

Mutant screens designed to isolate flightless flies have led to the identification of gene *loci* and important residues involved in IFM development and function [13, 14]. Since any perturbation in IFM was not detrimental to flies, hence such screen had been very beneficial in observing a large number of flies and scoring mutants uncovering larger genome regions. Two efficient methods were employed for isolating genetic lesions. Firstly, chemical mutagens like ethyl methanesulphonate (EMS) and ethylnitrosourea (ENU) are fed to flies to isolate the spectrum of different lesions dispersed in the genome to carry out saturation mutagenesis screen [15–17]. Secondly, P-element mobilization gets random insertions or deletions in the germline and then screening for mutants.

Another strategy called enhancer trap screening was employed to acquire knowledge of new genes expressing in IFM. An enhancer trap line utilized reporter gene under minimal promoter (insufficient to drive expression on its own) flanked by P-element sequence which could respond to transposase enzyme leading to mobilization and random insertion into the genome, thereby entrapping local gene enhancers [18]. The pattern (spatio-temporal) of enhancer activity if any in IFM could then be read by a reporter gene expression, and thus the role of the associated gene could be ascertained in muscle development. However, such strategy had limitations as they yielded only partial molecular information related to gene expression pattern only and hence were redesigned to give information related to more dynamic molecule, i.e. protein through the development of protein trap strategy [19]. Open resources of such *Drosophila* protein tagged with green fluorescent protein [20] could greatly help in visualizing their subcellular localization and build protein interactome which is very necessary for context to sarcomere organization and functioning.

1.4 IFM Mutant Phenotypes

IFM specification involves hierarchical gene expression which has to be switched on and off at the right developmental time to give rise to fully functional differentiated muscles. *twist* (*twi*) and *notch* (*N*) are two upstream genes expressed in myoblast which fuses to form IFM. However, constitutive expression of *twi* post fusion inhibits muscle differentiation, and similar phenotype is mimicked by *N* [21]. Another gene *vestigial* (*Vg*) was found to be crucial in imposing IFM fate as *vg null* leads to muscle degeneration through apoptosis and switching of IFM to direct flight muscle (another muscle in *Drosophila* thorax) like features [22]. Inhibition of small GTPase family of protein Rac1 led to early IFM defects and delay in myoblast fusion [23]. Similarly mutation in *stripe* [24] and ecdysone hormone-responsive gene *Broad complex* [25] led to improper IFM attachment to the epidermis. Motoneurons also influence muscle development as denervation of neuron associated with IFM diminished myoblast population though initial myoblast migration and fusion ensued normally. DVMs failed to develop eventually, and DLM either had smaller muscle mass or missing fibres [26]. Structural protein mutants have been invaluable in understanding the sarcomere organization. Null mutants as well as point mutations for IFM-specific protein isoforms of actin, myosin, tropomyosin, etc. enumerate meaningful insights on thin and thick filament assembly as well as regulation of muscle contraction [27, 29]. Assembly of disrupted Z band and resulting myofibrillar defects in kettin, alpha-actinin and muscle LIM protein mutants emphasizes the role of much more structural protein in proper sarcomere development [30, 32].

2 Materials

2.1 Tissue Dissection, Fixation, Mounting and Imaging: Reagents and Equipment

1. Liquid nitrogen kept in a small thermocol box with a lid.
2. Sylgard dissection plate (Fig. 3a). Sylgard 184 silicone elastomer from Sigma or Dow chemicals can be used according to the instruction manual to prepare your own dissecting plates. Briefly, the kit comprises of base and curing agent which is mixed in a 10 (base): 1 (curing agent) ratio and mixed manually in a disposable container. The mixture is finally poured in a small diameter glass or plastic petri dishes, partially filling it avoiding any air bubble and left for curing at room temperature for more than 2 days.
3. Fine forceps (91150-20), Vannas micro spring scissors straight (91500-09) and stainless steel minute pins 0.1 mm diameter (26002-10) from Fine Science Tools (Fig. 3b–d).
4. Set of stainless steel razor blades, petri dishes, glass slide, coverslips, tissue roll, paint brush and cavity block.
5. Microcentrifuge tubes (1.5 mL) for keeping samples.
6. Haemotoxylin and eosin (HE) stains.
7. Phosphate-buffered saline (PBS) pH 7.4 at room temperature.

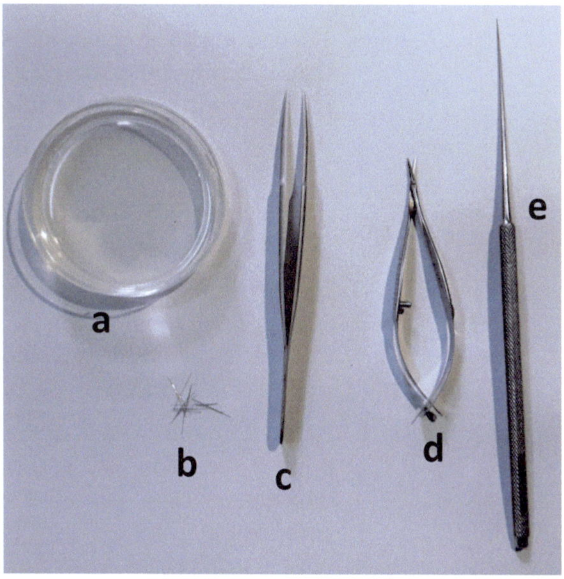

Fig. 3 (**a**) Sylgard plate. (**b**) Minutien pin. (**c**) Straight forceps. (**d**) Spring microscissor. (**e**) Pointed needle

8. 4% Paraformaldehyde (PF) in PBS. PF dissolves slowly and needs to be kept at 37 °C overnight. After complete dissolution aliquot the solution and freeze it in −20 °C. This can be used for over a month.
9. PBTx (PBS comprising 0.3% Triton X-100).
10. Glycerol.
11. Ethanol and its dilutions.
12. Methyl salicylate.
13. DPX mounting medium.
14. Nail polish.
15. Phalloidin-tetramethylrhodamine B isothiocyanate (*Phalloidin-TRITC*) (Sigma-Aldrich).
16. UltraCruz™ mounting medium (Santa Cruz Biotechnology).
17. Light microscope with polarized light optics (Olympus).
18. Zeiss Meta-V confocal microscope.

3 Methods

3.1 Microdissection of Pupa and Fixation

1. Take a pupa (20–50 h old) on a glass slide covered with a wet piece of tissue. By looking under the microscope carefully, make an incision at the posterior end where there is a small space in between the pupal case and developing pupa (Fig. 4a).
2. Remove the pupal case very slowly by rolling and making small incisions with fine forceps until we reach the anterior end.
3. Then scoop the pupae out using paint brush while still holding the remaining pupal case with forceps. Make sure that the pupa remains wet throughout this procedure (*see* **Note 6.4.1**).
4. Put the pupa on a Sylgard dissection plate filled with PBS, and pin it at three points, one at the edge of the head and others on the two winglike appendages of pupa with ventral side up (Fig. 4b).
5. Insert the scissor into the pupa from the abdominal end and make a small incision. Insert the scissor (make sure you do not touch the base of dorsal side), and cut the entire ventral body wall till the head.
6. Remove minute pin 2 and 3 and pin it in a new position by pulling the body wall and pinning it sideways, thus exposing developing IFM (situated roughly at upper half along the anterior-posterior axis).
7. IFM is situated at the dorsal inner side, and many layers of the fat body are above it making it impossible to observe the tissue (Fig. 4c).

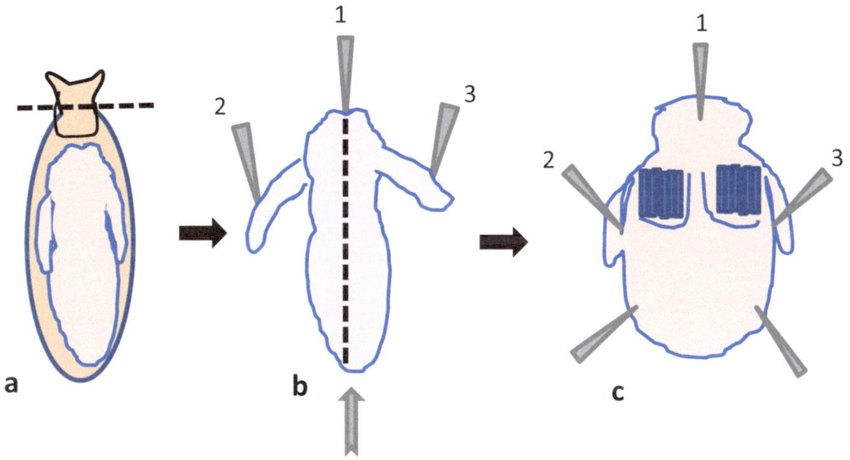

Fig. 4 Dissection of the pupa. (**a**) Removal of the pupal case by first breaking and inserting the forceps at the anterior end. (**b**) Pinning of the pupa, the ventral side facing up in Sylgard plate using a minutien pin at indicated positions. Midway cutting of the body wall. (**c**) Removal of the fat body reveals IFM in the upper quadrant

8. Carefully remove the fat bodies either with forceps or pointed needle and by gentle blowing with narrow straw pipe fitted with a 200 μL tip.
9. Wash the dissected pupae with PBS while it is still pinned in the Sylgard plate.
10. Fill the Sylgard plate with 4% PF for fixation at room temperature for 1 h in a rocker sliding at a very slow speed (*see* **Note 6.4.2**).
11. Remove the PF and wash with PBS and proceed accordingly for staining.

3.2 Adult Fly Dissection to Expose IFM

1. Take one anesthetized fly, and place it at one of the edges of a clean glass slide where a 50% glycerol solution has been smeared. Glycerol helps in sticking fly to the slide.
2. Avoid over-etherization of the fly and also over-usage of glycerol for sticking the fly on a glass slide.
3. The fly should be positioned in a way that its ventral side remains up with wings extended. Tap gently with a brush to the abdomen so that the fly remains flat on the surface.
4. Placing the fly at the corner of glass slide makes it convenient for dipping the fly in the liquid nitrogen while still holding the glass slide at the other end.
5. Dip the edge of the glass slide where the fly is placed into the box filled with liquid nitrogen for approximately 4–5 s.
6. Quickly dissect the fly with the help of razor while observing under the microscope along the ventral midline (Fig. 5).

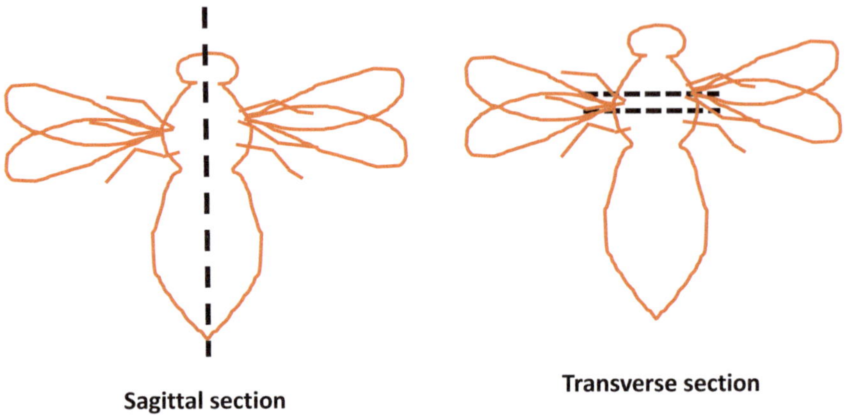

Fig. 5 Orientation of cutting the snap-frozen tissue in liquid nitrogen to expose IFM

This gives a sagittal view of the six DLM fibres running along the anterior-posterior axis.

7. Transverse section of the fly thorax can also be dissected by putting the razor perpendicular to the anterior-posterior axis and making two cuts (as denoted in the Fig. 5) to get a thin section.
8. Dissected flies are transferred to the microcentrifuge tube with the help of brush and fixed with 4% PF solution for 30 min. Gentle shaking can also be incorporated for proper fixation of tissues.
9. Care must be taken while aspirating solutions from microcentrifuge especially with transverse tissue sections.

3.3 Haemotoxylin-Eosin (HE) Staining

1. Fixed tissues of the pupa are now stained with eosin (0.5%) for 5 min followed by haematoxylin staining for another 1 min.
2. Subsequently dehydration of the tissues is done through alcohol series (50%, 70%, 90% and 100%), each for 1 h.
3. Use methyl salicylate to clear the pupal tissue for 1 h at room-temperature. This step is also known as tissue clearing. This step helps to make the tissue more transparent for imaging. Methyl salicylate has a high refractive index, thus percolating the tissue and equalizing the refractive index of the tissue resulting in less light scattering during imaging.
4. Finally stained tissues are mounted with DPX using coverslips and avoiding air bubbles.
5. Take images with upright microscope fitted with a digital camera (Carl Zeiss DM 1000) (Fig. 6a).

3.4 Polarized Light Imaging

1. For polarized light imaging, adult fly thorax obtained at Step 6 of Sect. 6.3.2 is serially dehydrated in alcohol series (50%,

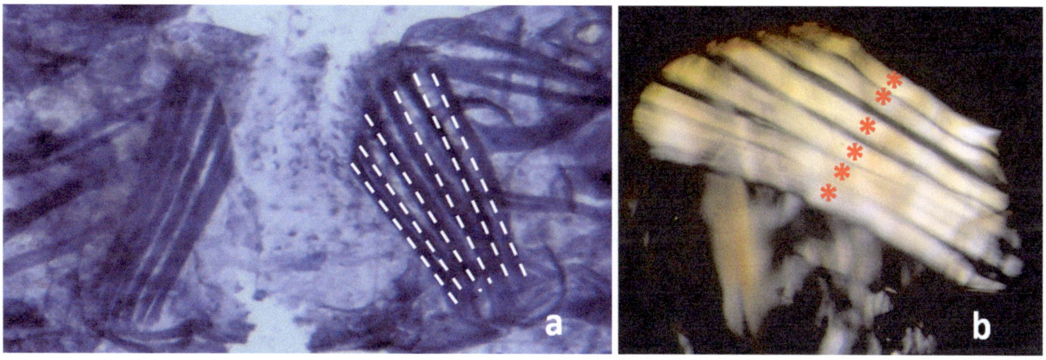

Fig. 6 (**a**) HE stained pupae showing DLM fibres. (**b**) Polarized light microscopy image of adult fly DLMs

70%, 90% and finally absolute alcohol) for a minimum of 2 h each (*see* **Note 6.4.3**).

2. After dehydration, ethanol is aspirated and left-over droplets are allowed to evaporate.
3. Methyl salicylate is then added to the tissues for at least 2 h before proceeding for mounting (at this stage tissue can be left at room temperature in methyl salicylate for more than a month).
4. The tissue is transferred to the cavity block along with methyl salicylate, and fly abdomen, head and wings are removed.
5. Bisected tissue can then be lifted with the help of a forceps and keep on the surface of a tissue paper to remove the excess of methyl salicylate.
6. Finally take a drop of DPX Mountant, and mount the sample in a way that IFM is facing towards the observer.
7. Air bubbles should be avoided while placing the coverslip, and the coverslips are secured at the edges by applying nail polish and allowed to air-dry.
8. Take images in stereomicroscope fitted with polarizer and analyser (Fig. 6b).

3.5 Confocal Microscopy

1. Pupa as well as adult fly tissues which are fixed from Sects. 6.3.1 and 6.3.2 can be utilized for immunohistochemistry (pinned pupa should be removed from the Sylgard plate and transferred to microcentrifuge tube for all the treatments).
2. Wash with PBTx thrice for 5 min each.
3. Prepare 1:200 dilution phalloidin-tetramethylrhodamine B isothiocyanate (*phalloidin-TRITC*) (Sigma-Aldrich) in PBS, and incubate the tissue at room temperature for 2 h or overnight at 4 °C (remember to cover the microcentrifuge with aluminium foil).
4. Wash with PBTx twice for 5 min with gentle shaking.

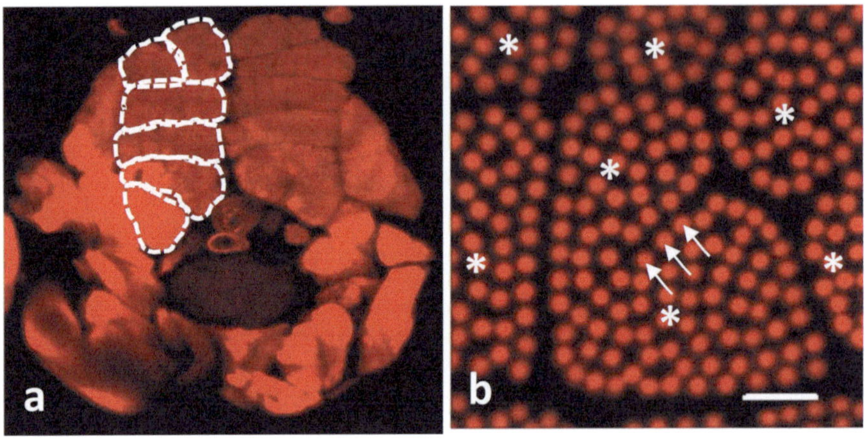

Fig. 7 Transverse view of DLM fibres. (**a**) Six bundles of muscle fibre of DLM in one lateral half of the thorax. (**b**) The single bundle consists of many myofilaments (marked by asterisks) which in turn are composed of many myofibrils (arrow point to single myofibril)

5. Mount in UltraCruz mounting medium which contains anti-fading agent, and seal the coverslip edge with nail paint.
6. Take the images under a confocal microscope with properly stained IFM in focus, and analyse with inbuilt software to assign scale bars, etc.
7. Myofilaments in a given muscle fibre in transverse tissue sections can be either counted manually or by using image J analysis software (Fig. 7a, b).

4 Notes

1. Pupal case removal requires much perseverance and with practice one can do it effortlessly.
2. Avoid using very old paraformaldehyde fixative solution as tissue fixation is prerequisite for immunohistochemistry.
3. Ensure proper tissue dehydration, and strictly adhere to the duration of incubation and dilutions of ethanol given in the protocol for polarized light microscopy.

Acknowledgements

We acknowledge DBT-BUILDER Project (BT/PR-9028/INF/22/193/2013) for lab equipment to Department of Life Sciences, Central University of Jharkhand and Indian Council of Medical Research (ICMR) for Senior Research Fellowship (SRF) grant to M.D.

References

1. Crossley AC (1978) The morphology and development of the *Drosophila* muscular system. In: Ashburner M, Wright TRF (eds) Genetics and biology of *Drosophila*. 2b. Acad. Press, New York, pp 499–560
2. Reedy MC, Beall C (1993) Ultrastructure of developing flight muscle in *Drosophila*. I. Assembly of myofibrils. Dev Biol 160:443–465. https://doi.org/10.1006/dbio.1993.1320
3. Peckham M, Molloy JE, Sparrow JC, White DCS (1990) Physiological properties of the dorsal longitudinal flight muscle and the tergal depressor of the trochanter muscle of *Drosophila melanogaster*. J Muscle Res Cell Motil 11(3):203–215
4. Fernandes J, Bate M, Vijayraghavan K (1991) Development of the indirect flight muscles of *Drosophila*. Development 113(1):67–77
5. Weitkunat M, Schnorrer F (2014) A guide to study *Drosophila* muscle biology. Methods 68(1):2–14. https://doi.org/10.1016/j.ymeth.2014.02.037
6. Bernstein SI, O'Donnell PT, Cripps RM (1993) Molecular genetic analysis of muscle development, structure, and function in *Drosophila*. Int Rev Cytol 143:63–152
7. Vigoreaux JO (2001) Genetics of the *Drosophila* flight muscle myofibril: a window into the biology of complex systems. BioEssays 23(11):1047–1063. https://doi.org/10.1002/bies.1150
8. Nongthomba U, Clark S, Cummins M, Ansari M, Stark M, Sparrow JC (2004) Troponin I is required for myofibrillogenesis and sarcomere formation in *Drosophila* flight muscle. J Cell Sci 117(9):1795–1805. https://doi.org/10.1534/genetics.106.056812
9. Firdaus H, Mohan J, Naz S, Arathi P, Ramesh SR, Nongthomba U (2015) A cis-regulatory mutation in troponin-I of *Drosophila* reveals the importance of proper stoichiometry of structural proteins during muscle assembly. Genetics 200(1):149–165. https://doi.org/10.1534/genetics.115.175604
10. Gunage RD, Dhanyasi N, Reichert H, VijayRaghavan K (2017) *Drosophila* adult muscle development and regeneration. Semin Cell Dev Biol 72:56–66. https://doi.org/10.1016/j.semcdb.2017.11.017
11. Dutta D, Anant S, Ruiz-Gomez M, Bate M, VijayRaghavan K (2004) Founder myoblasts and fibre number during adult myogenesis in *Drosophila*. Development 131(15):3761–3772. https://doi.org/10.1242/dev.01249
12. Ojima K, Lin ZX, Zhang ZQ, Hijikata T, Holtzer S, Labeit S, Sweeney HL (1999) Initiation and maturation of I-Z-I bodies in the growth tips of transfected. J Cell Sci 112:4101–4112
13. Homyk T, Sheppard DE (1977) Behavioral mutants of *Drosophila melanogaster*. I. Isolation and mapping of mutations which decrease flight ability. Genetics 87(1):95–104
14. Deak II (1977) Mutations of *Drosophila melanogaster* that affect muscles. Development 40(1):35–63
15. Cripps RM, Ball E, Stark M, Lawn A, Sparrow JC (1994) Recovery of dominant, autosomal flightless mutants of *Drosophila melanogaster* and identification of a new gene required for normal muscle structure and function. Genetics 137:151–164
16. An H, Mogami K (1996) Isolation of 88F actin mutants of *Drosophila melanogaster* and possible alterations in the mutant actin structures. J Mol Biol 260(4):492–505. https://doi.org/10.1006/jmbi.1996.0417
17. Cripps RM (2006) The contributions of genetics to the study of insect flight muscle function. In nature's versatile engine: insect flight muscle inside and out. Springer, Boston, pp 2–15
18. Bellen HJ (1999) Ten years of enhancer detection: lessons from the fly. Plant Cell 11(12):2271–2281. https://doi.org/10.1105/tpc.11.12.2271
19. Morin X, Daneman R, Zavortink M, Chia W (2001) A protein trap strategy to detect GFP-tagged proteins expressed from their endogenous loci in *Drosophila*. Proc Natl Acad Sci 98(26):15050–15055. https://doi.org/10.1073/pnas.261408198
20. Sarov M, Barz C, Jambor H, Hein MY, Schmied C, Suchold D, Stender B, Janosch S, Kj VV, Krishnan RT, Krishnamoorthy A (2016) A genome-wide resource for the analysis of protein localisation in *Drosophila*. elife 5:e12068. https://doi.org/10.7554/eLife.12068
21. Anant S, Roy S, Raghavan KV (1998) Twist and Notch negatively regulate adult muscle differentiation in *Drosophila*. Development 125(8):1361–1369
22. Bernard F, Lalouette A, Gullaud M, Jeantet A, Cossard R, Zider A, Ferveur J, Silber J (2003) Control of *apterous* by *vestigial* drives indirect flight muscle development in *Drosophila*. Dev Biol 260:391–403

23. Fernandes JJ, Atreya KB, Desai KM, Hall RE, Patel MD, Desai AA, Benham AE, Mable JL, Straessle JL (2005) A dominant negative form of Rac1 affects myogenesis of adult thoracic muscles in *Drosophila*. Dev Biol 285:11–27. https://doi.org/10.1016/j.ydbio.2005.05.040
24. Fernandes JJ, Celniker SE, VijayRaghavan K (1996) Development of the indirect flight muscle attachment sites in *Drosophila*: role of the PS integrins and the stripe gene. Dev Biol 176(2):166–184. https://doi.org/10.1006/dbio.1996.0125
25. Restifo LL, White K (1992) Mutations in a steroid hormone-regulated gene disrupt the metamorphosis of internal tissues in *Drosophila*: salivary glands, muscle, and gut. Roux Arch Dev Biol 201(4):221–234. https://doi.org/10.1007/BF00188753
26. Fernandes JJ, Keshishian H (2005) Motoneurons regulate myoblast proliferation and patterning in *Drosophila*. Dev Biol 277(2):493–505. https://doi.org/10.1016/j.ydbio.2004.09.038
27. Beall CJ, Sepanski MA, Fyrberg EA (1989) Genetic dissection of *Drosophila* myofibril formation: effects of actin and myosin heavy chain null alleles. Genes Dev 3:131–140
28. Kreuz AJ, Simcox A, Maughan D (1996) Alterations in flight muscle ultrastructure and function in *Drosophila* tropomyosin mutants. J Cell Biol 135(3):673–687
29. Barbas JA, Galceran J, Torroja L, Prado A, Ferrus A (1993) Abnormal muscle development in *hdp3* mutant of *Drosophila melanogaster* is caused by splicing defect affecting selected troponin-I isoforms. Mol Cell Biol 13:1433–1439
30. Hakeda S, Endo S, Saigo K (2000) Requirements of Kettin, a giant muscle protein highly conserved in overall structure in evolution, for normal muscle function, viability, and flight activity of *Drosophila*. J Cell Biol 148:101–114
31. Roulier EM, Fyrberg C, Fyrberg E (1992) Perturbations of *Drosophila* alpha-Actinin cause muscle paralysis, weakness, and atrophy but do not confer obvious nonmuscle phenotypes. J Cell Biol 116:911–922
32. Stronach BE, Renfranz PJ, Lilly B, Beckerle MC (1999) Muscle LIM proteins are associated with muscle sarcomeres and require dMEF2 for their expression during *Drosophila* myogenesis. Mol Biol Cell 10:2329–2342. https://doi.org/10.1091/mbc.10.7.2329

Chapter 7

Simple Histochemical Methods to Detect Cell Death in the Eye-Antennae Imaginal Disc of *Drosophila*

Swetapadma Sahu and Monalisa Mishra

Abstract

During early time point of development, a large number of cells are produced. As development progresses, many cells die to shape the body in its right form. *Drosophila melanogaster* has been extensively used to study various developmental mechanisms including cell death. It is an evolutionarily conserved mechanism needed for the cellular growth, development and maintenance of the organism. During development, cell death is initiated as an action of internal clock present within the body, but still many external factors can alter the time and frequency of cell death. The nucleic acids present within the body reflect the physiological condition of an organism. Various fluorescent dyes can detect the nucleic acid damage and ultimately the dead cells. Dyes that are commonly used to stain nucleic acid include acridine orange, propidium iodide, Hoechst and DAPI. All these dyes bind to the nuclei acid, and thus the live or dead stage of the cell can be visualized under a fluorescent microscope. The current study describes the staining techniques of all the four dyes in various *Drosophila* tissues.

Key words 4′, 6-diamidino-2-phenylindole (DAPI), Acridine orange, Hoechst, Propidium iodide

Abbreviation

AO	Acridine orange
DAPI	4′, 6-diamidino-2-phenylindole
PBS	Phosphate buffer saline
PBST	Phosphate buffer saline with Tween-20
PFA	Paraformaldehyde
PI	Propidium iodide

1 Introduction

Various physiological and environmental factors can damage the cell and thus initiate cell death [1]. During development, cell death is a natural phenomenon, to form the right structure. In this

Monalisa Mishra (ed.), *Fundamental Approaches to Screen Abnormalities in Drosophila*, Springer Protocols Handbooks, https://doi.org/10.1007/978-1-4939-9756-5_7,

context, various environmental factors may induce abnormal cell death. Numerous techniques have been developed to visualize healthy and dead cells. Fluorescence immunostaining is a powerful tool to identify live versus dead cells [2]. Using this technique, damages in the tissue can be identified [3]. The fluorophore binds to the cellular components and flourishes at a certain spectral range which can be visualized under a fluorescent microscope [4, 5]. Thus, various cell biological phenomena can be studied using immunostaining techniques [6]. In live cells, as the cell membrane is negatively charged, positively charged and some of the neutral molecules can pass through it, while negatively charged molecules can't pass through the cell membrane. The negatively charged dye can only enter into the cell if there is a breakage within the membrane [7]. Taking this property into account, negatively charged dyes are used to detect the damage in a cell or tissue.

Fluorescent dyes like Hoechst, diaminophenylindole (DAPI), propidium iodide (PI) and acridine orange (AO) get associated with nucleotides to produce fluorescence [8]. However, the mechanism via which they bind to the nucleic acid varies. Below we are describing the various ways for cell-dye interaction.

1. Intercalating agents: propidium iodide and ethidium bromide
2. Minor groove-binding agents: Hoechst dyes and DAPI
3. Random: acridine orange

1.1 Propidium Iodide

PI detects the apoptotic cells and is occasionally counterstained with different coloured fluorescent dyes [9, 10]. PI is impermeable for live cells but can penetrate easily through dead cells [11]. It has no specificity for nitrogenous bases. It binds only to DNA and RNA. One molecule of the dye can bind to 4–5 base pairs of DNA. PI binds to RNA after DNase treatment [12]. Binding of dye to nucleic acids enhances the fluorescence up to 20–30 times. An absorption and emission maximum of propidium iodide is 535 and 617 nm, respectively. A mercury arc or xenon lamp can excite PI [13].

1.2 Acridine Orange

Acridine orange is a cell-permeable cationic dye used to detect apoptosis [14]. It binds to both DNA or RNA with single and double strands [15]. The excitation and emission maxima are 502 and 525 nm, respectively. When it intercalates to double-stranded DNA, it emits green colour [16]. The excitation is at 460 nm and emission is 650 nm when it is associated with RNA and emits red colour [15]. One molecule of AO binds to three base pairs of DNA to flourish. In a single-stranded DNA/RNA, it interacts to the exposed phosphate backbone and emits a red colour [17, 18]. The emission spectra reflects the binding of dye to either with the single- and double-stranded DNA [19].

1.3 Hoechst

Hoechst is positively charged and can be easily assessable to live cells. It has a high affinity for AT-rich regions of DNA in fixed cells, and after binding, its fluorescence can enhance up to 30-fold. It has a less toxic effect than DAPI. The excitation and emission wavelength are 360 and 460 nm, respectively. It emits blue colour like DAPI and can be excited under UV filter in a fluorescence microscope [20].

1.4 DAPI

4′, 6-diamidino-2–phenylindole (DAPI) is a commonly used fluorophore which detects both live and dead cells [21]. After binding the fluorescence increases up to 20 times [22]. DAPI has chemical formula $C_{16}H_{15}N_5{\bullet}2HCl$ and its molecular weight is 350.25 g/mol. The excitation and emission wavelength are 340 and 488 nm, respectively [23]. The fluorescence blue colour can be detected under UV light. It can bind to double-stranded DNA strongly, and the excitation reaches to 360 nm, while emission shifts to 460 nm [23]. It binds to mostly the minor groove region of DNA double strand with higher specificity to AT-rich regions [24]. DAPI fades faster than other dyes [25]. The current protocol describes the staining of all the dyes which can mark cell death in eye-antennae imaginal disc.

2 Materials and Chemicals Required

Dissect the eye-antennae imaginal disc from the third inster larvae using 4% cold PFA (*see* **Note 5**).

This will be used throughout the experiment for the staining purpose.

2.1 Reagents

1. Acridine orange (HIMEDIA-GRM3087)
2. Hoechst (HIMEDIA-TC266)
3. DAPI (HIMEDIA-MB097)
4. Propidium iodide (HIMEDIA-TC252)
5. 1X Phosphate Buffered Saline(1X PBS)
6. 1X Phosphate Buffered Saline with 1% Tween-20 (PBST)
7. Para Formaldehyde (PFA) (HIMEDIA-MB059)
8. 1 N NaOH (HIMEDIA-MB095)
9. HCL
10. Glycerol (HIMEDIA-MB060)
11. Whatman-40

2.2 Materials and Equipment

1. Needle
2. Syringe (5 ml)
3. Glass slide (25 × 75 mm, ABDOS: G10101)

4. Petri plate (90 × 5 mm, Riviera)
5. Coverslip (18 mm, BLUE STAR)
6. Wax pencil (or small birthday candle)
7. Stereomicroscope (Leica)
8. Hot stir plate (TARSON Digital spinot: 6030)
9. Beaker (100 ml, Riviera-7,010024)
10. Stirring beads (TARSON-4117)
11. pH meter (EUTECH instruments-WZ-99855-16)

3 Method

3.1 Preparation of 1X PBS and 1X PBST

1. To prepare 100 ml of 1X PBS, add all the ingredients (Table 1) with 80 ml milliQ water and allow it to dissolve.
2. Adjust the pH of the solution to 7.2 by adding NaOH or HCl as per your requirement.
3. For making 100 ml of 1X PBST, mix all the components as given in the table (Table 1), and dissolve it in 80 ml of milliQ water.
4. After maintaining the pH, adjust the volume to 100 ml by adding milliQ water. Autoclave the solutions prepared prior to use [26] (*see* **Note 1–4**).

Table 1
(Recipe for PBS and PBST preparation)

	1X PBST (for 100 ml)		1X PBS (for 100 ml)	
	Chemical	**Amount (g)**	**Chemical**	**Amount (g)**
1	NaCl (HIMEDIA-GRM853)	0.8	NaCl	0.8
2	KCl (HIMEDIA-MB043)	0.02	KCl	0.02
3	Na_2HPO_4 (HIMEDIA- MB024)	0.14	Na_2HPO_4	0.14
4	KH_2PO_4 (HIMEDIA)	0.02	KH_2PO_4	0.02
5	Twin 20 (SIGMA-P9416)	0.2 ml		

3.2 Preparation of Fixative (4% PFA)

1. Heat 70 ml of 1X PBS in a glass beaker up to 60 °C on a hot plate with stirring beads (*see* **Note 1**).
2. Add 4 g of paraformaldehyde to it and stir continuously.
3. Add NaOH drop wise to make a completely clear solution.
4. Once this state is reached, cool and filter it.
5. Adjust the pH to 7.2 by adding HCl.
6. Now adjust the volume to 100 ml and make aliquots and finally store it in 4 °C (*see* **Note 5–7**).

3.3 Preparation of Mounting Media (20% Glycerol)

1. To make 20% glycerol, add 20 ml of glycerol to 80 ml of milliQ water, adjusting the volume to 100 ml, and slowly mix the components by pipetting. Store it in 4 °C (*see* **Note 4**).

3.4 Preparation of Working Concentration of Acridine Orange

1. Acridine orange (M.W. 369.96) can be stored at room temperature in its solid form.
2. To prepare the stock solution, dissolves 1 mg/ml of AO in milliq water.
3. After dilution, store it in 4 °C in dark or covered with aluminium foil.
4. The prepared solution can be used for months if handled carefully without exposure towards the light (*see* **Notes 8 and 9**).
5. To make the working solution of 1 μg/ml, we have to take 1 μl of stock solution and dilute it with 1 ml of milliQ water.
6. Keep it in the 4 °C freeze in dark condition.
7. Incubate the dissected tissue in a drop (20–30 μL) of 0.5 μg/mL acridine orange solution.

3.5 Preparation of Working Solution of DAPI/ Hoechst

1. Prepare 1 mg/ml of stock solution (*see* **Note 10**).
2. To prepare the stock solution, dissolve 1 mg of DAPI (M.W. 350.25)/Hoechst (M.W. 561.93) in 1 mL of milliQ water.
3. To make the working solution of 1 μg/ml, take 1 μl of stock solution and dilute it with 1 ml of 1X PBS (See notes 10 and 11)
4. Sonicate it to achieve the right concentration.
5. This is your working solution (*see* **Notes 10–13**).
6. Store it at 4 °C in the dark condition.
7. Hoechst can be diluted as the same method as DAPI.

3.6 Preparation of Working Concentration of Propidium Iodide

1. The dry form of PI can be stored at room temperature (*see* **Note 14**).
2. Dilute the PI (M.W. 668.39) with autoclaved water to make a 1 mg/ml stock solution (*see* **Note 15**).

3. After dilution, store it in 4 °C in dark and cover it with aluminium foil.
4. The prepared solution can be used for months if handled carefully without exposure to light.
5. To make the working solution of 1 μg/ml, we have to take 1 μl of stock solution and dilute it with 1 ml of milliQ water and keep it in a freeze in dark condition (*see* **Note 16**).

3.7 Fixation of Samples

1. Collect 10–15 third instar larvae from the food vials in a petriplate.
2. Wash them using 1X PBS.
3. Dissect the samples under a stereomicroscope by using a drop of cold 4% PFA to decrease the chances of cell damage.
4. Collect the imaginal discs for fixation in an 0.5 ml tube [27].
5. Fix the samples with 500 μl of 4% PFA in a 2 ml tube for 30 min at room temperature or overnight at 4 °C (*see* **Note 5**).
6. Discard the 4% PFA carefully from the tube.
7. Wash three times with 400 μl of 1X PBS to the samples.
8. Keep in a shaker for 5 min each wash at room temperature [28].
9. Samples need to be permeabilized for assessing the dyes. Post-fix the samples by keeping them with 400 μl of 1X PBST for 15–20 min [29] (Fig. 1).

3.8 Staining of Samples

1. Pipette out the PBST from the tube carefully.
2. Stain the samples using 40 μl of 1 μg/ml required stain.
3. Keep the tubes in dark for 5 min.
4. Remove the stain and wash the sample using PBS for two times, 5 min each wash.
5. In the case of acridine orange, the washing is done once because it is more sensitive (Fig. 2) (*see* **Notes 8 and 9**).

3.9 Preparation of Microscopic Slides for Staining

1. Take a glass slide and clean it using acetone.
2. To the centre of the slide, make a circle with wax to avoid spreading of samples to the borders of the slide during washing (*see* **Note 13**).
3. Put the stained samples along with 1X PBS at the centre of the wax circle.
4. Carefully remove the PBS by using a syringe (*see* **Note 6**).
5. Soak the excess amount of PBS using a triangle-shaped filter paper (Whatman-40) (*see* **Note 12**).
6. Do clean the slide carefully to prevent the loss of samples.

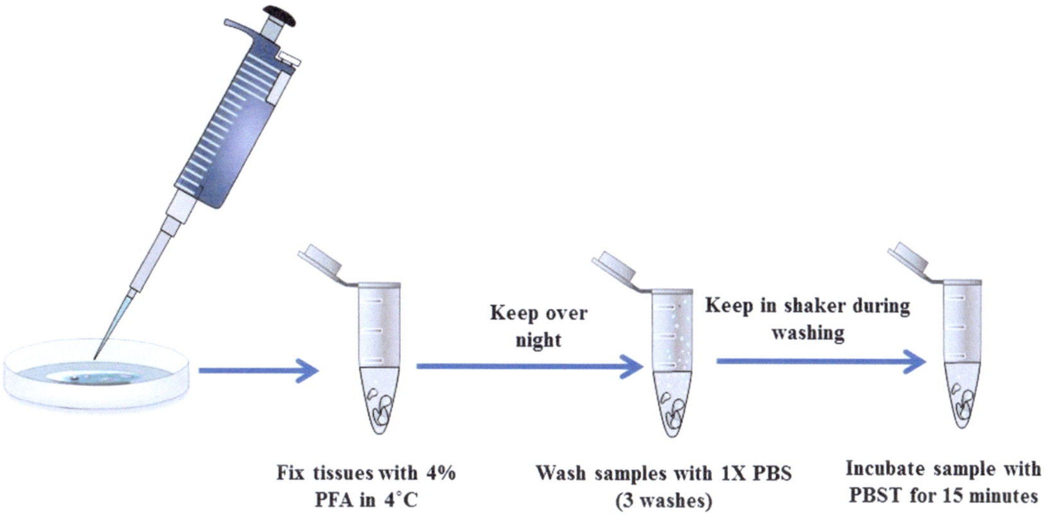

Fig. 1 Dissected discs are transferred from petriplate to tubes containing 4% PFA. After overnight incubation in 4 °C, tissues are washed with 1X PBS. Then they are fixed with 1XPBST for 15 min to make its membrane permeable

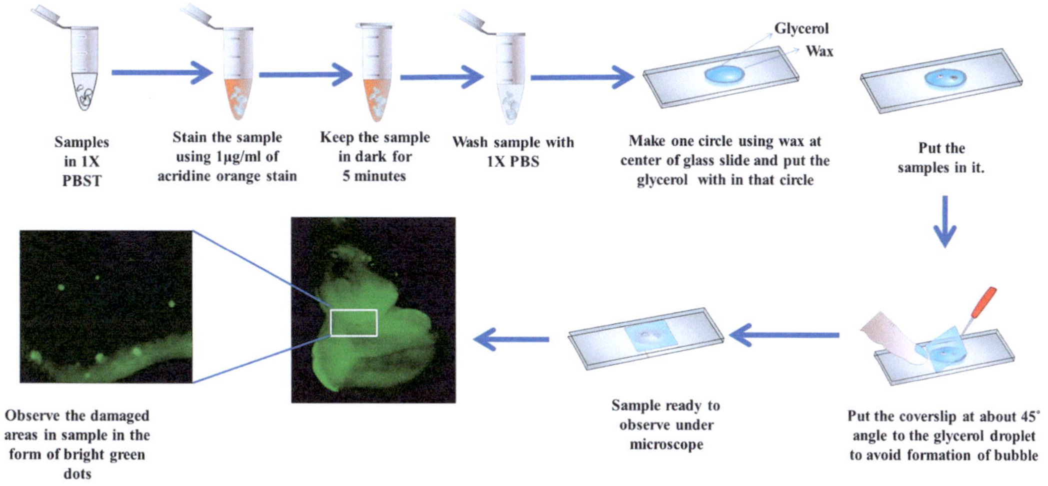

Fig. 2 Steps showing acridine orange staining of an eye-antennae imaginal disc

7. Place one drop (~ 20 μl) of 20% glycerol to the samples at the centre.
8. Now observe the position of the imaginal disc under the microscope to arrange the samples in a proper orientation.
9. Now take one clean coverslip and put it above the sample by placing it at 45° using a needle.

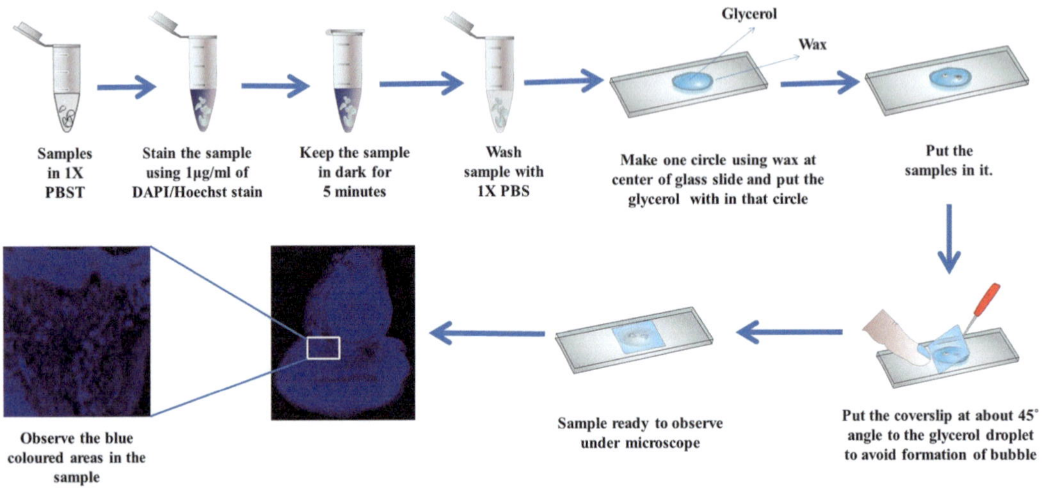

Fig. 3 Steps showing DAPI/Hoechst staining of an eye-antennae imaginal disc

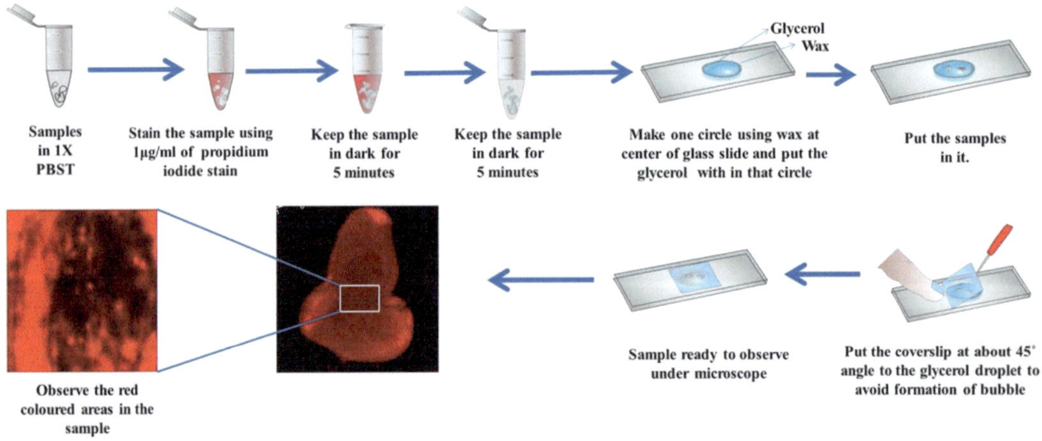

Fig. 4 Steps showing propidium iodide staining of an eye-antennae imaginal disc

10. Care should be taken that there is no air bubble near the sample since it interferes with the imaging.
11. Soak excess glycerol using triangle-shaped cut Whatman-40 or with a tissue paper.
12. Observe the slide under the microscope (Figs. 3 and 4) (*see* **Note 7**).

4 Note

1. Care must be taken during heating PBS so that there should not be over boiling.
2. The volume of the solution should not be adjusted before the maintenance of pH.

3. Prepared staining solutions must be stored in different aliquots to avoid light exposure.
4. Some chemicals are carcinogenic, so wearing gloves is recommended during its handling.
5. PFA fixation should be done at a cold temperature to avoid cell death.
6. Transfer the chemicals using a syringe to avoid damage of samples.
7. Store the prepared slides in the freeze till use.

(Acridine orange)

8. Acridine orange is sensitive towards light and undergoes photobleaching very soon, so excess exposure of light should be avoided before imaging the sample.
9. Aliquots must be prepared and each time a fresh aliquot of stain can be used to avoid light exposures.

(DAPI/ Hoechst)

10. Avoid contact with the skin since it may cause irritation.
11. The pH of chemicals must be maintained properly to get a good image.
12. Before staining, washing buffers must be blotted carefully; otherwise, it may affect the concentration of the stain resulting in improper imaging.
13. If non-specific stains are seen during imaging wash properly using PBS.

(Propidium iodide)

14. Propidium iodide is a mutagen, so it needs careful handling.
15. Use of gloves is recommended throughout the experiment.
16. To get a proper visualization of damages in DNA, proper concentration of PI must be used; otherwise, it may overstain the cytosol.

Acknowledgements

SS is thankful to DST/INSPIRE Fellowship/2016/IF160247 for financial support. MM lab is supported by Grant No. BT/PR21857/NNT/28/1238/2017, EMR/2017/003054, Odisha DBT 3325/ST(BIO)-02/2017.

References

1. Plemel JR, Caprariello AV, Keough MB, Henry TJ, Tsutsui S, Chu TH, Schenk GJ, Klaver R, Yong VW, Stys PK (2017) Unique spectral signatures of the nucleic acid dye acridine orange can distinguish cell death by apoptosis and necroptosis. J Cell Biol 216(4):1163–1181

2. Herzenberg LA, Sweet RG, Herzenberg LA (1976) Fluorescence-activated cell sorting. Sci Am 234(3):108–118
3. Neyfakh AA (1988) Use of fluorescent dyes as molecular probes for the study of multidrug resistance. Exp Cell Res 174(1):168–176
4. Lichtman JW, Conchello J-A (2005) Fluorescence microscopy. Nat Methods 2(12):910
5. Martin RM, Leonhardt H, Cardoso MC (2005) DNA labeling in living cells. Cytometry A 67(1):45–52
6. Kusuzaki K, Murata H, Matsubara T, Satonaka H, Wakabayashi T, Matsumine A, Uchida A (2007) Acridine orange could be an innovative anticancer agent under photon energy. In Vivo 21(2):205–214
7. Johnson I (1998) Fluorescent probes for living cells. Histochem J 30(3):123–140
8. von Bertalanffy L (1963) Acridine orange fluorescence in cell physiology, cytochemistry and medicine. Protoplasma 57(1–4):51–83
9. Nicoletti I, Migliorati G, Pagliacci M, Grignani F, Riccardi C (1991) A rapid and simple method for measuring thymocyte apoptosis by propidium iodide staining and flow cytometry. J Immunol Methods 139(2):271–279
10. Macklis JD, Madison RD (1990) Progressive incorporation of propidium iodide in cultured mouse neurons correlates with declining electrophysiological status: a fluorescence scale of membrane integrity. J Neurosci Methods 31(1):43–46
11. Cevik IU, Dalkara T (2003) Intravenously administered propidium iodide labels necrotic cells in the intact mouse brain after injury. Cell Death Differ 10(8):928
12. Krishan A (1975) Rapid flow cytofluorometric analysis of mammalian cell cycle by propidium iodide staining. J Cell Biol 66(1):188–193
13. Rieger AM, Hall BE, Schang LM, Barreda DR (2010) Conventional apoptosis assays using propidium iodide generate a significant number of false positives that prevent accurate assessment of cell death. J Immunol Methods 358(1–2):81–92
14. Chohan K, Griffin J, Carrell DT (2004) Evaluation of chromatin integrity in human sperm using acridine orange staining with different fixatives and after cryopreservation. Andrologia 36(5):321–326
15. Virant-Klun I, Tomazevic T, Meden-Vrtovec H (2002) Sperm single-stranded DNA, detected by acridine orange staining, reduces fertilization and quality of ICSI-derived embryos. J Assist Reprod Genet 19(7):319–328
16. Singh NP, McCoy MT, Tice RR, Schneider EL (1988) A simple technique for quantitation of low levels of DNA damage in individual cells. Exp Cell Res 175(1):184–191
17. Baisch H (1988) Different quiescence states of three culture cell lines detected by acridine orange staining of cellular RNA. Cytometry 9(4):325–331
18. Rigler R (1969) Acridine orange in nucleic acid analysis. Ann N Y Acad Sci 157(1):211–224
19. Raap A, Marijnen J, Vrolijk J, Van der Ploeg M (1986) Denaturation, renaturation, and loss of DNA during in situ hybridization procedures. Cytometry 7(3):235–242
20. Cai K, Yang J, Guan M, Ji W, Li Y, Rens W (2005) Single UV excitation of Hoechst 33342 and propidium iodide for viability assessment of rhesus monkey spermatozoa using flow cytometry. Arch Androl 51(5):371–383
21. Dmitrieva NI, Burg MB (2008) Analysis of DNA breaks, DNA damage response, and apoptosis produced by high NaCl. Am J Physiol-Renal Physiol 295(6):F1678–F1688
22. Lakowicz JR, Gryczynski I, Malak H, Schrader M, Engelhardt P, Kano H, Hell SW (1997) Time-resolved fluorescence spectroscopy and imaging of DNA labeled with DAPI and Hoechst 33342 using three-photon excitation. Biophys J 72(2 Pt 1):567
23. Gryczynski I, Malak H, Lakowicz JR (1996) Multiphoton excitation of the DNA stains DAPI and Hoechst. Bioimaging 4(3):138–148
24. Tanious FA, Veal JM, Buczak H, Ratmeyer LS, Wilson WD (1992) DAPI (4′, 6-diamidino-2-phenylindole) binds differently to DNA and RNA: minor-groove binding at AT sites and intercalation at AU sites. Biochemistry 31(12):3103–3112
25. Gallardo-Escarate C, Álvarez-Borrego J, Von Brand E, Dupré E, Del Rio-Portilla MA (2007) Relationship between DAPI-fluorescence fading and nuclear DNA content: an alternative method to DNA quantification? Biol Res 40 (1):29–40
26. Wilk R, Murthy SU, Yan H, Krause HM (2010) In situ hybridization: fruit fly embryos and tissues. Current Protocols Essential Laboratory Techniques 4(1):9.3. 1–9.3. 24
27. Chen F (2016) Preparation and immunofluorescence staining of the trachea in Drosophila larvae and pupae. Bio-protocol 6(9):e1797. https://doi.org/10.21769/BioProtoc.1797
28. Micchelli CA (2014) Whole-mount immunostaining of the adult Drosophila gastrointestinal tract. Methods 68(1):273–279
29. Bonaccorsi S, Giansanti MG, Cenci G, Gatti M (2012) Formaldehyde fixation of Drosophila testes. Cold Spring Harb Protoc 2012(8): pdb.prot067355

Chapter 8

Comet Assay to Detect the Severity of DNA Damage in *Drosophila*

Gyanaseni Dhar and Monalisa Mishra

Abstract

The comet assay is a rapid and sensitive method to detect DNA damages in individual cells. In this method, the electrophoretic migration of the DNA from the nucleus is achieved after the lysis of the cell in a thin layer of agarose. During electrophoresis, the loops containing a break lose their supercoiling nature and become free to extend towards the anode. The migration of DNA within the agarose results in a comet shape, which can be observed under a fluorescent microscope. The length of the comet tail compared to the head exhibits the number of DNA breaks. Based on the shape of the comet, the severity of the DNA damage can be measured. The size, shape and amount of DNA inside the 'comet' determine the severity of the DNA damage. The damage can be scored from the fluorescence images either by necked eye or by using different softwares. This assay is used to test the DNA damage, by various chemicals and infectious agents using *Drosophila* as a model system.

Key words Comet assay, Single cell gel electrophoresis, DNA strand breaks, Genotoxic compounds

Abbreviation

EDTA	Ethylenediaminetetraacetic Acid
EtBr	Ethidium bromide
HCl	Hydrochloric acid
LMA	Low melting point agarose
NaOH	Sodium hydroxide
NMPA	Normal melting point agarose
PBS	Phosphate-buffered saline

1 Introduction

Östling and Johanson have first described the concept of microgel electrophoresis in 1984 [1]. The principle is based upon the measurement of DNA single-stranded breaks, which occurs due to the relaxation of the supercoiling of DNA. Singh et al. modified the

Monalisa Mishra (ed.), *Fundamental Approaches to Screen Abnormalities in Drosophila*, Springer Protocols Handbooks, https://doi.org/10.1007/978-1-4939-9756-5_8,

idea in 1988 and introduced the use of alkaline condition to the media [2]. In this method, DNA gel electrophoresis gets incorporated with fluorescence microscopy, to visualize the migration of DNA strand in individual agarose embedded cells. Later, the method is modified by Ostling and Johanson in 1990 and termed as 'comet assay' (also referred as the single cell gel electrophoresis assay (SCG or SCGE assay)) after the emergence of the DNA from individual cells. The image appears like a comet and contains a head and a tail. The head is composed of high molecular weight of DNA, and the tail comprises of the end of migrating fragment. The tail movement measures both the DNA amount and distribution within the tail. If there is a break in the negatively charged DNA, the supercoiling of the DNA get relaxed, and it gets opened. The broken ends of the DNA migrate to the anode in the presence of an electric field [3]. In case of intact DNA, the free ends are lacking and thus prevent the migration of large-size fragment. Thus, based on DNA migration, the number of damaged DNA can be measured in a cell population. This mechanism makes comet assay, a simple, rapid and sensitive method to detect the DNA damage [4].

The comet assay has several advantages over other DNA damage assays. They include the following: (a) the DNA damage can be seen at the individual cell level, (b) less number of cells are required per sample, (c) the sensitivity of the DNA damage can be detected, and (d) this can be used for single cell population of both in vitro and in vivo [4–6]. The principle is based upon the speed of the migration of the DNA in various electric fields. The damaged DNA migrates faster than the intact DNA [7]. The samples are dissociated (in case of tissue) and are diluted with phosphate-buffered saline (PBS) to form the single cell suspension. The single cell suspension was mixed with low melting agarose (LMA). The cell mixed agarose gel was placed on a precoated slide to get a uniform layer [8]. A coverslip is applied to the uniformly spread precoated slide. In this way, the migration of the DNA from the cells embedded into the agarose will occur. Next, the slides are immersed into the lysis solution for the removal of proteins, nuclear and cellular membrane and the small-sized DNA molecules migrated away from the resident nucleus. Thus, the DNA unwinding and electrophoresis will occur with a specific pH. In case of negatively charged DNA break, the supercoiled DNA was relaxed and allows the broken end to migrate to the anode at the time of electrophoresis. In the case of undamaged DNA, (1) free ends are missing, and (2) it contains the large-size fragments. These two factors prevent the migration of the DNA. If the unwinding of DNA and electrophoresis occur at neutral pH -7, it facilitates the double-stranded break and crosses link detection. If it occurs at pH 12.1–12.4, it facilitates the detection of breakage of single- and double-stranded DNA and incomplete excision repair site. If the pH is more than 12.6, it indicates the alkali-labile sites with the addition to all categories of lesions [4, 9]. Finally, neutralization and staining with DNA binding dye performed before the analysis of the microscopic slide under the microscope. Under this process, a small portion of DNA extends to the anode and forms the electrophoretic track, which gives a shape of a 'comet'. Two different types of hypothesis were proposed for the formation of the comet tail [10]. (1)The first is based on the treatment of cell with free radicals; ionizing irradiation causes linear fragmentation of the DNA which either connects with the residual nucleoid or loses it. In alkaline condition, the fragments are single-stranded, and in a neutral condition, the fragments are double-stranded [11, 12]. (2) The second one proposes the nucleoid is composed of negatively supercoiled DNA loops which are attached to the nucleoid protein [13, 14]. If one single-stranded break appears in the loop, it gets relaxed, resulting in the disappearance of supercoil and migration of DNA [10]. The migration of the DNA results in the formation of the tail which can be visualized by fluorescence dyes. The DNA present in the tail is proportional to the ionizing radiation dose [10].

The comets are classified as five groups based on the basis of the appearance of the tail [15, 16]. First one is C0 which lacks a tail, C1–C3 with increasing order of tail length, and C4 represents the comet with high fragmented DNA. For the analysis, three parameters like tail length, tail moment and the amount of DNA present in the tail [8, 10, 17] are taken into consideration. The tail is measured as the distance from the head centre to the end of the tail.

1.1 Applications of the Comet Assay

The availability and simplicity of comet assay make it an indispensable tool for clinical and biomonitoring studies [10]. Potential mutagens like nanoparticles, physical factors, drug and a number of chemicals were tested using this technique [18, 19]. It is used in various in vivo experiments to check the safety of various chemicals, pharmaceuticals [20] and chemo protection ability of phytochemicals [15]. For ecological monitoring, comet assay is performed on various model organisms to detect the genotoxicity of the compound. In the marine environment, mussels are used as the model organism [5]. For soil, earthworm coelomocytes are used [21]. In humans, the comet assay is performed on white blood cells to check the effect of radiation [22], disease [23], smoking [24] and nutrition-induced DNA damage in human being. Various lipids present in the diets cause an alteration in oxidative DNA damage in lymphocytes [25] and thus used in the various diagnosis and forensic applications [8, 15].

1.2 Comet Assay Using *Drosophila*

Drosophila is well known to check the genotoxicity of various chemicals. In *Drosophila*, the brain ganglia, midgut cells, imaginal disc cell and haemocyte are targeted to check the in vivo genotoxicity [26, 27]. In mammals, lymphocytes are the main target cells to measure DNA damage. In the case of invertebrate, haemocytes in the hemolymph are used instead of lymphocytes [28, 29]. The haemocytes are highly sensitive towards genotoxic agents, and the processing methodology is very simple. Haemocytes comes in direct contact with the toxic materials which are circulating within the hemolymph. So, it is used as a target for evaluating the genotoxic assessment [30]. Among insects, the comet assay was first done in *Drosophila* by Gaivao [31]. In her Ph.D work, she checked the comet assay on imaginal disc and brain ganglia cells. Later, the comet assay from brain ganglia cells of third instar larvae was published by Bilbao et al. [32]. Later, comet assay was carried out on various tissues like midgut cells [33–35], haemocytes [26] and imaginal disc cells [36]. The current protocol describes the comet assay using *Drosophila* haemocytes as the target tissue.

2 Materials and Chemical Required

2.1 Lysis Solution Preparation

1. 2.5 M NaCl (HiMedia-GRM853)
2. 100 mM EDTA (HiMedia-TC115)
3. 10 mM Trizma base (HiMedia-TC072)
4. Triton X (HiMedia-MB031)
5. Distilled water
6. NaOH (HiMedia-MB095)
7. Measuring cylinder (100 mL)
8. PH meter

2.2 Electrophoresis Buffer

1. 10 N NaOH (HiMedia-MB095)
2. 200 mM EDTA (HiMedia-TC115)
3. Distilled water
4. Glass bottle (1000 mL)

2.3 Neutralization Buffer

1. 0.4 M Tris (HiMedia-TC072)
2. Distilled water
3. HCl
4. Glass bottle
5. pH meter

2.4 Staining Solution

1. EtBr (HiMedia-MB071)
2. Distilled water

2.5 Slide Preparation

1. Slide
2. 1% Normal melting point agarose (NMPA)
3. Distilled water
4. Microwave oven
5. Beaker (50 mL)
6. Coverslip

2.6 Hemolymph Collection

1. Third instar larva
2. 1.5 Eppendorf tube (TARSON)
3. 0.5 Eppendorf tube (TARSON)
4. Blade
5. Needle
6. Centrifuge

2.7 Comet Assay

1. Electrophoresis unit (TARSON)
2. 0.65% LMA (low melting point agarose)
3. PBS
4. Refrigerator (4 °C)
5. Slides
6. Graduated pipette
7. Distilled water
8. Fluorescence microscope

3 Methods

3.1 Lysis Solution Preparation

1. To prepare 100 mL of lysis solution, take 14.61 g of NaCl (2.5 M) in a measuring cylinder.
2. Add 3.72 g of EDTA (100 mM) to it.
3. Add 0.12 g of Trizma base (10 mM) to the measuring cylinder.
4. Add about 70 mL of water to the measuring cylinder and mixed it properly.
5. Add NaOH dropwise to adjust the pH 10.
6. Then add the rest amount of water to make the volume 100 mL.
7. At the time of experiment add freshly 1 mL of Triton X to the above solution.
8. Store it at refrigerator (4 °C).

3.2 Electrophoresis Buffer

1. Dissolve 200 g of NaOH in 500 mL of water to make the stock solution (stock 1).
2. Take 14.89 g of 200 mM EDTA in 200 mL of water, and prepare the stock solution (stock 2).

3. Take 30 mL of NaOH solution (from stock 1), and mix it with 5 mL of EDTA (from stock 2). Add water to this mixture to make the volume 1000 mL.
4. This mixture will serve as 1X electrophoresis buffer.

3.3 Neutralization Buffer

1. Add 4.84 g of 0.4 M Tris to ml 80 mL of water.
2. Adjust the pH 7.5 by adding HCl dropwise.
3. Add rest amount of water to adjust the volume 100 mL.
4. Store the buffer at room temperature.

3.4 Staining Solution

1. Add 50 mg of EtBr into 1 ml of water. This will serve as the stock solution.
2. Take 1–2 μl of EtBr from the above solution, and add 50 mL of water to it. This will serve as a staining solution.

3.5 Comet Assay Procedure

3.5.1 Agarose Preparation

1. For the preparation of 1% normal melting point agarose (NMPA), take 1gm of powdered agarose in a glass beaker.
2. Add 100 mL of distilled water to the glass beaker.
3. Place the beaker in the microwave oven at low power for 3–4 min (*see* **Note 1**).
4. Make sure that the agarose dissolved completely in water.

3.5.2 Agarose Precoated Slide

1. Dip the slides into 1% agarose solution (Fig. 2, **Step 1**).
2. Wipe one side to make it clean.
3. Allow agarose to air-dry to form a thin film on the slide.
4. Slides can be prepared ahead of time and can be stored with desiccant (*see* **Note 2**).

3.5.3 Collection of Sample (Hemolymph)

1. Collect 25–30 numbers of larvae and wash it with PBS (Fig. 1).

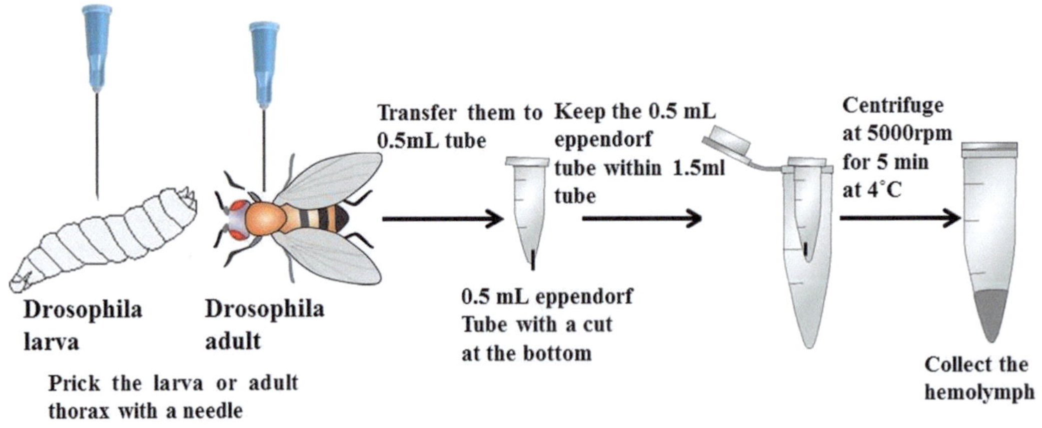

Fig. 1 Collection of hemolymph

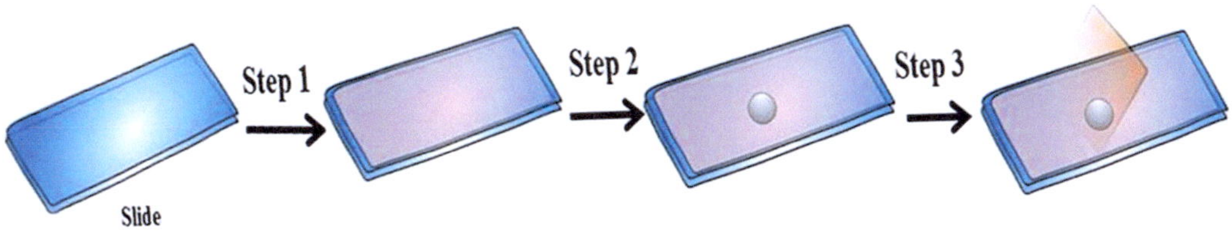

Fig. 2 Slide preparation and sample mounting. Step 1, coated with 1% NMPA overnight. Step 2, put hemolymph +0.65% LMA. Step 3, put coverslip on the hemolymph

2. Take washed larvae in 0.5 mL Eppendorf tube.
3. Make a cut at the bottom of the tube.
4. Place the Eppendorf within a 1.5 mL of Eppendorf tube.
5. Centrifuge at 4 °C with 5000 rpm for 5 min.
6. You will get the hemolymph in the 1.5 ml Eppendorf tube.
7. Suspend the hemolymph with 80 μl of 0.65% low melting point agarose (LMA) and 10 μl of cold PBS.
8. Rapidly mix the above mixture.
9. Pipet 100 μl of agarose mixed hemolymph (from the above step) onto the agarose-covered precoated slide (Fig. 2, **Step 2**).
10. Add it in a way that there will not be any bubble on the surface of the slide.
11. Add the coverslip and allow the agarose to dry (Fig. 2, **Step 3**).
12. Make sure that agarose is fully set before submerge it in the lysis solution.
13. After drying the coverslip was taken out.

3.5.4 Lysis and Electrophoresis

1. Take the lysis solution in the gel electrophoresis tank.
2. Place the slides horizontally lower to the gel electrophoresis tank.
3. Lyse samples for 1 h at 4 °C in the dark (Fig.3, **Step 1**).
4. After 1 h, carefully remove the slide.
5. Rinse the slide to remove salt and detergent.
6. Submerge the slides in fresh electrophoresis buffer for 15 min (Fig. 3, **Step 2**) (*see* **Notes 3, 4, 5**).
7. Conduct the electrophoresis for 25 min at 25 V and 300 mA (Fig. 3, **Step 3**).

3.5.5 Slide Staining

1. Remove the slides from the electrophoresis chamber, and rinse it with a neutralizing solution (Fig. 3, **Step 4**).

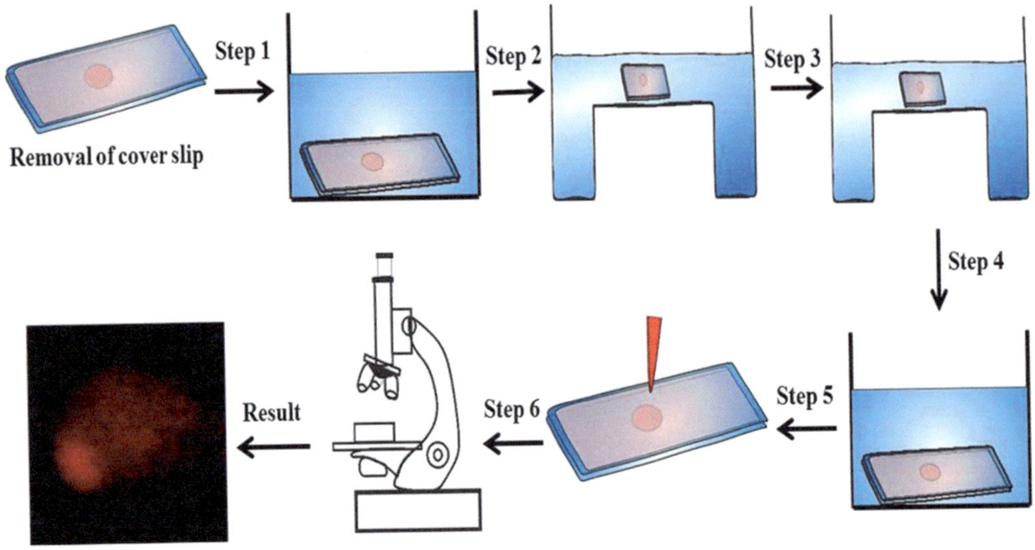

Fig. 3 Comet assay procedure. Step 1, dip the slide inside the lysis solution for 1 h. Step 2, slide submerged inside the electrophoresis buffer for 15 min. Step 3, electrophoresis occur for 20–30 min 25 V and 300 mA. Step 4, dip the slide inside the neutralizing solution for 10 min. Step 5, stain with EtBr. Step 6, analyse under fluorescence microscope

2. Place the slides in a staining solution containing EtBr for 20 min (Fig. 3, **Step 5**).
3. Next, rinse slides with distilled water to remove excess stain.

3.5.6 Slide Analysis

1. Observe the slides under fluorescence microscope (Fig. 3, **Step 6**).
2. Take various images from the slides.
3. Use image analysis software, to analyse individual images.
4. Parameters like total intensity (DNA content), tail length, the percentage of DNA damage in tail and tail moment can be monitored from the slide.

4 Notes

1. At the time of agarose melting, avoid vigorous boiling of the agarose, and make sure that the agarose dissolved completely.
2. Do not work in a humid environment. This will result to fall off the agarose from the slide.
3. Slides should be placed in the tank gently.
4. The slide contained in a gel tank should not be transferred.

5. At the time of lysis and electrophoresis, the agarose gel should not fall off from the slide.

Acknowledgements

GD is thankful to DST/INSPIRE Fellowship/2016/IF160243 for financial support. MM lab is supported by Grant No. BT/PR21857/NNT/28/1238/2017, EMR/2017/003054 and Odisha DBT 3325/ST(BIO)-02/2017.

References

1. Ostling O, Johanson KJ (1984) Microelectrophoretic study of radiation-induced DNA damages in individual mammalian cells. Biochem Biophys Res Commun 123(1):291–298
2. Singh NP, McCoy MT, Tice RR, Schneider EL (1988) A simple technique for quantitation of low levels of DNA damage in individual cells. Exp Cell Res 175(1):184–191
3. Olive PL, Banáth JP (2006) The comet assay: a method to measure DNA damage in individual cells. Nat Protoc 1(1):23
4. Kumaravel T, Vilhar B, Faux SP, Jha AN (2009) Comet assay measurements: a perspective. Cell Biol Toxicol 25(1):53–64
5. Dixon DR, Pruski AM, Dixon LR, Jha AN (2002) Marine invertebrate eco-genotoxicology: a methodological overview. Mutagenesis 17(6):495–507
6. Lee RF, Steinert S (2003) Use of the single cell gel electrophoresis/comet assay for detecting DNA damage in aquatic (marine and freshwater) animals. Mutat Res 544(1):43–64
7. Olive P, Banath J, Fjell C (1994) DNA strand breakage and DNA structure influence staining with propidium iodide using the alkaline comet assay. Cytometry 16(4):305–312
8. Shukla RK (2017) Forensic application of comet assay: an emerging technique. Forensic Sci Res 2(4):180–184
9. Miyamae Y, Iwasaki K, Kinae N, Tsuda S, Murakami M, Tanaka M, Sasaki YF (1997) Detection of DNA lesions induced by chemical mutagens using the single-cell gel electrophoresis (comet) assay.: 2. Relationship between DNA migration and alkaline condition. Mutat Res 393(1):107–113
10. Afanasieva K, Sivolob A (2018) Physical principles and new applications of comet assay. Biophys Chem 238:1
11. Horváthová E, Dušinská M, Shaposhnikov S, Collins AR (2004) DNA damage and repair measured in different genomic regions using the comet assay with fluorescent in situ hybridization. Mutagenesis 19(4):269–276
12. Shaposhnikov SA, Salenko VB, Brunborg G, Nygren J, Collins AR (2008) Single-cell gel electrophoresis (the comet assay): loops or fragments? Electrophoresis 29(14):3005–3012
13. Cook P, Brazell I (1975) Supercoils in human DNA. J Cell Sci 19(2):261–279
14. Cook PR, Brazell IA, Jost E (1976) Characterization of nuclear structures containing superhelical DNA. J Cell Sci 22(2):303–324
15. Collins AR (2004) The comet assay for DNA damage and repair. Mol Biotechnol 26(3):249
16. Collins AR, Oscoz AA, Brunborg G, Gaivao I, Giovannelli L, Kruszewski M, Smith CC, Štětina R (2008) The comet assay: topical issues. Mutagenesis 23(3):143–151
17. Lovell DP, Omori T (2008) Statistical issues in the use of the comet assay. Mutagenesis 23(3):171–182
18. Brendler-Schwaab S, Hartmann A, Pfuhler S, Speit G (2005) The in vivo comet assay: use and status in genotoxicity testing. Mutagenesis 20(4):245–254
19. Glei M, Schneider T, Schlörmann W (2016) Comet assay: an essential tool in toxicological research. Arch Toxicol 90(10):2315–2336
20. Hartmann A, Agurell E, Beevers C, Brendler-Schwaab S, Burlinson B, Clay P, Collins A, Smith A, Speit G, Thybaud V (2003) Recommendations for conducting the in vivo alkaline comet assay. Mutagenesis 18(1):45–51
21. Verschaeve L, Gilles J (1995) Single cell gel electrophoresis assay in the earthworm for the detection of genotoxic compounds in soils. Bull Environ Contam Toxicol 54(1):112–119
22. Somorovska M, Szabova E, Vodička P, Tulinska J, Barančoková M, Fabry R, Líšková A, Riegerova Z, Petrovska H, Kubova J (1999) Biomonitoring of genotoxic risk in workers in a rubber factory: comparison of

the comet assay with cytogenetic methods and immunology. Mutat Res 445(2):181–192

23. Collins AR, Kn R, Somorovská M, Petrovská H, Ondrušová A, Vohnout B, Fábry R, Dušinská M (1998) DNA damage in diabetes: correlation with a clinical marker. Free Radic Biol Med 25(3):373–377
24. Betti C, Davini T, Giannessi L, Loprieno N, Barale R (1994) Microgel electrophoresis assay (comet test) and SCE analysis in human lymphocytes from 100 normal subjects. Mutat Res 307(1):323–333
25. Jenkinson AM, Collins A, Duthie S, Wahle K, Duthie G (1999) The effect of increased intakes of polyunsaturated fatty acids and vitamin E on DNA damage in human lymphocytes. FASEB J 13(15):2138–2142
26. Carmona ER, Guecheva TN, Creus A, Marcos R (2011) Proposal of an in vivo comet assay using haemocytes of Drosophila melanogaster. Environ Mol Mutagen 52(2):165–169
27. Gaivão I, Sierra LM (2014) Drosophila comet assay: insights, uses, and future perspectives. Front Genet 5:304
28. Meister M (2004) Blood cells of Drosophila: cell lineages and role in host defence. Curr Opin Immunol 16(1):10–15
29. Irving P, Ubeda JM, Doucet D, Troxler L, Lagueux M, Zachary D, Hoffmann JA, Hetru C, Meister M (2005) New insights into Drosophila larval haemocyte functions through genome-wide analysis. Cell Microbiol 7(3):335–350
30. Villela IV, de Oliveira IM, da Silva J, Henriques JAP (2006) DNA damage and repair in haemolymph cells of golden mussel (Limnoperna fortunei) exposed to environmental contaminants. Mutat Res 605(1):78–86
31. Gaivão I (1999) Genotoxic evaluation of reactive oxygen species generatig compounds: a study in *Drosophila melanogaster*. Ph. D. thesis, Universidade de Trás-Os-Montes and Alto Douro, Vila Real
32. Bilbao C, Ferreiro JA, Comendador MA, LMa S (2002) Influence of mus201 and mus308 mutations of Drosophila melanogaster on the genotoxicity of model chemicals in somatic cells in vivo measured with the comet assay. Mutat Res 503(1–2):11–19
33. Mukhopadhyay I, Chowdhuri DK, Bajpayee M, Dhawan A (2004) Evaluation of in vivo genotoxicity of cypermethrin in Drosophila melanogaster using the alkaline Comet assay. Mutagenesis 19(2):85–90
34. Siddique HR, Chowdhuri DK, Saxena D, Dhawan A (2005) Validation of Drosophila melanogaster as an in vivo model for genotoxicity assessment using modified alkaline Comet assay. Mutagenesis 20(4):285–290
35. Sharma A, Shukla A, Mishra M, Chowdhuri DK (2011) Validation and application of Drosophila melanogaster as an in vivo model for the detection of double strand breaks by neutral Comet assay. Mutat Res 721(2):142–146
36. Verma A, Sengupta S, Lakhotia SC (2012) DNApol-ϵ gene is indispensable for the survival and growth of Drosophila melanogaster. Genesis 50(2):86–101

Chapter 9

Reactive Oxygen Species Detection of *Drosophila* Cells by Flow Cytometry

Ronnaly Rout, Srirupa Basu, and Monalisa Mishra

Abstract

Flow cytometry is an accurate and versatile technique to measure the fluorescence intensity of a large population of cells. In this chapter, we are using flow cytometer to detect the number of cells producing reactive oxygen species (ROS) in *Drosophila* under stress condition. ROS constitute a group of reactive molecules and free radicals derived from molecular oxygen. In this protocol, we are describing estimation of ROS from various types of *Drosophila* larval tissues, like imaginal disc, gut and freshly isolated hemolymph.

Key words Flow cytometer, Reactive oxygen species (ROS), 2′-7′-Dichlorodihydrofluorescein diacetate (DCF DA), 2′-7′-Dichlorofluorescein (DCF), *Drosophila* tissues, Hemolymph

1 Introduction

1.1 Production of ROS

All aerobic organisms undergo respiration utilizing molecular oxygen. It involves the transfer of electrons through a chain of carrier molecules in the mitochondrion to create an electrochemical gradient which furnishes ATP synthesis. Molecular oxygen finally accepts electrons and is reduced to form water [1]. However, some electrons leak from this pathway causing partial reduction of oxygen to generate free radicals which are reactive and unstable. Some of them are the singlet oxygen (O,) superoxide anion (O_2-), hydroxyl radicals (OH), various peroxides (ROOR′) and hydroperoxides (ROOH) including hydrogen peroxide (H_2O_2) [2]. All these molecules are collectively termed as reactive oxygen species (ROS). The mitochondria contribute for maximum production of ROS [3]. The damaging effect of ROS is controlled by cellular defence mechanism by synthesizing enzymes having antioxidative properties like catalase, superoxide dismutase (SOD), thioredoxin reductase and glutathione peroxidase (Gpx). In certain diseases like cancer, atherosclerosis, cardiovascular diseases, neurodegenerative diseases (Alzheimer's and Parkinson's) and diabetes [4–6], ROS is generated within the body. ROS resulted in increased oxidative

Monalisa Mishra (ed.), *Fundamental Approaches to Screen Abnormalities in Drosophila*, Springer Protocols Handbooks, https://doi.org/10.1007/978-1-4939-9756-5_9,

stress within the body [7]. In many cases ROS acts as a signalling molecule [8]. ROS provides an estimate of the redox status of the cell, and being highly reactive, it is essential to detect its level accurately during measurement [9].

1.2 Detection of ROS by DCFDA

ROS can be detected histologically, [10] by chemiluminescence [11] or by fluorescence measurements [12, 13]. Compared to fluorescence, the other methods have limitations like low sensitivity and specificity and time-consuming, and it often produces artefacts [14]. Among various fluorescent molecules, 2′,7′-dichlorodihydrofluorescein diacetate (DCFDA/H2DCFDA) is commonly used. This dye can freely enter into the cell and gets converted to 2′,7′-dichlorodihydrofluorescein (DCFH2) on oxidation by esterase enzyme. Later, it is oxidized to 2′,7-′-dichlorofluorescein (DCF) by ROS to emit a bright green fluorescence [15] which gets trapped intracellularly [16, 17]. All these chemical conversion steps are shown in Fig. 1. Various ROS like H_2O_2, NO_2 and peroxynitrite anions including glutathione concentration [18–20] can oxidize this dye [10] and thus provide the overall status of oxidative stress present within the cell.

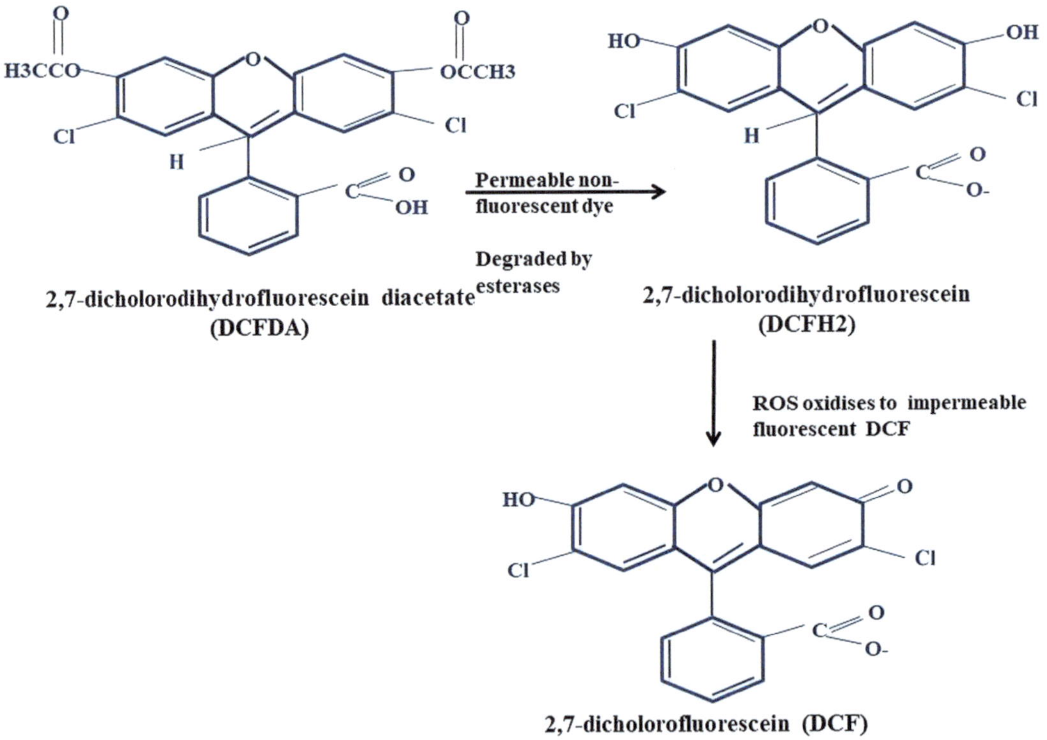

Fig. 1 Conversion of DCFDA dye to DCF by the action of ROS

This dye is excited at 488 nm wavelength, and the oxidation status can be checked by instruments like fluorimeter, confocal scanning microscopy or fluorescence microscopy and by flow cytometer. Among the different fluorescent assay methods, flow cytometry is the most accurate technique to quantify the intracellular fluorescence of cells which provides relative fluorescence units [21]. At a time thousands of cells are analysed, thus providing a precise fluorescence value. In addition, its sensitivity is high and can measure as low as 100 fluorescent cells. Flow cytometer has inbuilt optical detectors which receive emitted light scattered from cells after passing through specific band filters. Depending on the wavelength of emitted light, there are different band filters of which FL1 with 525 nm band pass filter detects only the green light. Therefore to detect DCF fluorescence, FL1 is selected which is the green channel used in this protocol [22].

1.3 Flow Cytometry for Analysing Drosophila Cells

Flow cytometry technique has been implemented in *Drosophila* tissue culture cell lines like S2, Kc and live cells from the organism [23]. Various methods are available for analysing *Drosophila* embryo [24], ovaries [25], larval imaginal discs [26] and haemocytes [27] by flow cytometer. Even first instar larvae using a fluorescent marker has been detected through flow cytometer [28]. It is well known that *Drosophila* haemocytes contribute for immune response by participating phagocytosis/encapsulation of pathogens, the release of antimicrobial peptides and reactive oxygen species (ROS) [29, 30]. Flow cytometry has been applied to freshly harvested *Drosophila* hemolymph cells for surface antibody reactivity [31]. In another study, hemolymph was analysed and sorted by using metabolite indicators like lectins, reporters (GFP and ß-galactosidase) and antibodies [27]. In this protocol, freshly harvested whole hemolymph from *Drosophila* larvae has been taken for ROS estimation.

Similarly, *Drosophila* larval tissues have also been analysed by flow cytometer for measurement of ROS. It has been reported that certain stress genes like hsp70, hsp83, hsp60 and hsp26 cause alteration of ROS production by adding volatile organic compounds like benzene, toluene and xylene [32]. Similarly, an orthologue gene Wwox in *Drosophila* is involved in aerobic metabolism and oxidative stress and thus modulates the ROS expression level [13]. The current protocol describes the analysis of ROS from various tissues by using flow cytometer.

2 Materials

1. 2′,7′-Dichlorodihydrofluorescein diacetate, acetyl ester (DCFDA) dye (Sigma-D6883)
2. Dimethylsulphoxide (DMSO) (HiMedia-MB058)
3. Trypsin-EDTA solution (Sigma-59418C)

4. Hydrogen peroxide of 30% strength (Merck-CK6C660813)
5. No. 5 Dumont forceps
6. 0.5 ml and 1.5 ml microcentrifuge tubes (Tarson-500015, 500016)
7. Petri plates (Borosil)
8. Paint brush
9. Pipettes 1 ml, 100 μl (Tarson)
10. Pipette tips of 100-1000 μl, 2-200 μl (Tarson-521020, 521010)
11. 100 μm cell strainer (HiMedia-TCP026)
12. Plastic tubes for flow cytometry (Tarson-850090)
13. Dissecting microscope
14. 12 Volt Orbital Shaker (Tarson)
15. Flow cytometry (Becton Dickinson Accuri C6)
16. Centrifuge (ThermoFisher)
17. 4 °C and −20 °C refrigerators (Samsung/Elanpro)
18. Incubator (37 °C, 5% CO2) (BR Biochem)
19. pH meter (Riviera, Eutech)

3 Method

3.1 Preparation of Reagents

1. Preparation of DCFDA stock: Prepare a stock solution of DCFDA (10 mM concentration) by diluting it with DMSO (*see* **Note 2**). Make the stock in an amber coloured 15 ml falcon tube. Store the solution in 4 °C.
2. Preparation of DCFDA working solution: Dilute the stock solution with 1X PBS to make the working solution of 10 μM concentration. Vortex the working solution for 15–30 s to disperse the dye evenly. This step should be in amber falcon tube (*see* **Notes 3 and 4**).
3. Preparation of 10X PBS: For preparing 10X PBS, take NaCl (40 g), KCL (1 g), Na_2HPO_4 (7.2 g) and K_2HPO_4 (1.2 g) in a measuring cylinder, and add 500 ml of distilled water. Adjust the pH to 7.4 and make volume up to 1000 ml.
4. Preparation of 1X PBS for dissection and isolation: Dilute the 10X PBS solution to working solution (1X) by adding milliQ water, and store the working solution at 4 °C.
5. Preparation of Trypsin-EDTA for tissue dissociation: Add 1X PBS to EDTA in a ratio of 1: 9 (EDTA 9 and PBS 1) to prepare the working solution. Make a working concentration of EDTA by adding 450 μl of Trypsin-EDTA with 50 μl of 1X PBS to make a total volume of 500 μl.

3.2 Preparation of Drosophila Sample

1. Fly strains: Wild-type *D. melanogaster* (*Oregon R*) is used in this study. Flies and larvae are grown on standard food vials having the composition corn meal powder, sucrose, yeast, agar, nepagin (methyl-p-hydroxy benzoate) and propionic acid and maintained at 24 ± 1 °C.
2. Collect the third instar larvae with a paint brush, and keep in 1XPBS at room temperature.
3. Wash the larvae properly to get rid of any food particle attached to the body.

3.3 Hemolymph Collection and Staining

1. Collect around 25–30 third instar larvae.
2. Wash the larvae with PBS and transfer them to a 0.5 ml centrifuge tube.
3. Make a longitudinal cut at the bottom of the centrifuge tube.
4. Rupture the larva cuticle with a microneedle. Do all these steps on ice (*see* **Note 5**). Transfer the ruptured larvae to the 0.5 ml tube. Put the cap on, and put it inside a 1.5 ml centrifuge tube. Take the cap off of the 1.5 ml centrifuge tube.
5. Centrifuge at 5000 rpm for 10 min at 4 °C.
6. Collect the hemolymph from the 1.5 ml centrifuge tube.
7. Take around 20 μl of hemolymph, and make up the volume to 100 μl by adding 1X PBS.
8. To the hemolymph, add 10 μl of 10 mM DCFDA dye at 37 °C in dark.
9. Incubate the dye with the hemolymph for 30 min.
10. After the onset of time, take the stained hemolymph in the flow cytometry tube and do the analysis immediately. All the steps involved in hemolymph collection and staining are shown in Fig. 2.

3.4 Dissection of Drosophila Tissues and Staining

1. Take 10–15 third instar larvae in a petri dish.
2. Dissect larval gut tissues, imaginal discs or any desired tissue of interest in 1XPBS cold PBS, and keep it in 1.5 ml centrifuge tube (*see* **Note 1**).
3. To the tissue add 0.5 ml of Trypsin-EDTA working solution.
4. Keep it for 1 h at 37 °C for incubation.
5. This step will give us a single cell suspension.
6. Pass the solution through a 100 μm cell strainer (HiMedia). This step will help to eliminate any clump or undigested tissue.
7. Centrifuge the tubes at 5000 rpm for 10 min at 25 °C.
8. Discard the supernatant.
9. Resuspend the cell pellet in 500 μl of 1XPBS.
10. Add 10 μl of 10 μM DCFDA to the resuspended cell pellet.

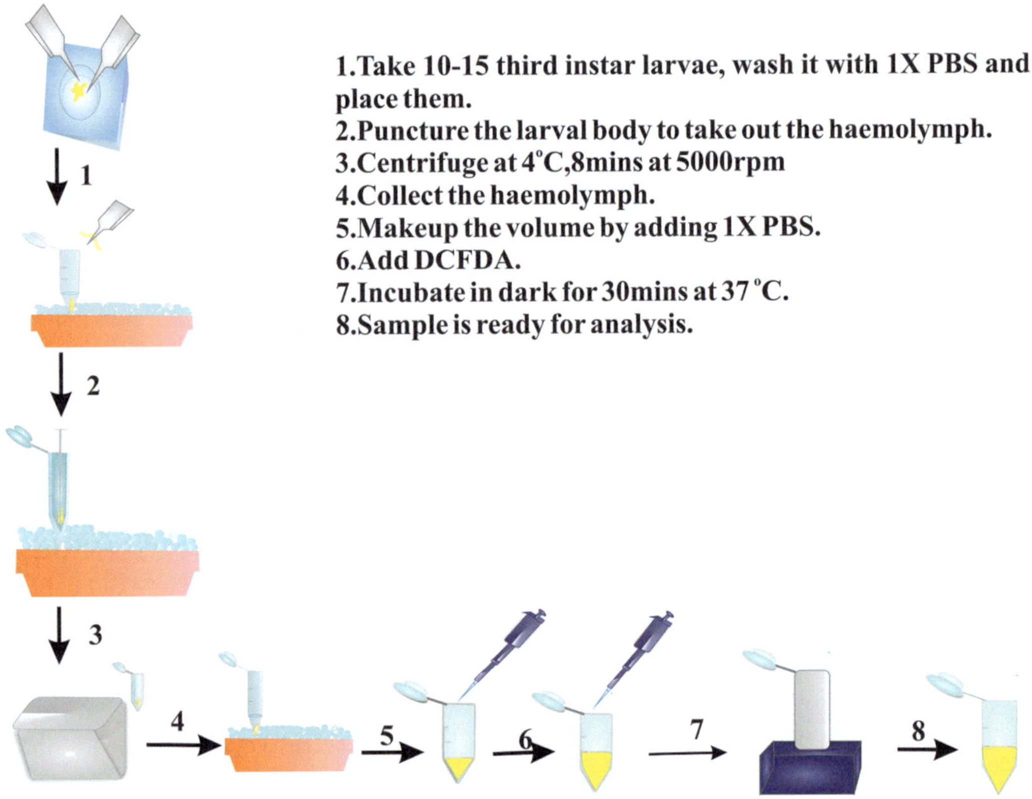

Fig. 2 Collection of hemolymph and staining with DCFDA dye

11. Incubate this step at 37 ° C for 30 min in dark.
12. To wash the stain after incubation, centrifuge at 5000 rpm for 10 min.
13. Decant the supernatant, and resuspend the pellet again in 500 μl PBS in flow cytometry tube.
14. Immediately analyse through flow cytometer. All the steps involved in tissue dissection and staining are shown in Fig. 3.

3.5 Positive and Negative Control

1. Take at least three sets of 1.5 ml centrifuge tubes.
2. The first one contains the unstained sample which can be hemolymph or tissue cells suspended in 0.5 ml 1XPBS without any dye. This is the blank sample tube or negative control.
3. The second tube is the control sample (*Oregon R*) stained with DCFDA dye.
4. The third tube is the positive control which is prepared by treating the control sample by adding H_2O_2 at 20 μM final concentration and then staining with DCFDA dye. (This is an optional step to induce ROS in the cells.)
5. Instead of **Step 4**, any mutant, treated, diseased or aged flies of your interest can be taken for the analysis of ROS.

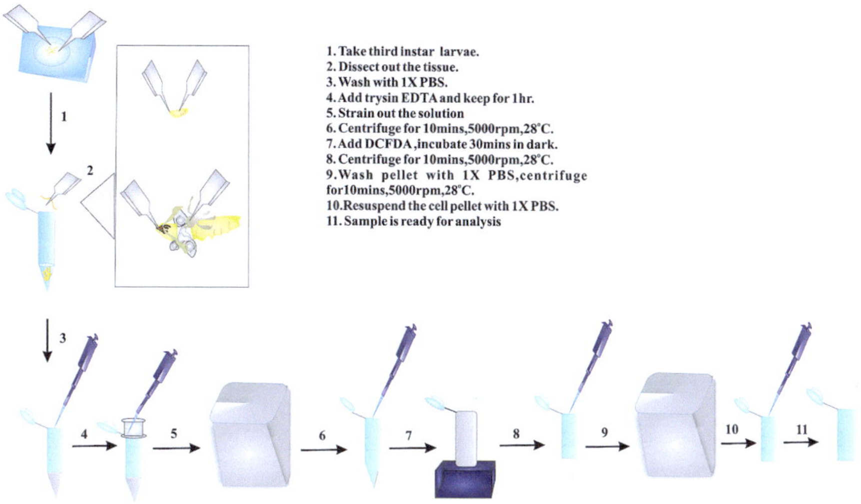

Fig. 3 Preparation of *Drosophila* tissue sample and staining with DCFDA dye

3.6 Steps for Flow Cytometer Acquisition

1. A flow cytometer has inbuilt different optical filters for detecting a wide range of fluorescent light. The light source excites the cells leading to emitted light which is converted to electrical pulses by optical detectors. The emitted light is sent to different detectors by using optical filters.
2. Various types of optical filters are inbuilt in the flow cytometer. The standard filters with its excitation emission range are the FL1 detector at (485/525 nm) wavelength, FL2 detector at (488/585 nm), FL3 at (488/670 nm) and FL4 at (640/675 nm) wavelength of light.
3. The FL1 can detect the fluorescence of the DCFDA dye since its excitation emission ranges from 485/525 nm wavelength. On degradation, DCF is formed which gives the green fluorescence; thus FL1 channel is selected for ROS analysis.
4. When laser light passes through the Eppendorf tube, it results in the scattering of light. The scattering of light can be classified into two types: (1) forward-scattered light (FSC) and (2) side-scattered light (SSC). The scattering of light depends on the particle size and internal granularity of the sample. FSC is proportional to cell surface area or size of the sample, whereas SSC is proportional to cell granularity or internal complexity.
5. The scattering of light is the source of the signal which will be detected by the flow cytometer.
6. The cell debris interferes with the scattering of light resulting in fewer signals than the intact cell. Thus, it is better to minimize the amount of cell debris in the sample of interest to be analysed.
7. The current study uses flow cytometer (BD Accuri) for detecting the signal. The equipment uses the C6 software for the analysis of the signal.

8. The software allows a workspace for the acquisition of the sample. The workspace is organized into four different options, namely, (1) data collection and acquisition, (2) multiple samples analysis, (3) statistics and (4) batch analysis.
9. Select the first option which is collect for data collection in the workspace template.
10. This step will allow you to acquire the sample displaying the dot plot which is a resultant of FSC versus SSC. The Y axis displays the SSC height, and the X axis displays the FSC height which is required to view the population of cells by scattering with similar SSC and FSC values.

3.7 Sample Collection

1. We can acquire the sample of our interest (gut, imaginal disc or hemolymph) at this step. First take the blank unstained (negative control) sample and acquire to set the dot plot. This will give you the background fluorescence/autofluorescence which is present in your sample itself.
2. Then acquire the control (*Oregon R*) sample stained with DCFDA dye.
3. Finally, acquire the (positive control sample DCFDA +H2O2) which we have used in our study (*see* **Note 6**).
4. Now compare the data with the control-stained sample (*see* **Note 7**).
5. Instead of step 3, you can use your treated sample of interest. Refer to **Step 5** of Sect. 3.5.
6. All the samples are acquired on the flow cytometer with at least 50,000 events per sample. However, the events vary from sample to sample. The channel you have selected is responsible for the event which displays the signal value.
7. Usually, healthy cells exhibit low ROS production due to cellular antioxidant defence mechanism. However, the treated cells or samples produce variation in the level of ROS.

3.8 Data Analysis

1. The flow cytometric data is stored in a standard format called as flow cytometer standard (FCS) format containing the sample information, instrument and analysis [33].
2. After saving the data, the cell population can be displayed in different formats like density plot or histogram plot while acquiring the sample.
3. To isolate the events of our specific population, we need to gate the plot by drawing a graphical boundary around the population of interest. Gating excludes the cell debris and allows analysing the data of events within the gate in subsequent plots (Fig. 4).

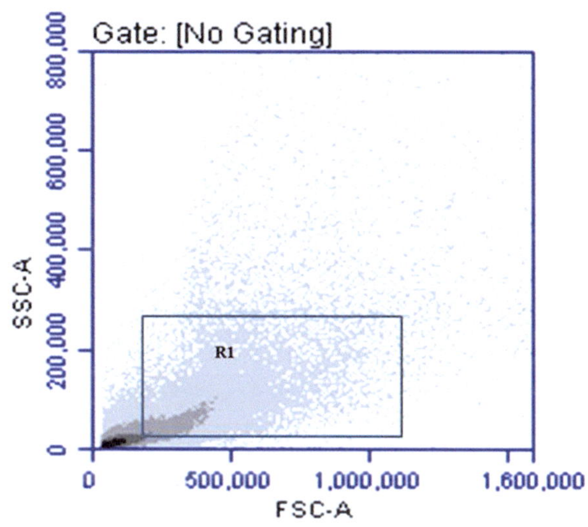

Fig. 4 Dot plot of hemolymph (SSC vs FSC)

4. A single-parameter histogram plot for DCFDA (FL1) is selected in the workspace for ROS analysis which displays a single peak as a homogenous cell population is taken for analysis.
5. The histogram is represented as the fluorescence value in the FL1 channel number in the horizontal X axis in the logarithmic scale and the number of events per channel number (FL1) in the vertical Y axis. Each event corresponds to its signal value in the FL1 channel.
6. Histogram markers are used which can be horizontal or vertical line bars used to specify a range of events for a single parameter. First the histogram peak of the negative control sample is selected, and a vertical marker is placed such that the VL representing unstained cell population corresponds to 100% and VR, the stained cells to designate positive events set at zero (Fig. 5A).
7. The stained control and the positive control sample histogram peak depict the events for DCFDA stained cells and represented as a percentage of positive cells (Fig. 5B, C).
8. The data are expressed as the relative shift due to a change in the fluorescence intensity of the dye. The greater the shift towards the right suggests a number of ROS molecules in the analysis of sample which is resulting in an increase in fluorescence.
9. You can also display the data in the histogram overlay plot by selecting the option from the analysis section of the workspace. This will allow you to compare all the samples in the same plot. The overlay plot (Fig. 6) depicts the peak for negative control (A), stained control (B) and positive control (C) samples in the hemolymph.

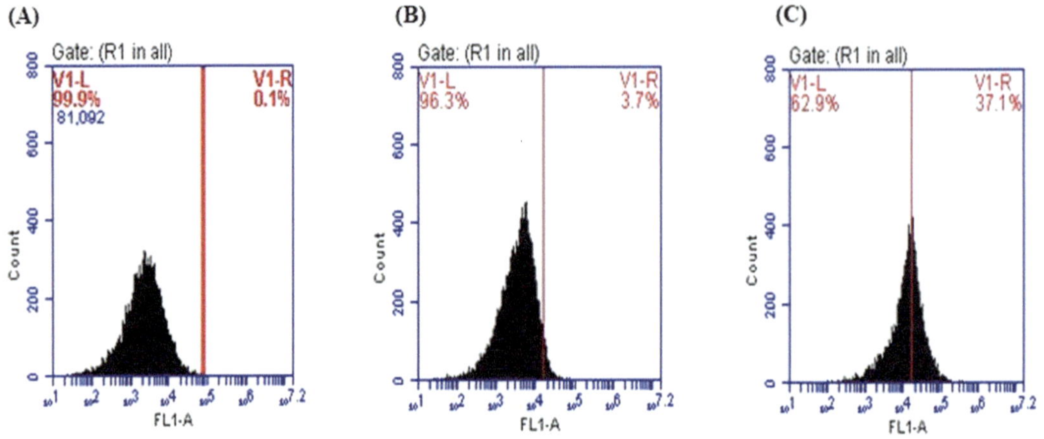

Fig. 5 Single-parameter histogram peak for DCFDA (FL1) of the (**A**) negative control (**B**), control-stained and (**C**) positive control sample

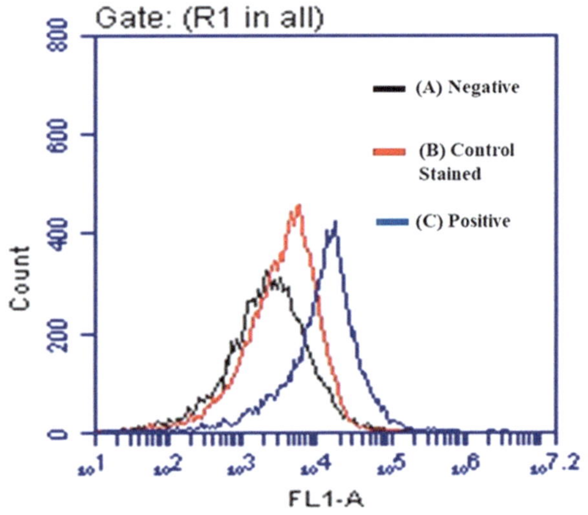

Fig. 6 Histogram overlay plot showing shifting of the peak in hemolymph due to DCFDA staining

10. Similarly for gut tissue cells and imaginal disc cells, the overlay graphs are represented in Fig. 7A, B.
11. All the samples are run in triplicates, and the mean values are calculated. The data generated from the plots can be exported for statistical analysis.
12. Subtract the mean fluorescence value of the blank sample from the mean fluorescence value of the stained control (*Oregon R*) sample. This will eliminate the background fluorescence and will give the fluorescence intensity value of only the control (*Oregon R*)-stained sample.

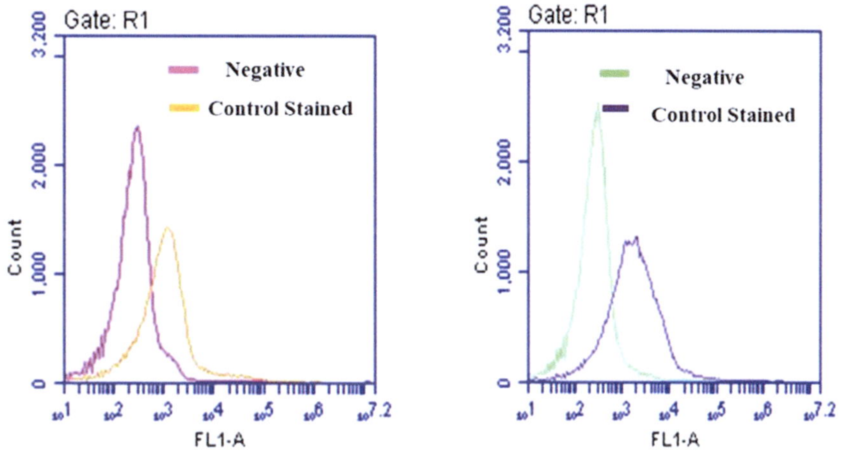

Fig. 7 Histogram overlay plot showing shifting of peak in (**A**) gut tissue cells and (**B**) imaginal disc cells due to DCFDA staining

13. The geometric mean of the fluorescence intensity can be compared between the control and treated sample. If the treated sample data mean is found to be greater than the control mean, it would be considered positive. The greater the difference between the two means denotes the number of positive molecules expressed per cell and the greater the positivity.

4 Note

1. The dissection and dissociation step should not exceed more than 2 h or else the tissue cells will undergo degradation and will give inaccurate ROS estimation.
2. The DCFDA should be dissolved only in anhydrous dimethylsulphoxide (DMSO) or dimethylformamide (DMF) or 100% ethanol.
3. Prepare the working solutions freshly, since it is air and light sensitive. Repeated freeze thaw cycle of the dye will affect the staining of the cells.
4. Use gloves when handling DMSO as it is toxic in nature. Specially avoid contact with skin, eyes or mucous membranes.
5. Hemolymph should be collected on ice to prevent melanisation.
6. Acquire the samples immediately or else it gets deteriorated.
7. Always deduct the background fluorescence shown by the negative control with the control sample.

Acknowledgements

We are thankful to Swetapadma Sahu and Gyanaseni Dhar for standardization of the experiment. RR is thankful to DST WOS-A for financial support. SB is thankful to Odisha DBT 3325/ST (BIO)-02/2017. MM lab is supported by Grant No. BT/PR21857/NNT/28/1238/2017, EMR/2017/003054 and Odisha DBT 3325/ST(BIO)-02/2017.

References

1. Hatefi Y (1985) The mitochondrial electron transport and oxidative phosphorylation system. Annu Rev Biochem 54(1):1015–1069
2. Halliwell B (1999) Antioxidant defence mechanisms: from the beginning to the end (of the beginning). Free Radic Res 31(4):261–272
3. Boveris A, Chance B (1973) The mitochondrial generation of hydrogen peroxide. General properties and effect of hyperbaric oxygen. Biochem J 134(3):707–716
4. Petersen RB, Nunomura A, Lee H-g, Casadesus G, Perry G, Smith MA, Zhu X (2007) Signal transduction cascades associated with oxidative stress in Alzheimer's disease. J Alzheimers Dis 11(2):143–152
5. Ozben T (2007) Oxidative stress and apoptosis: impact on cancer therapy. J Pharm Sci 96(9):2181–2196
6. Shults CW (2005) Antioxidants as therapy for Parkinson's disease. Antioxid Redox Signal 7(5–6):694–700
7. Halliwell B, Whiteman M (2004) Measuring reactive species and oxidative damage in vivo and in cell culture: how should you do it and what do the results mean? Br J Pharmacol 142(2):231–255
8. Owusu-Ansah E, Banerjee U (2009) Reactive oxygen species prime Drosophila haematopoietic progenitors for differentiation. Nature 461(7263):537
9. Cossarizza A, Ferraresi R, Troiano L, Roat E, Gibellini L, Bertoncelli L, Nasi M, Pinti M (2009) Simultaneous analysis of reactive oxygen species and reduced glutathione content in living cells by polychromatic flow cytometry. Nat Protoc 4(12):1790
10. Owusu-Ansah E, Yavari A, Banerjee U PROTOCOL EXCHANGE| Community CONTRIBUTED A protocol for in vivo detection of reactive oxygen species
11. Myhre O, Andersen JM, Aarnes H, Fonnum F (2003) Evaluation of the probes 2′,7′-dichlorofluorescin diacetate, luminol, and lucigenin as indicators of reactive species formation. Biochem Pharmacol 65(10):1575–1582
12. Xu C, Luo J, He L, Montell C, Perrimon N (2017) Oxidative stress induces stem cell proliferation via TRPA1/RyR-mediated Ca2+ signaling in the Drosophila midgut. Elife 6:e22441
13. Podratz JL, Lee H, Knorr P, Koehler S, Forsythe S, Lambrecht K, Arias S, Schmidt K, Steinhoff G, Yudintsev G (2017) Cisplatin induces mitochondrial deficits in Drosophila larval segmental nerve. Neurobiol Dis 97:60–69
14. Zhang Y, Dai M, Yuan Z (2018) Methods for the detection of reactive oxygen species. Anal Methods 10(38):4625–4638
15. Royall JA, Ischiropoulos H (1993) Evaluation of 2′,7′-dichlorofluorescin and dihydrorhodamine 123 as fluorescent probes for intracellular H2O2 in cultured endothelial cells. Arch Biochem Biophys 302(2):348–355
16. Mills EM, Takeda K, Yu Z-X, Ferrans V, Katagiri Y, Jiang H, Lavigne MC, Leto TL, Guroff G (1998) Nerve growth factor treatment prevents the increase in superoxide produced by epidermal growth factor in PC12 cells. J Biol Chem 273(35):22165–22168
17. Sundaresan M, Yu Z-X, Ferrans VJ, Irani K, Finkel T (1995) Requirement for generation of H2O2 for platelet-derived growth factor signal transduction. Science 270(5234):296–299
18. Gabriel C, Camins A, Sureda FX, Aquirre L, Escubedo E, Pallàs M, Camarasa J (1997) Determination of nitric oxide generation in mammalian neurons using dichlorofluorescin diacetate and flow cytometry. J Pharmacol Toxicol Methods 38(2):93–98
19. Possel H, Noack H, Augustin W, Keilhoff G, Wolf G (1997) 2,7-Dihydrodichlorofluorescein diacetate as a fluorescent marker for peroxynitrite formation. FEBS Lett 416(2):175–178

20. García-Ruiz C, Colell A, Marí M, Morales A, Fernández-Checa JC (1997) Direct effect of ceramide on the mitochondrial electron transport chain leads to generation of reactive oxygen species role of mitochondrial glutathione. J Biol Chem 272(17):11369–11377
21. Armstrong D (2010) Advanced protocols in oxidative stress II. Springer
22. Ameziane-El-Hassani R, Dupuy C (2013) Detection of intracellular reactive oxygen species (CM-H2DCFDA). Bio-protocol 3(1): e313
23. de la Cruz AFA, Edgar BA (2008) Flow cytometric analysis of Drosophila cells. In: Drosophila. Springer, pp 373–389
24. Cumberledge S, Krasnow MA (1994) Preparation and analysis of pure cell populations from Drosophila. In: Methods in cell biology, vol 44. Elsevier, pp 143–159
25. Bryant Z, Subrahmanyan L, Tworoger M, LaTray L, Liu C-R, Li M-J, van den Engh G, Ruohola-Baker H (1999) Characterization of differentially expressed genes in purified Drosophila follicle cells: toward a general strategy for cell type-specific developmental analysis. Proc Natl Acad Sci 96(10):5559–5564
26. Neufeld TP, de la Cruz AFA, Johnston LA, Edgar BA (1998) Coordination of growth and cell division in the Drosophila wing. Cell 93(7):1183–1193
27. Tirouvanziam R, Davidson CJ, Lipsick JS, Herzenberg LA (2004) Fluorescence-activated cell sorting (FACS) of Drosophila hemocytes reveals important functional similarities to mammalian leukocytes. Proc Natl Acad Sci 101(9):2912–2917
28. Krasnow MA, Cumberledge S, Manning G, Herzenberg LA, Nolan GP (1991) Whole animal cell sorting of Drosophila embryos. Science 251(4989):81–85
29. Lanot R, Zachary D, Holder F, Meister M (2001) Postembryonic hematopoiesis in Drosophila. Dev Biol 230(2):243–257
30. Hoffmann JA, Reichhart J-M (2002) Drosophila innate immunity: an evolutionary perspective. Nat Immunol 3(2):121
31. Kurucz E, Zettervall C-J, Sinka R, Vilmos P, Pivarcsi A, Ekengren S, Hegedüs Z, Ando I, Hultmark D (2003) Hemese, a hemocyte-specific transmembrane protein, affects the cellular immune response in Drosophila. Proc Natl Acad Sci 100(5):2622–2627
32. Singh MP, Reddy MK, Mathur N, Saxena D, Chowdhuri DK (2009) Induction of hsp70, hsp60, hsp83 and hsp26 and oxidative stress markers in benzene, toluene and xylene exposed *Drosophila melanogaster*: role of ROS generation. Toxicol Appl Pharmacol 235(2):226–243
33. Biosciences B (2000) Introduction to flow cytometry: a learning guide. Manual Part 1(1)

Chapter 10

Detection of Lipid Depot Within *Drosophila*

Nibedita Nayak and Monalisa Mishra

Abstract

Globally, diabetes and obesity are found to be the most challenging disease of the era. Although the complete treatment is not available, with the prevention and management, the disease can be treated. Various research are going on to understand the cause and consequence of obesity and diabetes using model organisms. Among all the other animal models, *Drosophila melanogaster* is emerging as an efficient model to study obesity and diabetes. In obesity-linked diabetes study, deposition of fat in the intestine, fat body and oenocytes can be checked by histological staining. To stain the fat depot within the body, Oil Red O, Sudan black and Nile red are used. The current protocol describes the localization of lipid in *Drosophila* tissue using various stains and also floating assay, a buoyancy-based screen assay to investigate fattyness in *Drosophila* larvae.

Key words Oil Red O, Sudan black, Nile red, Gut, Fat body, Floating assay

Abbreviations

PBS	Phosphate-buffered saline
PFA	Paraformaldehyde
PG	Propylene glycol

1 Introduction

Many diseases are caused due to abnormal metabolism [1]. Faulty metabolism includes imbalance in acid-base, calcium metabolism, DNA repair-deficiency, hyperlactatemia and iron metabolism. Due to irregular metabolism, diabetes, obesity, mitochondrial diseases, skin disease and brain-related disorders occur. Diabetes is a multi-organ disorder which causes obesity, amputations, blindness, renal failure, depression and cardiovascular diseases and premature death [2]. Although it is a critical disease, to date, appropriate treatment is still not available. Various animal models such as frogs, fish, rodents, *Caenorhabditis elegans*, *Saccharomyces cerevisiae* and

Monalisa Mishra (ed.), *Fundamental Approaches to Screen Abnormalities in Drosophila*, Springer Protocols Handbooks, https://doi.org/10.1007/978-1-4939-9756-5_10,

Drosophila are used to study diabetes to decipher the mechanism of the disease [3–11].

The diet of animals is interlinked with their metabolic state and diabetes [12, 13]. Thus, any alterations in dietary condition resulted in obesity and diabetes. A diet like sugar or fat disturbs the glucose homeostasis and thus results in obesity or diabetes [14–17]. *Drosophila melanogaster* has turned out to be an excellent and efficient model to explore the mechanism underlying the metabolic pathways. The triglyceride in normal diet breaks down into its smaller units known as monoacylglycerols, free fatty acids and free sterols. The smaller units absorbed by the intestine and later transferred to all body cells as a lipoprotein. The absorbed lipids, present in both larva and adult stage, can be utilized or stored as a fat within the gut or oenocytes. Thus, excess consumption of fat or sugar gets reflected in the storage tissues. The accumulated lipids in various tissues can be visualized by various lipophilic dyes. The lipophilic dyes such as Oil Red O, Sudan black and Nile red are commonly used to detect the lipids accumulation in the tissues.

1.1 Nile Red Staining

The Nile red is also known as 9-diethylamino-5H-benzo[a]phenoxazine-5-one. It is hydrophobic and lipophilic in nature, and its properties depend on the polarity of its internal milieu. Thus, the fluorescence property of Nile red is quenched when it comes in contact with the water as it is insoluble in it. However, the fluorescence intensity increases when it comes in contact with hydrophobic organic solvents and lipid droplets. Taking this property into account, Nile red is used to detect intracellular lipid droplet in unfixed tissues using fluorescent microscopy. Under fluorescent microscope wherever the lipid droplets (neutral lipid) are present that area appeared golden yellow [18–21]. Taking this mechanism into account, Nile red is used for staining various tissues [22–26].

The mechanism of how Nile red stains the gut is as follows: Nile red is soluble in a polar solvent. When the dye-solvent mixture applied to the tissue, the dye diffuses into the cells by interacting with the polar molecules present on the cell membrane. The dye inside the cell binds to the lipid globules (neutral lipids). The interaction between the dye-solvent complex and polar molecules of membrane exhibit fluorescence at higher wavelength, whereas the dye gives fluorescence at the lower wavelength when interacting with lipid globule. The dye-lipid complex, when excited at a wave length of 450–500 nm, absorbs blue-green photons and then emits a photon greater than 528 nm wavelength. This gives a fluorescence and the intensity is directly proportional to the amount of lipid.

1.2 Oil Red O

Oil Red O ('ORO') is a fat-soluble (lysochrome) diazo stain, which is used to visualize neutral lipids in fixed tissues. It mostly stains frozen sections; however, it also gives positive staining with paraffin sections. It stains lipids dark red which is easily visible under the fluorescent microscope at maximum absorption at 518 nm [27–30].

1.3 Sudan Black

It is a diazo dye used to stain neutral lipid depot in fixed tissue like Oil Red O, but unlike Oil Red O, the staining by Sudan black is less intense. Sudan black is nonfluorescence, lysochromic and basic in nature, and it stains both frozen tissue section and Parafilm tissue section as blue-black. The principle of working of this dye is: the basic component of the dye interacts with the acidic groups of the lipid globule in the tissues [30–32].

The applications of these stains on *Drosophila* to detect the lipid droplet are reported from various studies [31, 33–37]. The staining of these dyes with various tissues provides an effective way to examine the lipid depot in the fat body, oenocytes and gut. However, mishandling, time of incubation, the concentration of dye, temperature variation and microscope settings may give a false result. The current study describes the staining of these dyes in the gut and fat body of *Drosophila*.

1.4 Floating Assay

Diet-induced obesity arises due to breakdown of metabolism. Obesity results when the cells in the energy homeostasis fail to maintain balance in storing and utilizing the energy. Though the pathways involved in the conversion of circulating energy to stored forms (glycogen and triglycerides) and pathways mediating conversion of stored forms to utilized forms are well understood but the processes involved in balancing these interacting pathways still need much research. The regular diet of *Drosophila melanogaster* can be manipulated by adding high sugar or high fat to it, and the changes in energy homeostasis can be studied [3–41]. To understand the mechanism of storage and utilization of fat, the third instar larva of *Drosophila* serves as a potential model system. Thus, buoyancy-based screening assay developed to detect the deposition of fat in the *Drosophila* larvae by Ries et al. [42]. According to the principle, the fatty larvae will float in the solution (higher density), while the control larvae will be drowned. This is a simple method to perform research on obesity, genetics and therapeutic screening [42, 43].

2 Materials and Chemicals Required

1. Female and male *Drosophila* (Oregon R)
2. Fly food material (Yeast powder (Subala Yeast, SYD-003/10), sucrose (Fisher Scientific-qualigens-28106), cornmeal (from local market), agar agar type 1 (HiMedia-GRM666), milliQ water)
3. Diethyl ether (HiMedia-AS047)
4. Nile red dye (HiMedia-MB256)

5. Acetone (SRL-31566)
6. Oil Red O dye (HiMedia-TC256)
7. Propylene Glycol (PG) (HiMedia-AS066)
8. Sudan black (HiMedia-GRM260)
9. Glycerol (HiMedia-MB060)
10. Four percent paraformaldehyde (PFA)
11. 1X phosphate-buffered saline (PBS)
12. 1N NaOH (HiMedia- MB095)
13. HCL (HiMedia-RM5955)
14. Ethanol
15. Coconut oil
16. Cotton plug
17. Glass vials (25 × 100 mm)
18. Eppendorf tube (1 ml)
19. Glass slide (25 × 75 mm, ABDOS-G10101)
20. Coverslip (18 mm, Blue Star)
21. Beaker (Rivera)
22. Hot plate
23. Stirring beads
24. pH meter
25. Stopwatch
26. Pipette
27. Dissection microscope/stereomicroscope
28. Fluorescence microscope

3 Preparation of Chemicals

3.1 Preparation of 1X PBS

1. To prepare 100 ml of 1X PBS, add all the ingredients (as mentioned in Table 1) with 80 ml milliq water, and allow it to dissolve.
2. Adjust the pH of the solution to 7.2 by adding HCl dropwise.
3. After maintaining the pH, adjust the volume to 100 ml by adding milliQ water. Autoclave the solutions prepared prior to use [44].

3.2 Preparation of 4% PFA

1. Heat 70 ml of 1X PBS in a glass beaker up to 60 °C on a hot plate with stirring beads.
2. Add 4 gram of paraformaldehyde to it and stir continuously.
3. Add NaOH dropwise to make a completely clear solution.

Table 1
Recipe for PBS preparation

	1X PBS (for 100 ml)	
	Chemical	**Amount (g)**
1	NaCl	0.8
2	KCl	0.02
3	Na_2HPO_4	0.14
4	KH_2PO_4	0.02

4. Once this state is reached, cool and filter it.
5. Adjust the pH to 7.2 by adding HCl.
6. Now adjust the volume to 100 ml.
7. Aliquot it and finally store it in 4 °C.

3.3 Preparation of 20% Glycerol

1. To make 20% glycerol, add 20 ml of glycerol to 80 ml of milliQ water, adjust the volume to 100 ml, and slowly mix the components by pipetting. Store it in 4 °C.

3.4 Preparation of Working Concentration of Nile Red

1. Nile red can be stored at room temperature in its solid form.
2. To prepare, the stock solution dissolves 1 mg/ml of Nile red in acetone (*see* **Note 4**).
3. After dilution store it in 4 °C in dark or covered with aluminium foil (*see* **Note 5**).
4. The prepared solution can be used for months if handled carefully without exposure to the light (*see* **Note 5**).
5. Dilute the Nile red/acetone stock solution in PBS to get a working solution of 0.5 μg/ml (*see* **Note 4**).

3.5 Preparation of Working Concentration of Oil Red O

1. Oil Red O can be stored at room temperature in its solid form.
2. Oil Red O is freshly prepared every time prior to the staining experiment.
3. Add 5 mg of Oil Red O to 1 ml of propylene glycol, and heat the solution at 95–100 °C (*see* **Notes 4 and 6**).
4. Filter the warm solution using Whatman paper (25 μm), and allow it to stand overnight at room temperature in dark, and use it in the next day.

3.6 Preparation of Working Concentration of Sudan Black

1. Sudan black can be stored at room temperature in its solid form.
2. To make a working solution of Sudan black (0.5%), dissolve 5 mg in 1 ml of 75% ethanol (*see* **Note 4**).
3. Store the solution in 4 °C.

4 Methodology

4.1 Nile Red Stain

1. Make standard fly food for experimental set-up of control and treated (*see* **Note 1**).
2. Transfer the male and female flies in the ratio 2:3 or 5:7 by anaesthetizing the flies with ether (*see* **Note 2**).
3. After 3–4 days, collect the third instar larva, and wash them in 1X PBS to remove food attach to their body (*see* **Note 3**).
4. Dissect the gut and fat body of third instar larva under the stereomicroscope.
5. Then incubate the tissues in Nile red working solution (20–30 μL) and 20% glycerol (10 μL) for 1 h in dark (*see* **Note 5**).
6. After the incubation period, wash the samples with 1X PBS again to remove excess stain and mount the sample in the glass slide and put a coverslip.
7. Take the images under a fluorescence microscope (*see* **Note 7**).
8. The number of lipid droplet can be counted, and the size of the lipid droplet can be measured in ImageJ software (Fig. 1) (*see* **Note 12**).
9. After eclosion collects the adult flies, dissect the gut and fat body and follow the same procedure as that for the larva.

4.2 Oil Red O Stain

1. Make standard fly food for experimental set-up of control and treated (*see* **Note 1**).
2. Transfer the male and female flies in the ratio 2:3 or 5:7 by anaesthetizing flies with ether (*see* **Note 2**).
3. After 3–4 days, collect the third instar larva, and wash them in PBS to remove food attached to their body (*see* **Note 3**).
4. Dissect the gut and fat body of third instar larva under the stereomicroscope, and fix them using cold 4% formaldehyde for 30 min.
5. Wash the tissues twice in PBS and twice in 100% PG.
6. Then keep the tissues in Oil Red O working solution, and incubate for 1 h at 60 °C.
7. After the incubation period, wash the sample with 85% PG twice.
8. Then again wash the samples with PBS to remove excess stain and mount the sample in the glass slide and put a coverslip.
9. Take the images under a fluorescent microscope (*see* **Note 7**).
10. The number of lipid droplet can be counted, and the size of the lipid droplet can be measured in ImageJ software (Fig. 2) (*see* **Note 12**).

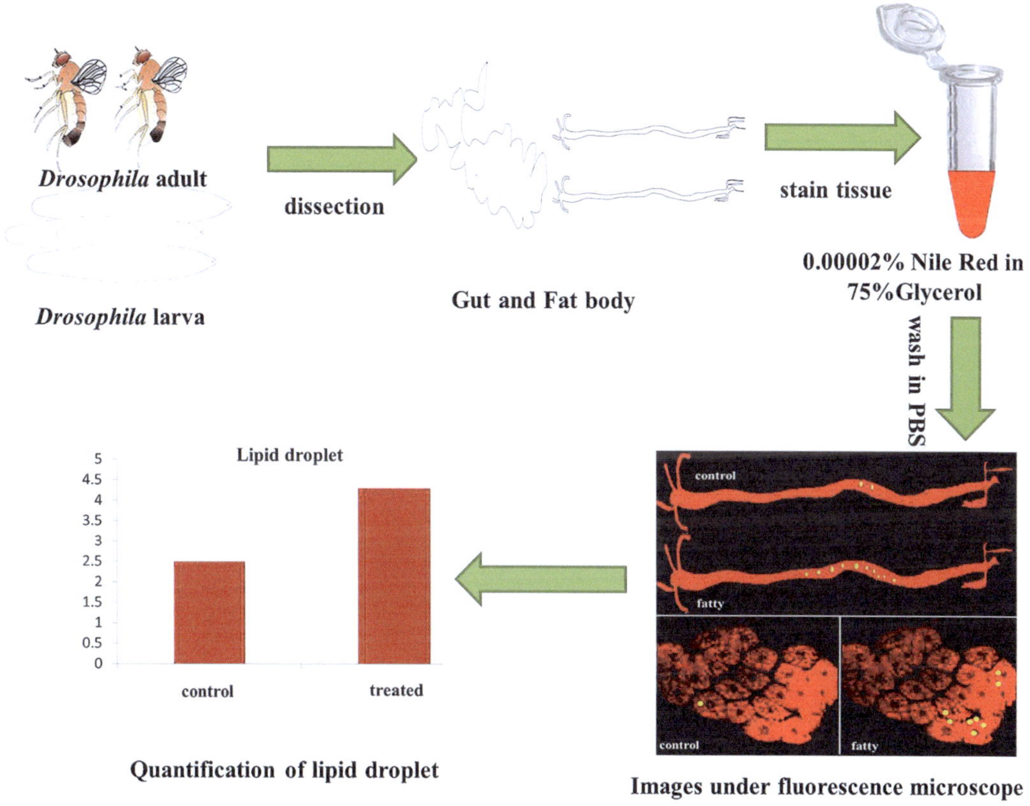

Fig. 1 Nile red stain for assessment of lipid in *Drosophila* gut and fat body

11. After eclosion collect the adult flies, dissect the gut and fat body and follow the same procedure as that for the larva.

4.3 Sudan Black

1. Make standard fly food for experimental set-up of control and treated (*see* **Note 1**).
2. Transfer the male and female flies in the ratio 2:3 or 5:7 by anaesthetizing flies with ether (*see* **Note 2**).
3. After 3–4 days, collect the third instar larva, and wash them in PBS to remove food attached to their body (*see* **Note 3**).
4. Dissect the fat body of third instar larva under the stereomicroscope, and fix it using cold 4% formaldehyde for overnight (*see* **Note 8**).
5. Wash the tissues twice in PBS and twice in 50% ethanol.
6. Then keep the tissues in the 20–30 μl of working solution of the stain, and incubate for 2 min.
7. After the incubation period, the sample undergoes serial washing with 50% ethanol followed by 25% ethanol and lastly with PBS to remove excess stain and mount the sample in the glass slide with glycerol and put a coverslip.

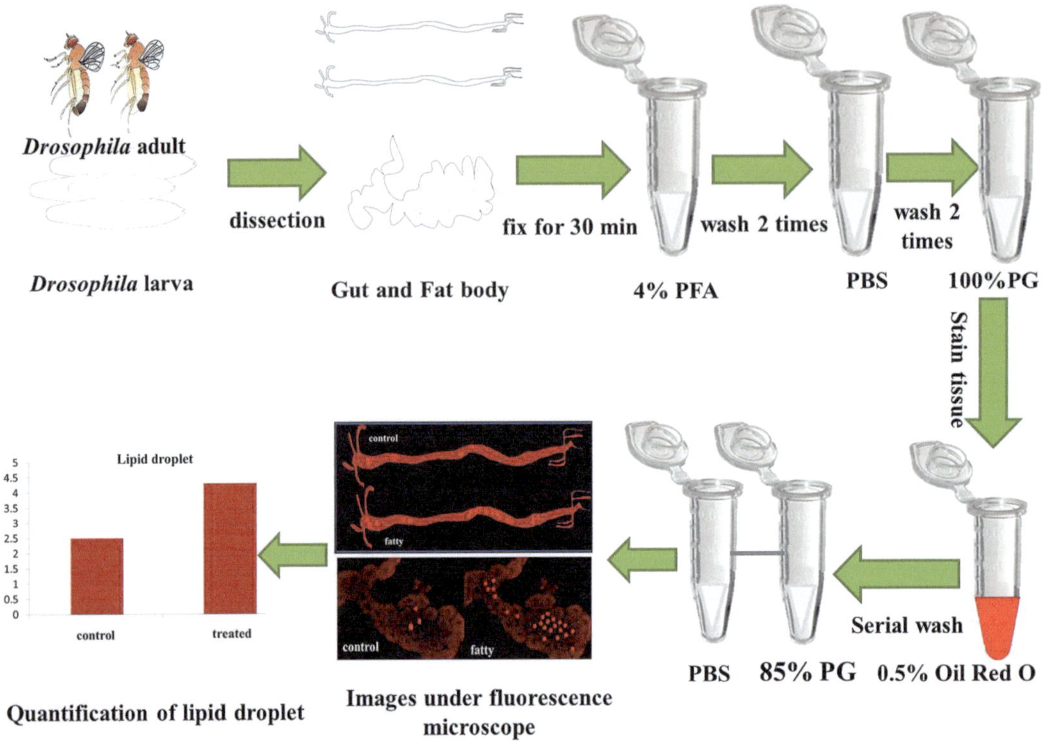

Fig. 2 Oil Red O stain for assessment of lipid in *Drosophila* gut and fat body

8. Take the images under a bright field microscope.
9. The number of lipid droplet can be counted, and the size of the lipid droplet can be measured using ImageJ software (Fig. 3) (*see* **Note 12**).
10. After eclosion collect the adult flies, dissect the fat body and follow the same procedure (**Steps 1–9**) as that for the larva.

4.4 Floating Assay

1. Make standard fly food for experimental set-up of control and treated (*see* **Note 1**).
2. Add coconut oil (high fat diet) to treated set-up (*see* **Notes 9 and 10**).
3. Transfer the male and female flies in the ratio 2:3 or 5:7 by anaesthetizing flies with ether (*see* **Note 2**).
4. After 3–4 days, collect the third instar larva, and wash them in PBS to remove food attach to their body (*see* **Note 3**).
5. Ten to 15 third instar larvae can be taken to 10 ml vials containing 8–30% sucrose dissolved in PBS (*see* **Note 11**).
6. To maintain equilibrium, mix the larvae, and let it be stable for 3–5 min.

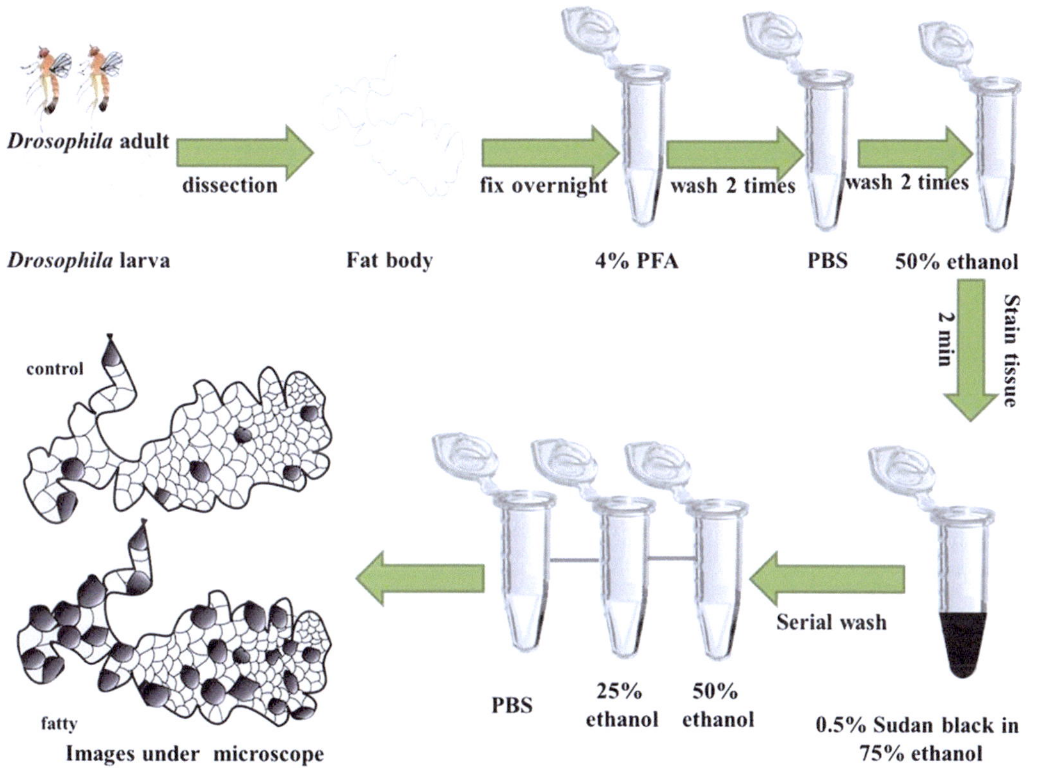

Fig. 3 Sudan black staining of *Drosophila* fat body to detect fat depot

7. According to the principle, the larvae those are fatty will float in the solution (higher density), while the control larvae will be found to be drowned in the vial at the bottom (Fig. 4) (*see* **Note 12**).

5 Notes

1. The fly should be maintained in optimal environmental conditions like temperature, light and humidity.
2. During anaesthetizing fly, an ear bud has to be used. After dipping inside ether excess of ether should be squeezed by pressing the bud at the tip of ether tube. Excess dose of ether may lead to the death of fly. So once they are anaesthetized, we need to keep them 20–30 min to be active for the next dose (if required).
3. Washing of larva and adult flies after removing them from food is essential because food may interfere with the assay.
4. The concentration of stain should be made carefully as higher concentration may lead to excessive staining.
5. As the dye is light sensitive, the dye should be always kept in dark for proper use.

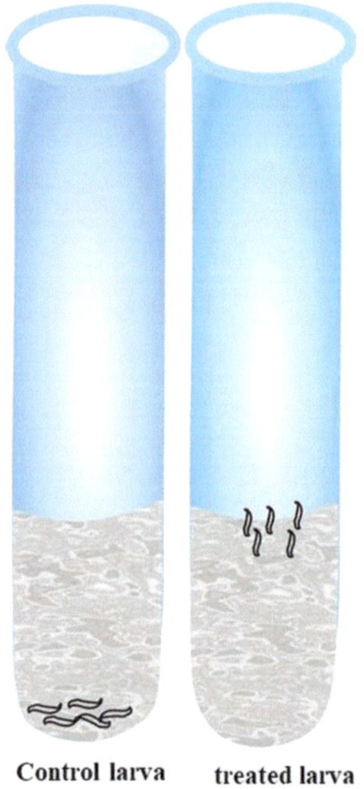

Fig. 4 Floating assay indicating control larva drowning and fatty larva floating in the sucrose solution

6. Do not heat the Oil Red O solution above 100 °C as overheating may cause excess background staining.
7. In the fluorescence microscope, the exposure time of the light should be fixed to maintain uniformity and eliminate the chance of error.
8. Some sample can be fixed with paraffin wax also, and the sectioning should be thin for proper staining.
9. This assay can be performed with larva treated with high sugar diet.
10. When coconut oil is used as a high fat diet then before transferring flies, care should be taken that oil should be properly soaked in food otherwise flies may stick to oil and die.
11. The sucrose solution should be made using PBS only.
12. The experiment has to be repeated 5–10 times to acquire statistically significant data.

Acknowledgements

NN is thankful to DST/INSPIRE Fellowship/2016/IF160257 for financial support. MM lab is supported by Grant No. BT/PR21857/NNT/28/1238/2017, EMR/2017/003054 and Odisha DBT 3325/ST(BIO)-02/2017.

References

1. Huynh K, Schneider M, Gareau M (2016) Altering the gut microbiome for cognitive benefit? In: The gut-brain axis. Elsevier, pp 319–337
2. Stratton IM, Adler AI, Neil HAW, Matthews DR, Manley SE, Cull CA, Hadden D, Turner RC, Holman RR (2000) Association of glycaemia with macrovascular and microvascular complications of type 2 diabetes (UKPDS 35): prospective observational study. BMJ 321(7258):405–412
3. Hunziker W, Kiener TK, Xu J (2009) Vertebrate animal models unravel physiological roles for zonula occludens tight junction adaptor proteins. Ann N Y Acad Sci 1165(1):28–33
4. Burggren WW, Warburton S (2007) Amphibians as animal models for laboratory research in physiology. ILAR J 48(3):260–269
5. Lieschke GJ, Currie PD (2007) Animal models of human disease: zebrafish swim into view. Nat Rev Genet 8(5):353
6. Chianese R, Chioccarelli T, Cacciola G, Ciaramella V, Fasano S, Pierantoni R, Meccariello R, Cobellis G (2011) The contribution of lower vertebrate animal models in human reproduction research. Gen Comp Endocrinol 171(1):17–27
7. Botstein D, Chervitz SA, Cherry M (1997) Yeast as a model organism. Science 277(5330):1259–1260
8. Kaletta T, Hengartner MO (2006) Finding function in novel targets: *C. elegans* as a model organism. Nat Rev Drug Discov 5(5):387
9. Ségalat L (2007) Invertebrate animal models of diseases as screening tools in drug discovery. ACS Chem Biol 2(4):231–236
10. Pandey UB, Nichols CD (2011) Human disease models in *Drosophila melanogaster* and the role of the fly in therapeutic drug discovery. Pharmacol Rev:pr. 110.003293
11. Álvarez-Rendón JP, Salceda R, Riesgo-Escovar JR (2018) *Drosophila melanogaster* as a model for diabetes type 2 progression. BioMed Res Int 2018
12. Buescher JL, Musselman LP, Wilson CA, Lang T, Keleher M, Baranski TJ, Duncan JG (2013) Evidence for transgenerational metabolic programming in Drosophila. Dis Models Mech 011924
13. Matzkin LM, Johnson S, Paight C, Markow TA (2013) Preadult parental diet affects offspring development and metabolism in *Drosophila melanogaster*. PLoS One 8(3):e59530
14. Pasco MY, Léopold P (2012) High sugar-induced insulin resistance in Drosophila relies on the lipocalin Neural Lazarillo. PLoS One 7(5):e36583
15. Musselman L, Fink J, Narzinski K, Ramachandran P, Hathiramani S, Cagan R, Baranski T (2011) A high-sugar diet produces obesity and insulin resistance in wild-type Drosophila. Dis Model Mech 4:842–849
16. Birse RT, Choi J, Reardon K, Rodriguez J, Graham S, Diop S, Ocorr K, Bodmer R, Oldham S (2010) High-fat-diet-induced obesity and heart dysfunction are regulated by the TOR pathway in Drosophila. Cell Metab 12(5):533–544
17. Reed LK, Williams S, Springston M, Brown J, Freeman K, DesRoches CE, Sokolowski MB, Gibson G (2010) Genotype-by-diet interactions drive metabolic phenotype variation in *Drosophila melanogaster*. Genetics
18. Greenspan P, Mayer EP, Fowler SD (1985) Nile red: a selective fluorescent stain for intracellular lipid droplets. J Cell Biol 100(3):965–973
19. Grönke S, Mildner A, Fellert S, Tennagels N, Petry S, Müller G, Jäckle H, Kühnlein RP (2005) Brummer lipase is an evolutionary conserved fat storage regulator in Drosophila. Cell Metab 1(5):323–330
20. Bertozzini E, Galluzzi L, Penna A, Magnani M (2011) Application of the standard addition method for the absolute quantification of neutral lipids in microalgae using Nile red. J Microbiol Methods 87(1):17–23
21. Huang G-H, Chen G, Chen F (2009) Rapid screening method for lipid production in alga based on Nile red fluorescence. Biomass Bioenergy 33(10):1386–1392
22. Lee SJ, Yoon B-D, Oh H-M (1998) Rapid method for the determination of lipid from the green alga *Botryococcus braunii*. Biotechnol Tech 12(7):553–556
23. Chen W, Zhang C, Song L, Sommerfeld M, Hu Q (2009) A high throughput Nile red method for quantitative measurement of neutral lipids in microalgae. J Microbiol Methods 77(1):41–47
24. Chen W, Sommerfeld M, Hu Q (2011) Microwave-assisted Nile red method for in vivo quantification of neutral lipids in microalgae. Bioresour Technol 102(1):135–141
25. Doan T-TY, Obbard JP (2011) Improved Nile red staining of Nannochloropsis sp. J Appl Phycol 23(5):895–901

26. Pick U, Rachutin-Zalogin T (2012) Kinetic anomalies in the interactions of Nile red with microalgae. J Microbiol Methods 88(2):189–196
27. Aceves-Piña EO, Quinn WG (1979) Learning in normal and mutant Drosophila larvae. Science 206(4414):93–96
28. Chien C-H, Chen W-W, Wu J-T, Chang T-C (2011) Label-free imaging of Drosophila in vivo by coherent anti-Stokes Raman scattering and two-photon excitation autofluorescence microscopy. J Biomed Opt 16(1):016012
29. Bader CA, Shandala T, Carter EA, Ivask A, Guinan T, Hickey SM, Werrett MV, Wright PJ, Simpson PV, Stagni S (2016) A molecular probe for the detection of polar lipids in live cells. PLoS One 11(8):e0161557
30. Tennessen JM, Barry WE, Cox J, Thummel CS (2014) Methods for studying metabolism in Drosophila. Methods 68(1):105–115
31. Fernando-Warnakulasuriya GJ, Wells MA (1988) Isolation and characterization of lipophorin from *Drosophila melanogaster* larvae. Arch Insect Biochem Physiol 8(4):243–248
32. Butterworth FM, Bodenstein D, King RC (1965) Adipose tissue of *Drosophila melanogaster*. I. An experimental study of larval fat body. J Exp Zool 158(2):141–153
33. Grönke S, Beller M, Fellert S, Ramakrishnan H, Jäckle H, Kühnlein RP (2003) Control of fat storage by a Drosophila PAT domain protein. Curr Biol 13(7):603–606
34. Welte MA, Gross SP, Postner M, Block SM, Wieschaus EF (1998) Developmental regulation of vesicle transport in Drosophila embryos: forces and kinetics. Cell 92(4):547–557
35. Gutierrez E, Wiggins D, Fielding B, Gould AP (2007) Specialized hepatocyte-like cells regulate Drosophila lipid metabolism. Nature 445(7125):275
36. Seecof RL, Dewhurst S (1974) Insulin is a Drosophila hormone and acts to enhance the differentiation of embryonic Drosophila cells. Cell Differ 3(1):63–70
37. Pho D, Pennanec'h M, Jallon J (1996) Purification of adult *Drosophila melanogaster* lipophorin and its role in hydrocarbon transport. Arch Insect Biochem Physiol 31(3):289–303
38. Rosini TC, Silva ASRD, Moraes CD (2012) Diet-induced obesity: rodent model for the study of obesity-related disorders. Rev Assoc Med Bras 58(3):383–387
39. Al-Anzi B, Sapin V, Waters C, Zinn K, Wyman RJ, Benzer S (2009) Obesity-blocking neurons in Drosophila. Neuron 63(3):329–341
40. Kunte AS, Matthews KA, Rawson RB (2006) Fatty acid auxotrophy in Drosophila larvae lacking SREBP. Cell Metab 3(6):439–448
41. Musselman LP, Fink JL, Narzinski K, Ramachandran PV, Hathiramani SS, Cagan RL, Baranski TJ (2011) A high-sugar diet produces obesity and insulin resistance in wild-type Drosophila. Dis Models Mech 007948
42. Reis T, Van Gilst MR, Hariharan IK (2010) A buoyancy-based screen of Drosophila larvae for fat-storage mutants reveals a role for Sir2 in coupling fat storage to nutrient availability. PLoS Genet 6(11):e1001206
43. Liu J, Li T, Yang D, Ma R, Moran T, Smith W (2012) Synphilin-1 alters metabolic homeostasis in a novel Drosophila obesity model. Int J Obes 36(12):1529
44. Wilk R, Murthy SU, Yan H, Krause HM (2010) In situ hybridization: fruit fly embryos and tissues. Curr Protocol Essent Lab Tech 4(1):9.3. 1–9.3. 24

Chapter 11

Estimation of Oxidative Stress and Survivorship in *Drosophila*

Nibedita Nayak and Monalisa Mishra

Abstract

Various exogenous factors like temperature, sound, pressure, microbe infection and chemicals can produce oxidative stress within the body. Similarly, endogenous factors like metabolism, age, genetic defects and disease also cause stress. When the body is under stress, it generates reactive oxygen species (ROS). However, the concentration of ROS varies according to the stimuli. ROS acts as a signaling molecule and alters various developmental, physiological and metabolic pathways. Hence, accurate methods are in need to detect the concentration of ROS within the body. There are numerous ways to measure different types of ROS either using spectrophotometry or using of chemiluminescence or fluorescence probe or using electron spin resonance. Being a small animal, the more accurate method of ROS detection will be helpful to detect the effect of various factors on the body. The current chapter describes the methods for detection of ROS using the biochemical and histological method. We are also describing the survivorship assay as a stress indicator in the model organism, *Drosophila*.

Key words Reactive oxygen species (ROS), *Drosophila*, DCFH-DA, NBT, Survivorship assay

Abbreviations

DCF-DA	Dichlorodihydrofluorescein diacetate
GAA	Glacial acetic acid
H_2O_2	Hydrogen peroxide
NBT	Nitroblue tetrazolium
PBS	Phosphate-buffered saline
PBST	Phosphate-buffered saline with Tween 20
PFA	Paraformaldehyde

1 Introduction

The reactive oxygen species (ROS) includes oxygen radicals like hydroxyl radicals (-OH), hydrogen peroxide (H_2O_2), nitric oxide, singlet oxygen and superoxide anions (O_2^-) [1]. ROS is generated

Monalisa Mishra (ed.), *Fundamental Approaches to Screen Abnormalities in Drosophila*, Springer Protocols Handbooks, https://doi.org/10.1007/978-1-4939-9756-5_11, © Springer Science+Business Media, LLC, part of Springer Nature 2020

either by energy transfer reactions or as a byproduct of different metabolic pathways [2]. To combat the excessive amount of ROS, every organism has an antioxidant defence system [3]. However, oxidative outburst (rapid rise in ROS level) [4] occurs when the balance between the production and degradation of ROS is disturbed by various internal and external stimuli [5, 6]. The function of ROS depends on the concentration. Low concentration of ROS is beneficial promoting cell survival and proliferation, whereas a moderate concentration may arrest cell cycle, while a high concentration is detrimental by activating apoptosis [7–9].

Generally, ROS have the ability to destroy any DNA, lipids and proteins. These free radical ROS cause oxidative stress or oxidative degradation which in turn can play a key role in inducing ageing and even age-related disorders [10–12]. To date, many reports deduced that ROS can be a major factor contributing to the neurodegenerative disease such as Alzheimer's by destroying vital molecules of the brain; it is also responsible for the development of other diseases such as cancer, atherosclerosis and diabetes [13–18]. There are cases where ROS plays a significant role in the onset of many diseases, thus methods are needed to combat the onset of disorders by treating with various antioxidants [13, 15, 19–21]. Thus it is indispensable to quantify ROS-mediated oxidative damage. Not only the concentration but locations of production, potentiality to degrade or diffuse and appropriate preference for the reaction are other features of ROS, which can affect the behaviour of the cells. Thus, biological importance of ROS can only be determined, if the exact concentration and location in the cell are known. Furthermore, the instability, mutual interference and short lifetime of ROS put up a hindrance to accomplish this objective. Numerous methods such as chromatography, genetically encoded fluorescent protein-based assays, fluorescent and chemiluminescent probe-based assay and spectrophotometry methods are available [22–24].

1.1 Fluorescence-Based Methods

The nonfluorescent probes are oxidant sensitive often gets oxidized by ROS and converted to fluorescence probes. The most widely used probes are Amplex Red (impermeable in the cell), dihydroethidium and dichlorodihydrofluorescein. The probe permeable in the cell exhibits the oxidative stress in the cell due to stimuli. In *Drosophila* model, the most widely used fluorescent probe to determine ROS is dichlorodihydrofluorescein diacetate (DCFH-DA). This dye is good enough to indicate the cellular oxidative stress by measuring the intracellular redox level directly. The DCFH-DA/DCFH behave differently in cells as the cells have a different level of permeability barrier to the probe and also a difference in the activity of esterase as a result the equilibrium concentrations of the probe in the cells differ. Thus, before setting up any experiments, checking the activity of DCFH-DA/DCFH in different cells is prerequisite. Principle of working is based on the oxidation of DCFH-DA. The probe is permeable to cell, and upon entry it is converted to nonfluorescent $DCFH^-$ (intermediated free radical) when hydrolysed by the esterase. The $DCFH^-$ easily diffuses inside where it is oxidized by non-specific ROS to be converted into green fluorescent DCF probe. The fluorescence intensity of the DCF, is directly proportional to oxidative stress [25].

In the presence of a transition metal only, the DCFH is sensitive to H_2O_2 (hydrogen peroxide) or cytochrome c [26]. The DCFH dye is more sensitive towards $ONOO^-$ than any other free radical. DCF-DA, a dehydrogenation probe, is also sensitive to ROS, but it has a wide range of sensitivity towards ROS as compared to DCFH-DA probe. Cells are permeable to DCF-DA, and it works on the same principle as that of DCFH-DA dye. The probe DCF-DA can be oxidized by H_2O_2, nitrogen oxide (NO) and another superoxide to form DCF [27, 28]. Thus, it can be inferred that DCFH-DA is only sensitive to H_2O_2 or peroxynitrite and thus is a more single radical-specific dye. DCF-DA is responsive to a wide range of ROS and therefore is a non-specific dye. It can recognize only a particular radical. The fluorescence intensity obtained is not directly proportional to the measurement of H_2O_2 as the data can be misleading by other radicals such as HOCl, $ONOO^-$, ONOOH and NO_2^-, which can efficiently convert DCFH to DCF [29]. Moreover, fluorescence intensity is misinterpreted when dye intermediate radical ($DCFH^-$) interacts with oxygen to produce surplus oxide (O_2^-) radicals adding up to the intensity produced by H_2O_2 [30]. The occurrence of factors, for instance, cytochrome c, ferrous ion, hematin, HRP (horseradish peroxidase) and peroxidases, can oxidize DCFH-DA to DCFH even when a cell lacks H_2O_2 [31, 32]. Precisely, DCFH-DA fails to detect a single radical in the cell due to inability in discriminating the ROS but can only deduce the cellular oxidative stress condition of cells upon receiving the stimuli.

1.2 Spectrophotometry Methods

The most conventional method used for detection of ROS is the spectrophotometry methods. It is the best way for semi-quantification of ROS. The working principle is based not only on the radicals and redox substances interaction but also on the difference in absorbance of substrate and product at a different wavelength. Out of cytochrome c reduction assay, aconitase inactivation assay, boronate assay and diaminobenzidine (DAB) assay, the most widely used spectroscopy method to detect ROS in *Drosophila* model is NBT (nitroblue tetrazolium) assay. Intracellular ROS level can be determined by NBT assay. The assay is based on colorimetric principle. The cells enclosing ROS when incubated with yellow-coloured nitroblue tetrazolium (Y-NBT) become permeable to cells and then get reduced to blue formazan deposits which can easily get dissolved in any organic solvent, and observance can be taken in the microplate reader. NBT assay is an easy and sensitive method to quantify even the slightest of the ROS in the cells [33, 34].

1.3 Survival Assay

The development, growth, survival and body size of an organism are influenced by the food consumed [35]. Whenever compounds or chemical is added to regular food of fly, it has an impact on growth, metabolism and survivability of the organism by generating ROS (reactive oxygen species). ROS acts as a signalling molecule and induces a defect in gene/metabolite [36–40]. Consequently, the assessment of these compounds on the fly can be done by the survival assay [41, 42].

Hence, this chapter provides comprehensive information to determine ROS and stress in *Drosophila* model which will be beneficial in disease pathology research.

2 Materials and Chemicals Required

1. Female and male *Drosophila* (*Oregon R*)
2. Fly food material (yeast powder (Subala Yeast, SYD-003/10), sucrose (Fisher Scientific-qualigens-28106), cornmeal (from local market), agar agar type 1 (HiMedia-RM666), milliQ water)
3. Diethyl ether (HiMedia-AS047)
4. 1X PBS
5. 1X PBST
6. 4% PFA
7. DCFH-DA dye (Sigma- D6883)
8. NBT (HiMedia-MB107)
9. Glacial acetic acid (GAA) (SRL- 30106)
10. Hydrogen peroxide (H_2O_2) (HiMedia-PCT-1511)
11. Whatman filter paper
12. Cotton plug
13. Fine sterilized needle
14. Glass vials (25 × 100 mm)
15. Eppendorf tube (1 ml)
16. Glass slide
17. Coverslip (18 mm, Blue Star)
18. Hot stir plate
19. Beaker

20. Stirring beads
21. pH meter
22. Stopwatch
23. Pipette
24. Dissection microscope/stereomicroscope
25. Microplates
26. Centrifuge
27. Spectrophotometer or microplate reader

3 Preparation of Chemicals

3.1 Preparation of 1X PBS and 1X PBST

1. To prepare 100 ml of 1X PBS, add all the ingredients (Table 1) with 80 ml milliQ water and allow it to dissolve.
2. Adjust the pH of the solution to 7.2 by adding NaOH or HCl as per your requirement.
3. For making 100 ml of 1X PBST, mix all the components as given in the table 1, and dissolve it in 80 ml of milliQ water.
4. Adjust the pH to 7.2.
5. After maintaining the pH, adjust the volume to 100 ml by adding milliQ water. Autoclave the solutions prepared prior to use [43].

3.2 Preparation of 4% PFA

1. Heat 70 ml of 1X PBS in a glass beaker up to 60 °C on a hot plate with stirring beads.
2. Add 4 g of paraformaldehyde to it, and stir continuously.
3. Add NaOH dropwise to make a completely clear solution.
4. Once this state is reached, cool and filters it.
5. Adjust the pH to 7.2 by adding HCl.

Table 1
Recipe for PBS and PBST preparation

	1X PBST (for 100 ml)		1X PBS (for 100 ml)	
	Chemical	Amount (g)	Chemical	Amount (g)
1	NaCl (HiMedia-GRM853)	0.8	NaCl	0.8
2	KCl (HiMedia-MB043)	0.02	KCl	0.02
3	Na_2HPO_4 (HiMedia-MB024)	0.14	Na_2HPO_4	0.14
4	KH_2PO_4 (HiMedia)	0.02	KH_2PO_4	0.02
5	Tween 20 (Sigma-P9416)	0.2 ml		

6. Now adjust the volume to 100 ml and make aliquots and finally store it in 4 °C.

3.3 Preparation of Working Concentration of DCFH-DA

1. DCFH-DA dye should be stored at −20 °C.
2. To prepare, the stock solution dissolves 4.85 g/ml of DCFH-DA in methanol (1 mM) (*see* **Note 4**).
3. After dilution store it in −20 °C in dark or covered with aluminium foil.
4. The prepared solution can be used for months if handled carefully without exposure towards the light (*see* **Note 5**).
5. Dilute the DCFH-DA stock solution in methanol to get a working solution of 10 μM concentration (*see* **Note 4**).
6. Keep it in dark condition at −20 °C (*see* **Note 5**).

3.4 Preparation of Working Concentration of NBT

1. NBT should be stored at normal room temperature.
2. To prepare, the stock solution dissolves 10 mg/ml of NBT in water (12.2 mM) (*see* **Note 4**).
3. After dilution store it in −20 °C in dark or covered with aluminium foil.
4. The prepared solution can be used for months if handled carefully without exposure towards the light (*see* **Note 5**).
5. Dilute the NBT stock solution in water to get a working solution of 1.6 mM concentration (*see* **Note 4**).
6. Keep it in dark condition at −20 °C (*see* **Note 5**).

4 Methodology

4.1 DCFH-DA Stain

1. Make standard fly food for experimental set-up of control and treated (*see* **Note 1**).
2. Transfer the male and female flies in the ratio 2:3 or 5:7 by anaesthetizing flies with ether (*see* **Note 2**).
3. After 3–4 days, collect the third instar larva, and wash them in PBS to remove food attach to their body (*see* **Note 3**).
4. Dissect the gut of third instar larva under the stereomicroscope, and then fix it in fixative reagent (4% PFA) in an Eppendorf tube at 4 °C overnight.
5. Next day, to remove excess PFA, wash the sample in PBS for two times, each for 5 min duration.
6. It is followed by washing of sample in PBST for three times for 5 min each.
7. Stain and incubate the sample with DCFH-DA dye (10 μM concentration) for 30 min in dark (*see* **Note 5**).

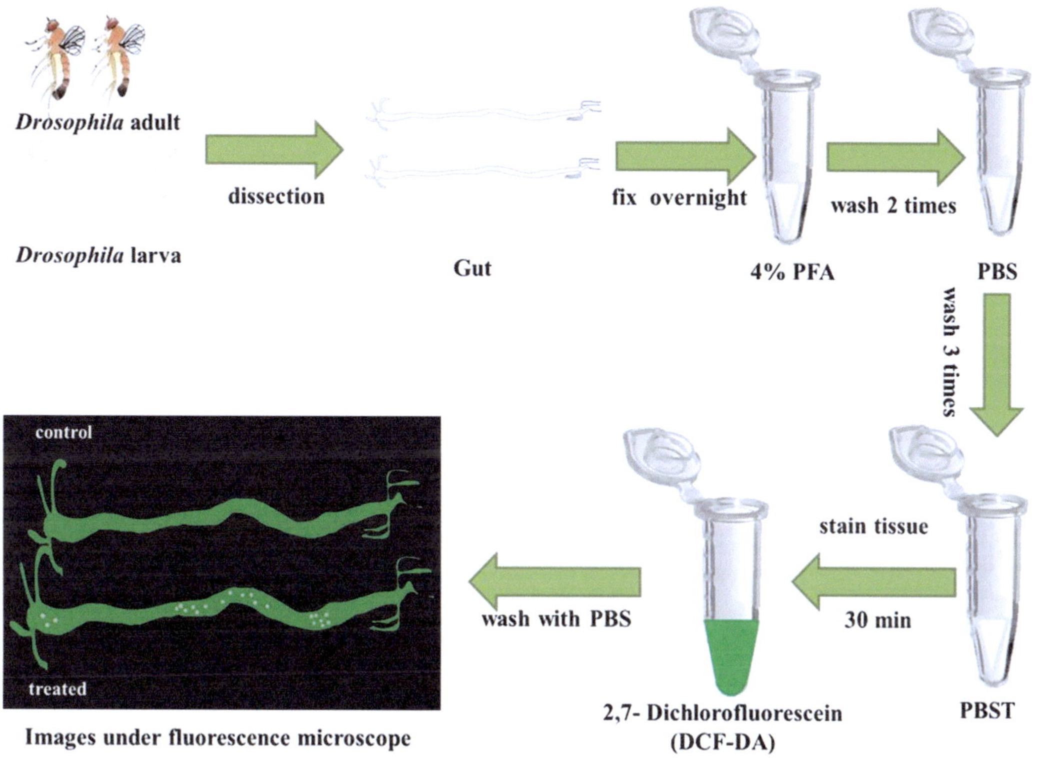

Fig. 1 Steps showing DCFH-DA staining of a sample

8. After the incubation period, wash the samples with PBS again to remove excess stain and mount the sample in the glass slide and put a coverslip.
9. Take the images under a fluorescence microscope (Fig. 1) (*see* **Notes 6 and 13**).
10. After eclosion collect the adult flies, dissect the gut and follow the same procedure as that for the larva.

4.2 NBT Assay

1. Make standard fly food for experimental set-up of control and treated (*see* **Note 1**).
2. Transfer the male and female flies in the ratio 2:3 or 5:7 by anaesthetizing flies with ether (*see* **Note 2**).
3. After 3–4 days, collect the third instar larva, and wash them in PBS to remove food attach to their body (*see* **Note 3**).
4. Collect the larva in the 0.5 ml Eppendorf tube which has a small hole made manually with a pin and then place that Eppendorf tube on the 1 ml Eppendorf tube (*see* **Note 10**).
5. Prick the larva near the mouth region only with the needle by keeping the Eppendorf on the ice (*see* **Notes 7, 8 and 9**).

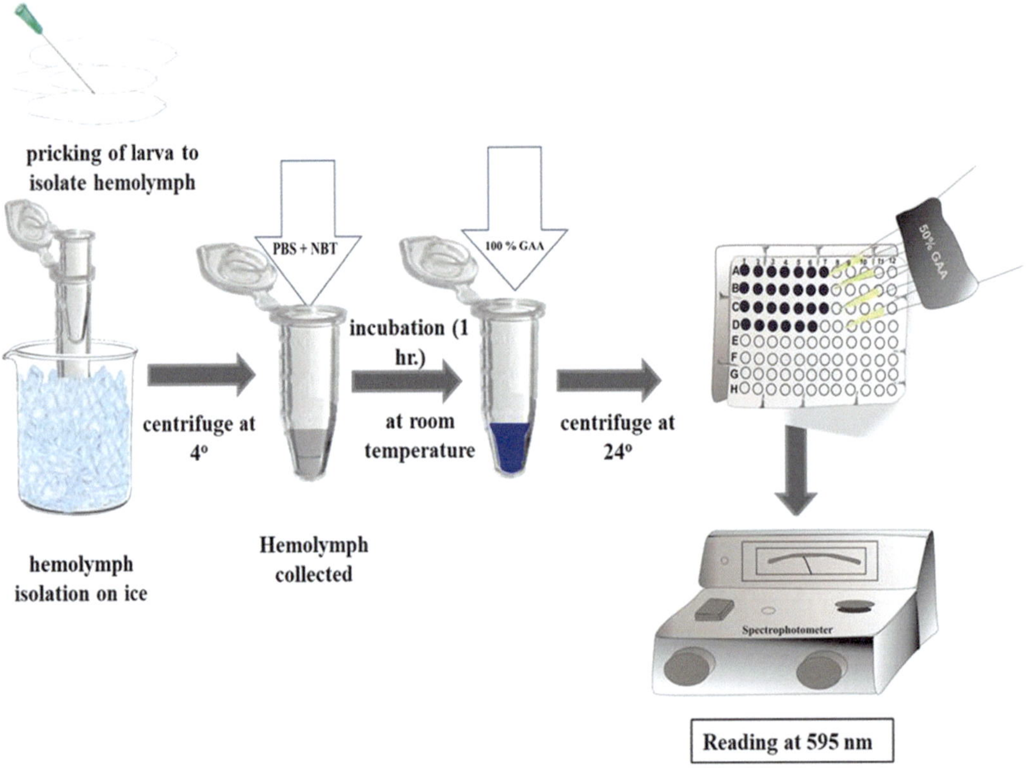

Fig. 2 Steps involved in NBT assay

6. Centrifuge the Eppendorf tubes (exact set-up) at 4500 rpm for 5 min at 4 °C to isolate the hemolymph in the 1 ml Eppendorf tube.
7. Take 5 μl of hemolymph, and add 10 μl of 1 X PBS followed by 15 μL of 1.6 mM NBT solution to it.
8. Incubate for 1 h at room temperature in dark (*see* **Note 5**), and then add an equal volume of 100% of GAA to it in order to stop the reaction (*see* **Note 11**).
9. Again centrifuge for 5 min at 24 °C at 4500 rpm, and to it add 150 μl of 50% of GAA (*see* **Note 11**).
10. Finally take the absorbance in the microplate reader at 595 nm wavelength (Fig. 2) (*see* **Note 13**).
11. After eclosion collect the adult flies, keep the flies in same Eppendorf tube set-up as of larva and prick with needle in the thorax region and follow the same procedure as that for the larva.

4.3 Survival Assay

1. Make standard fly food for experimental set-up of control and treated (*see* **Note 1**).

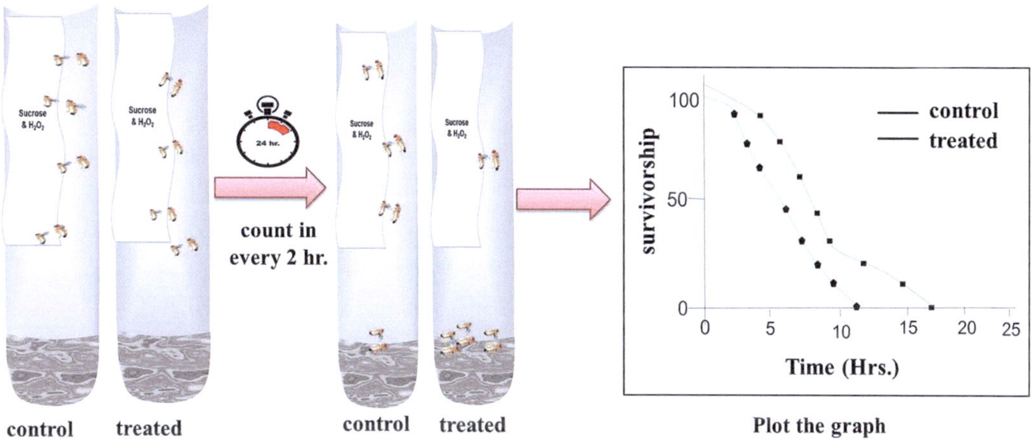

Fig. 3 The protocol of starvation assay or survivorship assay

2. Transfer the male and female flies in the ratio 2:3 or 5:7 by anaesthetizing flies with ether (*see* **Note 2**).
3. After 10–12 days, collect the adult flies of both control and treated.
4. Prepare a mixture of sucrose (6%) and H_2O_2 (9%) (*see* **Note 12**).
5. In the two glass vials containing agar agar type 1, place the Whatman filter paper soaked in the prepared mixture.
6. Then in one vial, transfer the control flies while in other the treated flies, and close the opening of the vial with a cotton plug.
7. Observe the flies in an interval of every 2 h, and note the number of alive and dead flies (Fig. 3).
8. The observation continues till all flies are dead.
9. Plot the graph in GraphPad Prism software (*see* **Note 13**).

5 Notes

1. The fly should be maintained in optimal environmental conditions like temperature and light and humidity.
2. During anaesthetizing fly, an ear bud has to be used. After dipping inside ether, excess of ether should be squeezed by pressing the bud at the tip of ether tube. Excess dose of ether may lead to the death of fly, so once they are anaesthetized, we need to keep them 20–30 min to be active for the next dose (if required).
3. Washing of larva and adult flies after removing them from food is essential because food may interfere with the assay.

4. The concentration of stain should be made carefully as higher concentration may lead to excessive staining.
5. As the dye is light sensitive, the dye should be always kept and used in dark.
6. In the fluorescence microscope, the exposure time of the light should be fixed to maintain uniformity and eliminate the chance of error.
7. Hemolymph collection should be done on ice to prevent melanization.
8. The needle used for pricking should be sterilized.
9. Rupturing of the larva body and adult should be done carefully to prevent mixing of hemolymph with body fluid.
10. The hole made in the 0.5 ml tube should be appropriate because larger hole may lead to draining of tissue to the 1 ml tube, while if the hole is small, then hemolymph could not be extracted.
11. GAA addition is two times in the procedure, but the concentration differs; thus, the addition of GAA should be done carefully.
12. The mixture should be kept in dark as H_2O_2 is sensitive to light.
13. The experiment has to be repeated five to ten times to acquire statistically significant data.

Acknowledgements

NN is thankful to DST/INSPIRE Fellowship/2016/IF160257 for financial support. MM lab is supported by Grant No. BT/PR21857/NNT/28/1238/2017, EMR/2017/003054 and Odisha DBT 3325/ST(BIO)-02/2017.

References

1. Klotz L-O (2002) Oxidant-induced signaling: effects of peroxynitrite and singlet oxygen. Biol Chem 383(3–4):443–456
2. Foyer CH (1994) Oxygen metabolism and the regulation of photosynthetic electron transport. In: Causes of photooxidative stress and amelioration of defense systems in plants. CRC Press, Boca Raton, pp 1–42
3. Alscher RG, Donahue JL, Cramer CL (1997) Reactive oxygen species and antioxidants: relationships in green cells. Physiol Plant 100(2):224–233
4. Apostol I, Heinstein PF, Low PS (1989) Rapid stimulation of an oxidative burst during elicitation of cultured plant cells: role in defense and signal transduction. Plant Physiol 90(1):109–116
5. Malan C, Greyling MM, Gressel J (1990) Correlation between CuZn superoxide dismutase and glutathione reductase, and environmental and xenobiotic stress tolerance in maize inbreds. Plant Sci 69(2):157–166
6. Prasad TK, Anderson MD, Martin BA, Stewart CR (1994) Evidence for chilling-induced oxidative stress in maize seedlings and a regulatory role for hydrogen peroxide. Plant Cell 6(1):65–74

7. D'Autréaux B, Toledano MB (2007) ROS as signalling molecules: mechanisms that generate specificity in ROS homeostasis. Nat Rev Mol Cell Biol 8(10):813
8. Droge W (2002) Free radicals in the physiological control of cell function. Physiol Rev 82(1):47–95
9. Verbon EH, Post JA, Boonstra J (2012) The influence of reactive oxygen species on cell cycle progression in mammalian cells. Gene 511(1):1–6
10. Lamb NJ, Gutteridge JM, Baker C, Evans TW, Quinlan GJ (1999) Oxidative damage to proteins of bronchoalveolar lavage fluid in patients with acute respiratory distress syndrome: evidence for neutrophil-mediated hydroxylation, nitration, and chlorination. Crit Care Med 27(9):1738–1744
11. Halliwell B (1999) Oxygen and nitrogen are pro-carcinogens. Damage to DNA by reactive oxygen, chlorine and nitrogen species: measurement, mechanism and the effects of nutrition. Mutat Res/Genet Toxicol Environ Mutagen 443(1):37–52
12. Sohal RS, Mockett RJ, Orr WC (2002) Mechanisms of aging: an appraisal of the oxidative stress hypothesis1, 2. Free Radic Biol Med 33(5):575–586
13. Halliwell B, Halliwell B, Halliwell B, Gutteridge J (2006) Role of free radicals in the neurodegenerative diseases: therapeutic implications for antioxidant treatment. Amyotroph Lateral Scler 7(sup1):67–67
14. Allan Butterfield D (2002) Amyloid β-peptide (1-42)-induced oxidative stress and neurotoxicity: implications for neurodegeneration in Alzheimer's disease brain. A review. Free Radic Res 36(12):1307–1313
15. Liu Q, Raina AK, Smith MA, Sayre LM, Perry G (2003) Hydroxynonenal, toxic carbonyls, and Alzheimer disease. Mol Asp Med 24(4–5):305–313
16. Chowienczyk P, Brett S, Gopaul N, Meeking D, Marchetti M, Russell-Jones D, Änggård E, Ritter J (2000) Oral treatment with an antioxidant (raxofelast) reduces oxidative stress and improves endothelial function in men with type II diabetes. Diabetologia 43(8):974–977
17. Halliwell B (2000) Lipid peroxidation, antioxidants and cardiovascular disease: how should we move forward? Cardiovasc Res 47(3):410–418
18. Parthasarathy S, Santanam N, Ramachandran S, Meilhac O (2000) Potential role of oxidized lipids and lipoproteins in antioxidant defense. Free Radic Res 33 (3):197–215
19. Lee C-K, Allison DB, Brand J, Weindruch R, Prolla TA (2002) Transcriptional profiles associated with aging and middle age-onset caloric restriction in mouse hearts. Proc Natl Acad Sci 99(23):14988–14993
20. Galli RL, Shukitt-Hale B, Youdim KA, Joseph JA (2002) Fruit polyphenolics and brain aging: nutritional interventions targeting age-related neuronal and behavioral deficits. Ann N Y Acad Sci 959(1):128–132
21. Steinberg D, Witztum JL (2002) Is the oxidative modification hypothesis relevant to human atherosclerosis? Do the antioxidant trials conducted to date refute the hypothesis? Circulation 105(17):2107–2111
22. Burns JM, Cooper WJ, Ferry JL, King DW, DiMento BP, McNeill K, Miller CJ, Miller WL, Peake BM, Rusak SA (2012) Methods for reactive oxygen species (ROS) detection in aqueous environments. Aquat Sci 74(4):683–734
23. Kuznetsov AV, Kehrer I, Kozlov AV, Haller M, Redl H, Hermann M, Grimm M, Troppmair J (2011) Mitochondrial ROS production under cellular stress: comparison of different detection methods. Anal Bioanal Chem 400(8):2383–2390
24. Yang HW, Hwang KJ, Kwon HC, Kim HS, Choi KW, Oh KS (1998) Detection of reactive oxygen species (ROS) and apoptosis in human fragmented embryos. Hum Reprod 13(4):998–1002
25. Wang H, Joseph JA (1999) Quantifying cellular oxidative stress by dichlorofluorescein assay using microplate reader1. Free Radic Biol Med 27(5–6):612–616
26. Karlsson M, Kurz T, Brunk UT, Nilsson SE, Frennesson CI (2010) What does the commonly used DCF test for oxidative stress really show? Biochem J 428(2):183–190
27. Rao K, Murali K, Padmanabhan J, Kilby DL, Cohen HJ, Currie MS, Weinberg JB (1992) Flow cytometric analysis of nitric oxide production in human neutrophils using dichlorofluorescein diacetate in the presence of a calmodulin inhibitor. J Leukoc Biol 51(5):496–500
28. Possel H, Noack H, Augustin W, Keilhoff G, Wolf G (1997) 2, 7-Dihydrodichlorofluorescein diacetate as a fluorescent marker for peroxynitrite formation. FEBS Lett 416(2):175–178
29. Zielonka J, Kalyanaraman B (2008) "ROS-generating mitochondrial DNA mutations can regulate tumor cell metastasis"—a critical commentary. Free Radic Biol Med 45(9):1217–1219

30. Folkes LK, Patel KB, Wardman P, Wrona M (2009) Kinetics of reaction of nitrogen dioxide with dihydrorhodamine and the reaction of the dihydrorhodamine radical with oxygen: implications for quantifying peroxynitrite formation in cells. Arch Biochem Biophys 484(2):122–126
31. Qian SY, Buettner GR (1999) Iron and dioxygen chemistry is an important route to initiation of biological free radical oxidations: an electron paramagnetic resonance spin trapping study. Free Radic Biol Med 26(11–12):1447–1456
32. Tarpey MM, Wink DA, Grisham MB (2004) Methods for detection of reactive metabolites of oxygen and nitrogen: in vitro and in vivo considerations. Am J Phys Regul Integr Comp Phys 286(3):R431–R444
33. Pick E, Charon J, Mizel D (1981) A rapid densitometric microassay for nitroblue tetrazolium reduction and application of the microassay to macrophages. J Reticuloendothel Soc 30(6):581–593
34. Rook G, Steele J, Umar S, Dockrell H (1985) A simple method for the solubilisation of reduced NBT, and its use as a colorimetric assay for activation of human macrophages by - γ-interferon. J Immunol Methods 82(1):161–167
35. Güler P, Ayhan N, Koşukcu C, Onder BŞ (2015) The effects of larval diet restriction on developmental time, preadult survival, and wing length in *Drosophila melanogaster*. Turk J Zool 39(3):395–403
36. Philbrook NA, Winn LM, Afrooz AN, Saleh NB, Walker VK (2011) The effect of TiO2 and Ag nanoparticles on reproduction and development of Drosophila melanogaster and CD-1 mice. Toxicol Appl Pharmacol 257(3):429–436
37. Pappus SA, Ekka B, Sahu S, Sabat D, Dash P, Mishra M (2017) A toxicity assessment of hydroxyapatite nanoparticles on development and behaviour of *Drosophila melanogaster*. J Nanopart Res 19(4):136
38. Sabat D, Patnaik A, Ekka B, Dash P, Mishra M (2016) Investigation of titania nanoparticles on behaviour and mechanosensory organ of *Drosophila melanogaster*. Physiol Behav 167:76–85
39. Piper MD, Skorupa D, Partridge L (2005) Diet, metabolism and lifespan in Drosophila. Exp Gerontol 40(11):857–862
40. Driver CJ, Cosopodiotis G (1979) The effect of dietary fat on longevity of *Drosophila melanogaster*. Exp Gerontol 14(3):95–100
41. Jovanović B, Cvetković VJ, Mitrović TL (2016) Effects of human food grade titanium dioxide nanoparticle dietary exposure on *Drosophila melanogaster* survival, fecundity, pupation and expression of antioxidant genes. Chemosphere 144:43–49
42. Mihajilov-Krstev T, Jovanović B, Jović J, Ilić B, Miladinović D, Matejić J, Rajković J, Đorđević L, Cvetković V, Zlatković B (2014) Antimicrobial, antioxidative, and insect repellent effects of *Artemisia absinthium* essential oil. Planta Med 80(18):1698–1705
43. Wilk R, Murthy SU, Yan H, Krause HM (2010) In situ hybridization: fruit fly embryos and tissues. Curr Protoc Essent Lab Tech 4(1):9.3. 1–9.3. 24

Chapter 12

Biochemical Estimation to Detect the Metabolic Pathways of *Drosophila*

Sumit Mukherjee and Monalisa Mishra

Abstract

Regulation of the metabolic homeostasis is a prerequisite for all the animals for their growth and development. Thus, every organism developed strategies to maintain a balance between energy intake and expenditure. The imbalance in metabolism leads to a number of diseases such as obesity, diabetes, cardiovascular diseases and cancer. Besides the existence of structural variation between the organisms, the metabolic pathways are evolutionary conserved. Thus, the study of the metabolic pathways in a model system will help us to understand the role of metabolism in the onset of a disease. *Drosophila melanogaster* is evolving as an important model to study various metabolic disorders of humanbeing. This organism possesses physiological and metabolic similarity with mammals. Thus, various metabolic components like glucose, trehalose, protein and lipid associated with the fat-sensitive pathways as well as glucose-induced disorder can be checked using the fly model. The current chapter briefly employs various methods to quantify the metabolites from the larval as well as adult *Drosophila* tissues.

Key words *Drosophila* metabolism, Glucose estimation, Trehalose, Triglyceride assay, Protein estimation

1 Introduction

All organisms maintain metabolic homeostasis for their growth and development. Thus, for proper growth, a balance between energy intake with energy expenditure is essential [1]. Any imbalance in metabolic homeostasis resulted in metabolic disorders like obesity, diabetes, cardiovascular disease and cancer. Presently, murine models, such as mice, rat and *C. elegans*, are used to study the metabolic disorders of a human being [2–5]. Besides that *Drosophila* is evolving as a model to study various metabolic disorders [6]. *Drosophila* share similarity in the metabolic pathways and organ composition with mammals [7]. It shares structural and functional similarity with vertebrate intestine [8, 9]. The fat body is analogous to adipose tissue, and a specialized cell known as oenocytes is the

Monalisa Mishra (ed.), *Fundamental Approaches to Screen Abnormalities in Drosophila*, Springer Protocols Handbooks, https://doi.org/10.1007/978-1-4939-9756-5_12,

analogous structure of vertebrate liver [7, 10, 11]. The digestive system possesses copper cells which share functional similarities with parietal cells [12]. Additionally, the short life span, genetic manipulation, fully sequenced genome and 75% similarity with human diseased genes make *Drosophila* as an ideal model to study metabolic disorders [13, 14]. The conserved genetic similarity implies that the basic mechanisms of these diseases are conserved throughout evolution [15].

During the metabolic disorder, the levels of carbohydrates, protein and lipid are altered resulting in hyperglycaemia and hyperlipidaemia. Thus, these parameters can be used to detect the metabolic status of the fly [16]. The circulating sugars in *Drosophila* occur mainly in two forms: (1) glucose and (2) trehalose. Trehalose is the abundant sugar in larvae; however, both trehalose and glucose are present in the adult fly. Fly uses carbohydrate for flight, muscle contraction, various biochemical reactions, and storage of energy as glycogen [16]. The free glucose level in flies is quantified by colorimetric enzymatic assay or glucose oxidase (GO) assay [17]. In this method, glucose oxidation is catalyzed by glucose oxidase to generate gluconic acid and hydrogen peroxide (H_2O_2). H_2O_2 reacts with o-dianisidine, a compound used to generate oxidized o-dianisidine, which is orange in colour. Finally, H_2SO_4 is used to terminate the reaction which gives a pink colour that can be measured at an absorbance of 540 nm. Trehalose is the principle circulatory sugar in the fruit fly apart from glucose and can be estimated using the same method [18]. However, in trehalose estimation, trehalase is used that digest trehalose to derive free glucose, which can be measured by using glucose oxidase reagent. This free glucose level can then be compared to the background glucose level to quantify the amount of trehalose converted to glucose.

For lipid estimation and extraction, the methods that are used [19, 20], detects the total fatty acids, diacylglycerol and triglycerides, by detecting the level of free glycerol generated by cleavage of triacylglycerol (TAG) with lipoprotein lipase [19, 21, 22]. Thin-layer chromatography (TLC) is also used for lipid estimation [23]. It is important to note the assays that are used for lipid estimation also generate glycerol from mono- and diacylglycerides along with TAG, so the presence of any form of glycerolipid needs to be checked by thin-layer chromatography (TLC) [24]. Although eye pigment (measured at 540 nm wavelength) of adult flies may interfere with the assay results, the method we are describing for lipid quantification is unable to detect the eye pigment.

For accurate and rapid estimation of protein, Bradford assay is used [25–27]. This method is simple, rapid and more accurate than Lowry method. In this method, the interference of the nonprotein components is reduced. Additionally, the Coomassie Blue G250 is used as a chromophore for the reaction. The dye binds to the protein molecules in acidic conditions resulting conversation of brown to blue colour. Coomassie exists in four ionic forms, of which three ionic forms predominate in the acidic solution. The anionic blue form binds to the protein and has maximum absorbance at 595 nm. Thus the protein concentration is measured from the binding of the dye to protein, and the absorbance is measured at 595 nm. This method detects basic amino acids, lysine, arginine and histidine since Coomassie interacts with amino acids to form a protein-dye complex. With this protocol a protein amount between 1 and 10 μg can be measured.

The amount of protein can be further estimated by using BCA protein assay [28, 29]. The principle lying behind the assay is that Cu^{+2} is reduced to Cu^{+1} by protein molecules in alkaline solution and displays a purple colour in the presence of bicinchoninic acid. The colour intensity is measured at an absorbance maximum of 562 nm. Four amino acids including cysteine, tryptophan and tyrosine cause the reduction of copper. However, the peptide backbones also contribute to form colour that helps to minimize the colour variation that may arise due to the difference in protein. BCA is relatively stable in alkaline conditions. It is a single step assay. BCA assay has a major disadvantage of susceptibility to interference by various chemicals such as metal chelators (EGTA, EDTA), reducing agents (dithiothreitol and β-marcaptoethanol) and buffers of high ionic strength. This interference can be overcome by diluting the samples.

The current protocol describes carbohydrate, lipid and protein in various developmental stages of *Drosophila*.

2 Materials

2.1 Glucose Estimation

1. Early third instar larvae (or adult flies).
2. 0-sized paint brush.
3. Micropestle.
4. Microcentrifuge tube (1.5 ml).
5. Phosphate-buffered saline (PBS): Add 8 g NaCl (137 mM; HiMedia-MB023), 200 mg KCl (2.7 mM; HiMedia-MB043), 1.44 g Na_2HPO_4 (10 mM; HiMedia-TC051) and 240 mg KH_2PO_4 (1.8 mM; HiMedia-TC011) to 800 ml milliQ water. Adjust the pH to 7.4 and make the volume up to 1000 ml.
6. Anhydrous glucose (HiMedia-MB037).
7. Glucose oxidase (Sigma-G3660).

8. O-Dianisidine (Sigma-D2679).
9. 12N sulphuric acid (H_2SO_4) (HiMedia-RM6244): To prepare 10 ml of 12N H_2SO_4, dilute 3.33 ml 36N H_2SO_4 with 6.67 ml milliQ water.
10. Plate reader (BIO-BASE).
11. 96-well plate (Tarson-980040).
12. Mini cooler (Tarson-525030).
13. Centrifuge (ThermoFisher).
14. Heat block (Abdos)

2.2 Trehalose Estimation

1. Early third instar larvae (or adult flies).
2. 0-sized paint brush.
3. Micropestle.
4. Microcentrifuge tube (1.5 ml).
5. Phosphate-buffered saline (PBS).
6. Trehalase buffer (TB): 5 mM tris buffer (pH 6.6), 2.7 mM KCl, 137 mM NaCl.
7. Trehalase stock (TS): Dilute 3 μl porcine trahalase (Sigma-T8778) in 1 ml trehalase buffer.
8. Anhydrous glucose (HiMedia-MB037).
9. Trehalose (HiMedia-TC177).
10. Glucose oxidase (Sigma-G3660).
11. O-Dianisidine (Sigma-D2679).
12. 12N sulphuric acid (H_2SO_4) (HiMedia-RM6244): To prepare 10 ml of 12N H_2SO_4, dilute 3.33 ml 36N H_2SO_4 with 6.67 ml milliQ water.
13. Plate reader (BIO-BASE).
14. 96-well plate (Tarson-980040).
15. Mini cooler (or dry ice).
16. Centrifuge (ThermoFisher).
17. Heat block (Abdos).

2.3 Triglyceride Assay

1. Early third instar larvae (or adult fly)
2. Phosphate-buffered saline (PBS)
3. Tween-20 (HiMedia-MB067)
4. Triolein equivalent glycerol standard (Sigma-G7793)
5. Triglyceride reagent (Sigma-T2449)
6. Free glycerol reagent (Sigma-F6428)
7. Parafilm M (Tarson)

8. 96-well plate (Tarson-980040)
9. Plate reader (BIO-BASE)
10. Mini cooler

2.4 Protein Estimation (Bradford Method)

1. Third instar larvae (or adult fly).
2. Phosphate-buffered saline (PBS).
3. NB lysis buffer: Dissolve 58.45 mg EDTA (2 mM), 605.7 mg tris base (50 mM) and 876.6 mg NaCl (150 mM) in 100 ml 1X PBST.
4. Bovine serum albumin (BSA) (HiMedia-MB083).
5. Coomassie Brilliant Blue G-250 (HiMedia-MB092).
6. Methanol (HiMedia-AS059).
7. Lysis buffer.
8. Phosphoric acid (H_3PO_4) (HiMedia-AS011).
9. Micropestle.
10. Mini cooler.
11. 96-well plate (Tarson-980040).
12. Plate reader (BIO-BASE).

2.5 Protein Estimation by Bicinchoninic Acid (BCA)

1. Third instar larvae (or adult fly)
2. Phosphate-buffered saline (PBS)
3. NB lysis buffer
4. Bovine serum albumin (BSA)
5. BCA assay kit (Thermo-23227)
6. Microfuge tube (1.5 ml)
7. Micropestle
8. Mini-cooler
9. 96-well plate (Tarson-980040)
10. Plate reader (BIO-BASE)

3 Method

3.1 Glucose Estimation

3.1.1 Preparation of the Glucose Oxidase Reagent

1. Take the vial of o-dianisidine dihydrochloride (5 mg) and keep it in 4 °C.
2. Add 1.0 ml milliQ water and mix the solution on a rocker for 10 min.
3. Store the solution in an amber-coloured tube to avoid light and store it at 4 °C.

4. For the preparation of the working reagent, add 0.8 ml of the o-dianisidine to 39.2 ml of glucose oxidase reagent to prepare a 40 ml solution of glucose oxidase/peroxidase working reagent. Mix the solutions thoroughly by inversion.
5. Store the solution at 4 °C.

3.1.2 Preparation of the Glucose Standards

1. Prepare a glucose stock solution (1 mg/ml).
2. For the standard solutions, dilute 16 μl of glucose stock (1 mg/ml) in 84 μl of PBS (final volume 100 μl).
3. Dilute the solution twofold in PBS (50 μl of 0.16 mg/ml solution + 50 μl of PBS to prepare 0.08 mg/ml, 50 μl of 0.08 mg/ml solution + 50 μl PBS to prepare 0.04 mg/ml and so on) to make 0.01, 0.02, 0.04, 0.08 and 0.16 mg/ml glucose standards.

3.1.3 Measurement of Free Glucose in Fly Sample

1. Take 25 early third instar larvae (or 10 adult flies) from the food vial with the help of a paint brush.
2. Wash them in 1 x PBS (Fig. 1). Adult flies can be directly transferred immediately after anaesthetizing them, no need to wash with PBS (*see* **Note 4.1.1**).

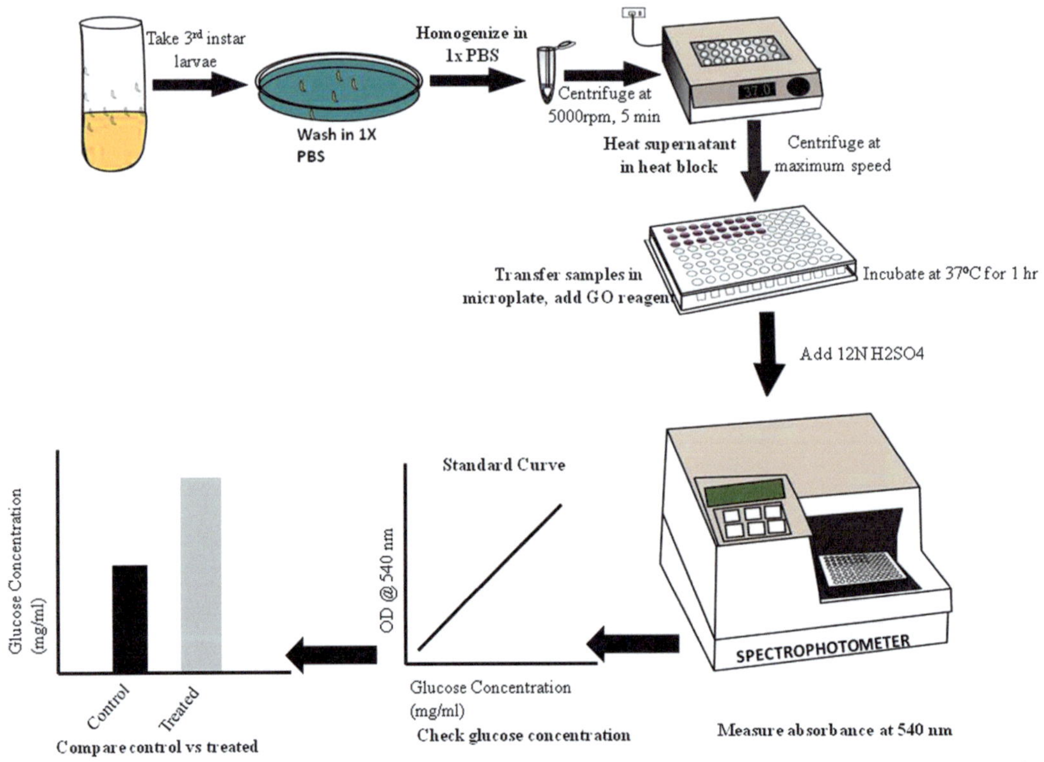

Fig. 1 Schematic diagram of glucose estimation by glucose oxidase assay

3. Collect the larvae (or adult) in a 1.5 ml tube.
4. Place the tubes on a minicooler (dry ice can also be used).
5. Add 100 μl of 1x PBS to the Eppendorf tube, and crush the larvae (or adult) with a micropestle on ice (Fig. 1) (*see* **Note 4.1.2**).
6. Centrifuge the homogenate at 5000 rpm at 4 °C for 5 min to separate the cell debris.
7. Transfer the supernatant to another 1.5 ml tube, and heat in a dry bath at 70°C for 10 min (Fig. 1) (*see* **Note 4.1.3**).
8. Centrifuge the tube at 12000 rpm for 3 min at 4 °C. Take supernatant in fresh 1.5 ml tube. Store the sample at −80 °C till the preparation of the standards.
9. Take the glucose standards (30 μl) along with 1x PBS (blank) in a microplate.
10. Add the samples (i.e. control and treated), 30 μl to the next row of the microplate (in case of larvae dilute the samples 1:4 with 1X PBS).
11. Next, add 100 μl of glucose oxidase to individual well. Seal the plate with Parafilm and keep at 37 °C for 1 h (*see* **Note 4.1.4**).
12. After the incubation, add 100 μl of 12N H_2SO_4 to each well to stop the reaction.
13. Measure absorbance at 540 nm. Note the values of the standards and the samples (Fig. 1).
14. Plot a standard curve and determine the concentration of free glucose.

3.2 Trehalose Estimation

3.2.1 Preparation of Glucose Oxidase Reagent

1. Take a preweighed vial and add 5 mg of o-dianisidine dihydrochloride to it.
2. Add 1.0 ml of milliQ water to it.
3. Mix the solution on a rocker. Avoid light exposure of the solution and keep at 4 °C.
4. For the preparation of the working reagent, take 0.8 ml of the o-dianisidine, and add 39.2 ml of glucose oxidase reagent to it. Mix for 10 min on a rocker.
5. Store the solution at 4 °C.

3.2.2 Preparation of Glucose Standards

1. Take 16 μl of glucose stock solution (1 mg/ml) in 84 μl TB to prepare glucose standard of concentration 0.16 mg/ml.
2. Perform a series of twofold serial dilution of 0.16 mg/ml solution with TB (50 μl of 0.16 mg/ml + 50 μl of TB for 0.08 mg/ml standard) to make 0.01, 0.02, 0.04 and 0.08 mg/ml standard solutions of glucose.

3.2.3 Preparation of the Trehalose Standards

1. Prepare a trehalose stock (1 mg/ml) by diluting 5 mg of trehalose in 5 ml of milliQ water.
2. For the standard solution, add 16 μl of trehalose stock with 34 μl TB and 50 μl TS (total 100 μl) to prepare for 0.16 mg/ml standard. Serially dilute the 0.16 mg/ml standard with a mixture of TB and TS in 1:1 ratio (for 0.08 mg/ml, 50 μl of 0.16 mg/ml + 25 μl TB + 25 μl TS) to make 0.01, 0.02, 0.04 and 0.08 mg/ml standard solution of trehalose.

3.2.4 Measurement of Trehalose Content

1. Take 25 early third instar larvae (or 10 adult flies) from the food vial.
2. Wash them in 1X PBS (adult flies do not need washing).
3. Transfer the larvae (or adult) to 1.5 ml microfuge tube.
4. Add 100 μl of trehalase buffer (TB) in each tube, and homogenize the larvae (or adult) with micropestle on ice (Fig. 2) (*see* **Note 4.2.1**).
5. Centrifuge the homogenized samples at 5000 rpm at 4 °C for 5 min.

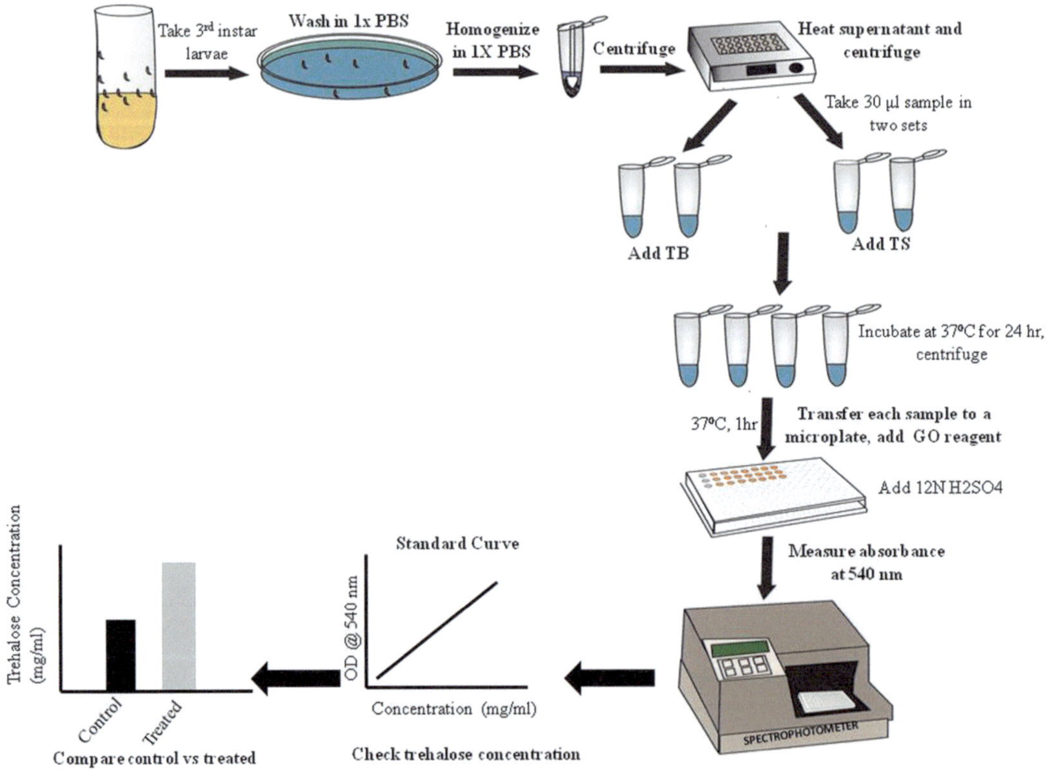

Fig. 2 Steps for estimation of trehalose by glucose oxidase assay

6. Heat supernatant at 70 °C for 10 min using a heat block and centrifuge at 12000 rpm at 4 °C for 3 min (Fig. 2).
7. Transfer supernatant in a tube, and store at −80 °C till the preparation of the standards.
8. Add 30 μl of glucose standards or TB blank to 30 μl of TB in 1.5 ml microfuge tubes.
9. Divide the trehalose standards in two sets of 30 μl each into 1.5 ml microfuge tubes. Add 30 μl of TB to one set and in the other set, add 30 μl TS (*see* **Note 4.2.2**).
10. Take 30 μl of each of the control and treated fly samples in two 1.5 ml tubes. Add 30 μl of TB to the first set to determine the background level of free glucose. Add 30 μl TS to the second set of the tube to convert trehalose to free glucose (Fig. 2).
11. Keep the samples including the standards and incubate at 37 °C for 24 h.
12. After the incubation, centrifuge all the samples at 12000 rpm for 3 min.
13. Take 30 μl of the sample into each well of a microplate, and add 100 μl of glucose oxidase reagent to individual well (Fig. 2).
14. Seal the plate with Parafilm and keep at 37 °C for 1 h.
15. Stop the reaction by adding 100 μl of 12N H_2SO_4 to each well including the blanks.
16. Take the absorbance at 540 nm and note the values.
17. Calculate free glucose in each sample from the standard curve. To measure the trehalose, subtract the absorbance of free glucose without TS, from the absorbance of the samples treated with trehalase. The concentration of trehalose is determined from the trehalose standard curve.

3.3 Triglyceride Assay

3.3.1 Preparation of Glycerol Standards

1. Add 40 μl of standard glycerol solution to 60 μl of PBST (to make total volume of 100 μl) to make 1.0 mg/ml triolein equivalent standard.
2. Perform a series of serial dilutions with PBST (50 μl of 1 mg/ml standard + 50 μl of PBST for 0.5 mg/ml standard) to make 0.125, 0.25 and 0.50 mg/ml standards.

3.3.2 Triglyceride Assay

1. Take 25 early third instar larvae (or 10 adult flies), and rinse them in a petri plate using 1X PBS (Fig. 3) (*see* **Note 4.3.1**).
2. Snap freeze the adults at 0 °C for homogenization.
3. Homogenize the larvae or adult rapidly with 100 μl of PBST (0.2% Tween-20 in 1X PBS) with a micropestle (Fig. 3).
4. Do the homogenization on ice (*see* **Note 4.3.2**).

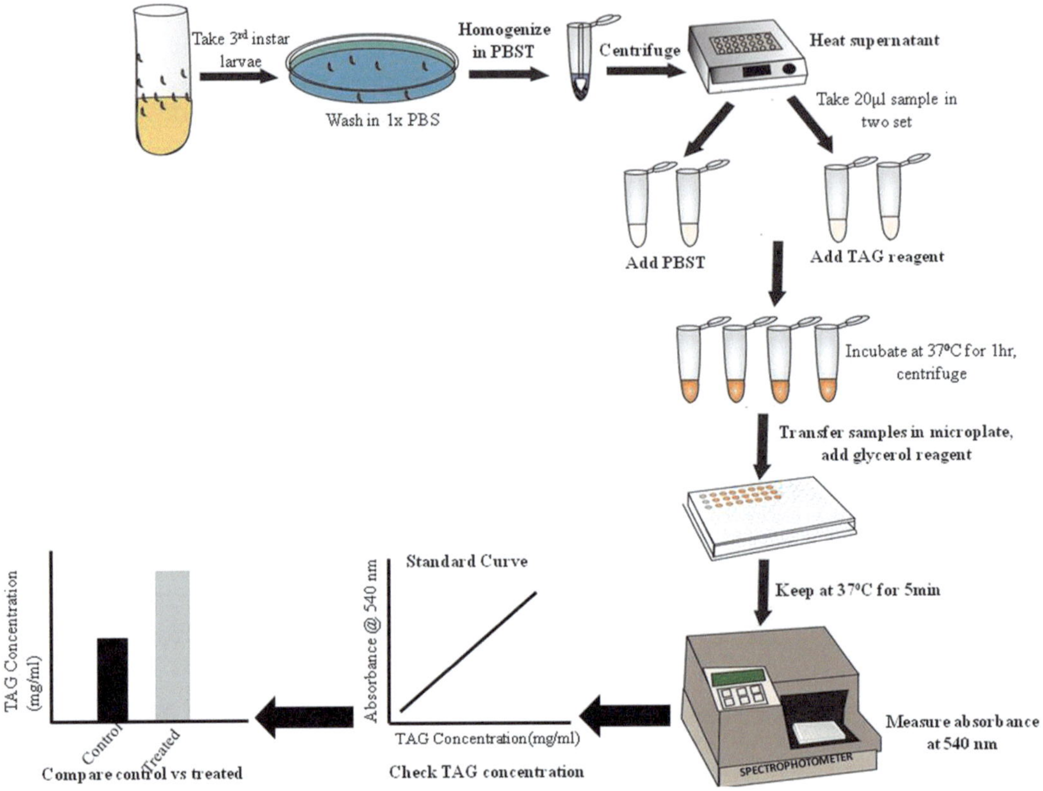

Fig. 3 Colorimetric assay to determine triglyceride content

5. Centrifuge the homogenate at 5000 rpm at 4 °C for 5 min to separate the cell debris. Transfer the supernatant to a new Eppendorf tube.
6. Heat the supernatant at 70 °C for 10 min (*see* **Note 4.3.3**). Store the heat inactivated samples in −80 °C till use. Do not centrifuge the heat inactivated samples (*see* **Note 4.3.4**).
7. Take two sets of 1.5 ml tube, and in each of them, add 20 μl of PBST blank, the glycerol standards and fly samples in individual tubes. Add 20 μl of PBST to one set (the samples to be used for measuring free glycerol). In the other set of tubes, add 20 μl of TAG reagent (the lipase in the TAG reagent will digest the triglyceride content of the samples to generate free glycerol) (Fig. 3).
8. Incubate all the samples at 37 °C for 1 h.
9. After the incubation, centrifuge at 12000 rpm for 3 min at room temperature. Take 30 μl sample in a 96-well plate.
10. Add 100 μl free glycerol reagent to individual well and mix properly (Fig. 3).

11. Seal the plate and incubate at 37 °C for 5 min.
12. Measure the absorbance at 540 nm and note the values (Fig. 3).
13. To measure the triacylglyceride concentration, the absorbance of free glycerol in the untreated sample without TAG reagent is subtracted from the concentration of total glycerol of samples treated with TAG reagent. Calculate the TAG concentration from the triolein-equivalent standard curve.

3.4 Estimation of Protein by Bradford Method

3.4.1 Preparation of Bradford Reagent

1. Add 50 mg of Coomassie Blue G250 in 50 ml of methanol.
2. Add 100 ml 85% phosphoric acid (H_3PO_4) to it and mix thoroughly.
3. Pour this solution slowly into 850 ml of distilled H_2O, and keep it for 10–15 min to dissolve the dye.
4. Filter the solution using Whatman filter paper to remove precipitates, and store in a brown bottle at room temperature.

3.4.2 Preparation of BSA Standards

1. Prepare a BSA stock solution (1 mg/ml).
2. Serially dilute the samples with distilled water to make the standards (0.1, 0.2, 0.4, 0.6, 0.8 mg/ml).

3.4.3 Protein Estimation in Drosophila

1. Take 25 third instar larvae (or 10 adult flies), and wash them in 1X PBS to remove the food particles (*see* **Note 4.4.1**).
2. Homogenize the animals in 100 μl NB lysis buffer with the help of a micropestle (*see* **Note 4.4.2**).
3. Perform centrifugation of the samples at 10000G for 20 s at 4 °C.
4. Perform centrifugation of the samples twice at 20000G at 4 °C for 3 min (Fig. 4).
5. Take a fresh tube and transfer the supernatant. Centrifuge this supernatant for two more times at 20000G at 4 °C for 3 min.
6. Prepare the standard solutions of BSA.
7. Transfer 5 μl of the standard solution or fly sample in a microplate. Take distilled water for the blank or control.
8. Add 250 μl Bradford reagent to individual well and mix properly (Fig. 4).
9. Incubate at dark for 5–60 min. Determine absorbance at 595 nm and note the values (*see* **Note 4.4.3**).
10. Plot the values in a standard curve, and measure protein concentration in the fly samples.

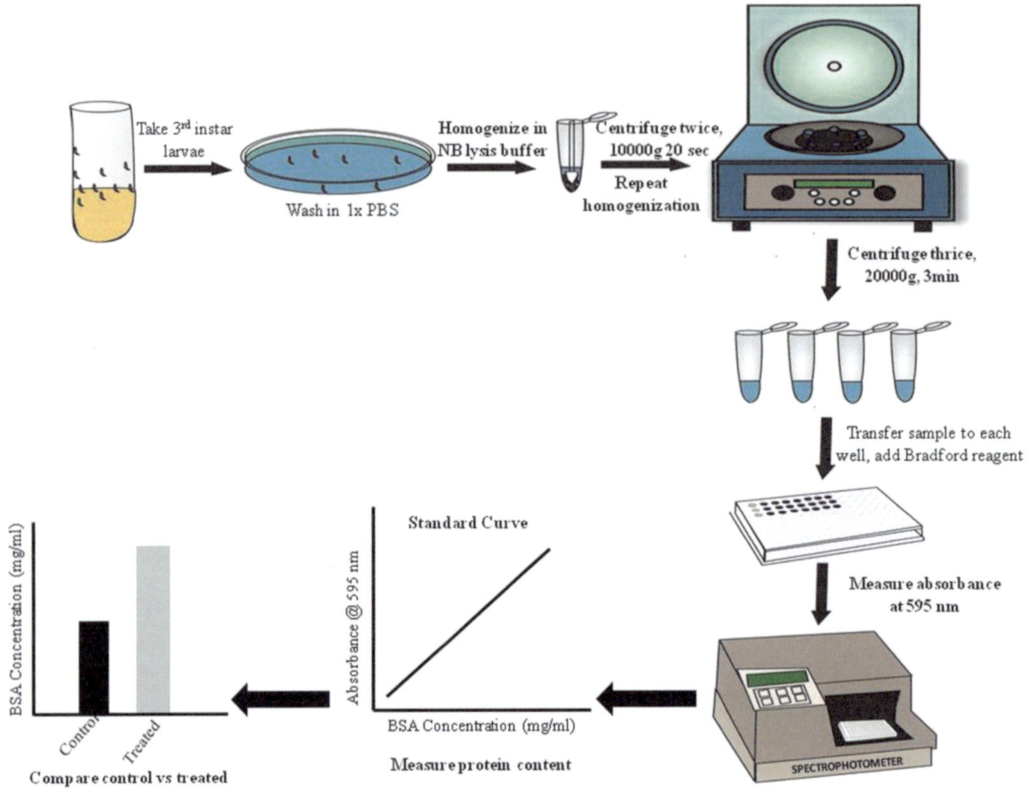

Fig. 4 Schematic diagram showing protein estimation by Bradford assay

3.5 Protein Estimation by Bicinchoninic Acid (BCA) Assay

3.5.1 Preparation of BSA Standards

1. Prepare a BSA stock solution (1 mg/ml).
2. Serially dilute the samples with distilled water to make the standards (0.1, 0.2, 0.4, 0.6, 0.8 mg/ml).

3.5.2 Estimation of Protein by BCA Assay

1. Take 25 third instar larvae (or 10 adult flies) from the food vial. Wash them with PBS to remove the food particles from the larval body. Adult flies do not need the washing step (*see* **Note 4.5.1**).
2. Take the larvae (or adult) in 1.5 ml tubes, and homogenize in 100 μl of NB lysis buffer by using a micropestle (Fig. 5).
3. Homogenize the sample on ice.
4. The homogenate was centrifuged three times at 25000g for 3 min (Fig. 5).
5. Transfer supernatant in a fresh 1.5 ml tube (*see* **Note 4.5.2**).
6. Prepare the BSA standards.

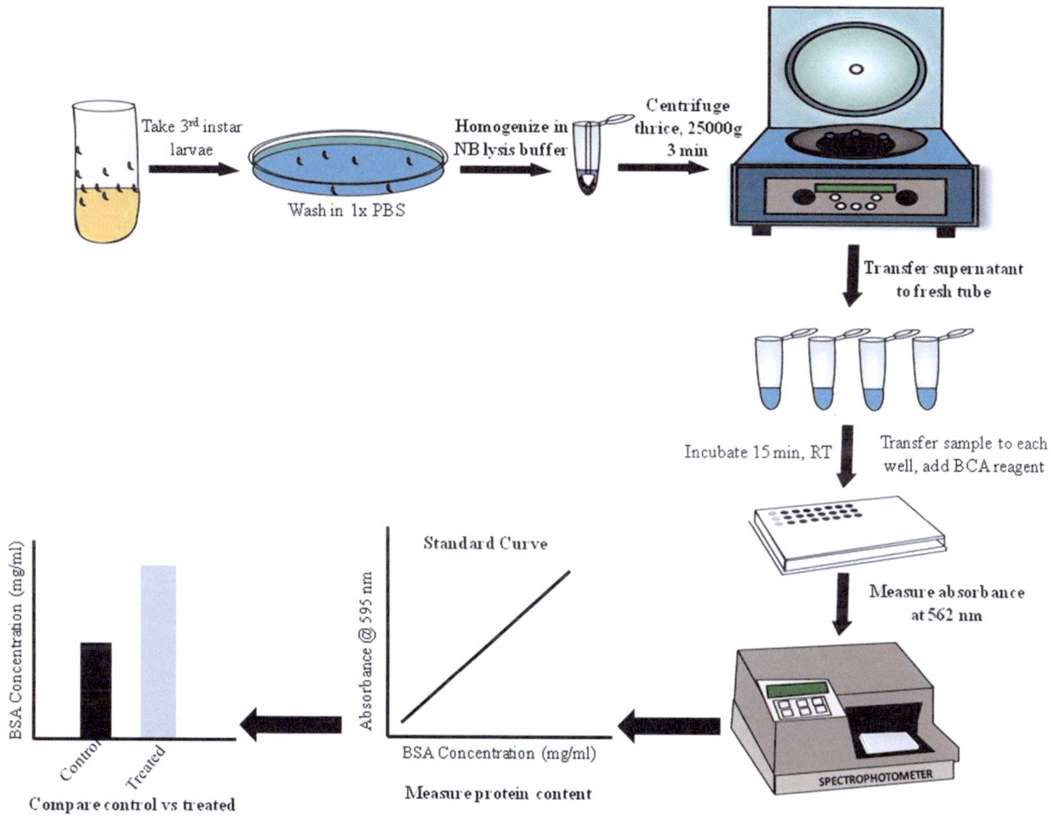

Fig. 5 Protein estimation by bichinconinic acid assay kit

7. Take 10 μl of the standard solutions and the sample (5 μl supernatant in 5 μl PBS) to individual wells of a microplate.
8. Add 200 μl of BCA reagent to it and mix properly (*see* **Note 4.5.3**).
9. Keep the samples for 15 min at dark and measure absorbance at 562 nm (Fig. 5).
10. Note the absorbance values, and plot them in the standard curve to obtain the protein concentration of the fly samples.

4 Notes

4.1 Glucose Estimation

1. Wash the larvae properly with PBS to remove the food particles from the cuticle.
2. Keep the homogenized samples on ice in order to prevent the breakdown of internal stored glucose such as glycogen.
3. Perform heat inactivation of the samples to inhibit enzyme actions that might lead to glucose breakdown.
4. Seal the 96-well plate to avoid evaporation of the samples.

4.2 Trehalose Estimation

1. Transfer the homogenate on ice as early as possible as the endogenous enzymes may degrade trehalose to glucose which interferes with the final results.
2. Make two separate sets of reactions, one treated with trehalase, the enzyme that degrades trahalose to glucose, and the other is untreated to measure the background glucose level.

4.3 Triglyceride Assay

1. Wash the larvae 3–4 times with PBS to remove food particles from the body.
2. Homogenization should be performed on ice as the stored glycerolipids might degrade into free glycerol by the endogenous enzymes and thus may interfere with the final results.
3. Move the samples rapidly for heat treatment.
4. After the heat treatment, the samples should not be centrifuged as lipids are moderately insoluble in PBST.

4.4 Estimation of Protein by Bradford Method

1. Snap freeze the animals rapidly in order to prevent degradation of proteins.
2. During homogenization, try to use fewer strokes of the pestle with the samples to reduce friction.
3. Do not keep the samples for a long time as the absorbance increases over time. Do not keep the samples for more than 1 h.

4.5 Protein Estimation by Bicinchoninic Acid (BCA)

1. Rinse the larva 3–4 times with PBS to remove the remnants of food particles from the larval body. Snap-freeze the adult flies.
2. Do not heat treat the samples for quantification of protein, as heating may degrade the protein and hinder the result.
3. Mix the reagents with the samples by gentle pipetting without creating air bubbles. Presence of air bubbles may cause interference in absorbance.

Acknowledgements

SM is thankful to MHRD for financial support. MM lab is supported by Grant No. BT/PR21857/NNT/28/1238/2017, EMR/2017/003054, Odisha DBT 3325/ST(BIO)-02/2017.

References

1. Galgani J, Ravussin E (2009) Energy metabolism, fuel selection and body weight regulation. Int J Obes 32(S7):S109
2. Moreno-Arriola E, Cárdenas-Rodríguez N, Coballase-Urrutia E, Pedraza-Chaverri J, Carmona-Aparicio L, Ortega-Cuellar D (2014) *Caenorhabditis elegans*: a useful model for studying metabolic disorders in which oxidative stress is a contributing factor. Oxidative medicine and cellular longevity 2014
3. Goldsworthy ME, Potter PK (2014) Modelling age-related metabolic disorders in the mouse. Mamm Genome 25(9–10):487–496
4. Kennedy AJ, Ellacott KL, King VL, Hasty AH (2010) Mouse models of the metabolic syndrome. Dis Model Mech 3(3–4):156–166

5. de Artinano AA, Castro MM (2009) Experimental rat models to study the metabolic syndrome. Br J Nutr 102(9):1246–1253
6. Bharucha KN (2009) The epicurean fly: using *Drosophila melanogaster* to study metabolism. Pediatr Res 65(2):132
7. Leopold P, Perrimon N (2007) Drosophila and the genetics of the internal milieu. Nature 450(7167):186
8. Liu X, Hodgson JJ, Buchon N (2017) Drosophila as a model for homeostatic, antibacterial, and antiviral mechanisms in the gut. PLoS Pathog 13(5):e1006277
9. Apidianakis Y, Rahme LG (2011) *Drosophila melanogaster* as a model for human intestinal infection and pathology. Dis Model Mech 4(1):21–30
10. Baker KD, Thummel CS (2007) Diabetic larvae and obese flies—emerging studies of metabolism in Drosophila. Cell Metab 6(4):257–266
11. Gutierrez E, Wiggins D, Fielding B, Gould AP (2007) Specialized hepatocyte-like cells regulate Drosophila lipid metabolism. Nature 445(7125):275
12. Dubreuil RR (2004) Copper cells and stomach acid secretion in the Drosophila midgut. Int J Biochem Cell Biol 36(5):742–752
13. Jennings BH (2011) Drosophila–a versatile model in biology & medicine. Mater Today 14(5):190–195
14. Roberts DB (2006) *Drosophila melanogaster*: the model organism. Entomol Exp Appl 121(2):93–103
15. Owusu-Ansah E, Perrimon N (2014) Modeling metabolic homeostasis and nutrient sensing in Drosophila: implications for aging and metabolic diseases. Dis Model Mech 7(3):343–350
16. Tennessen JM, Barry WE, Cox J, Thummel CS (2014) Methods for studying metabolism in Drosophila. Methods 68(1):105–115
17. Bergmeyer J, Grassl M (1974) Methods of enzymatic analysis, 3rd edn, vol I
18. Chen Q, Ma E, Behar KL, Xu T, Haddad GG (2002) Role of trehalose phosphate synthase in anoxia tolerance and development in *Drosophila melanogaster*. J Biol Chem 277(5):3274–3279
19. Hildebrandt A, Bickmeyer I, Kühnlein RP (2011) Reliable Drosophila body fat quantification by a coupled colorimetric assay. PLoS One 6(9):e23796
20. Rietveld A, Neutz S, Simons K, Eaton S (1999) Association of sterol-and glycosylphosphatidylinositol-linked proteins with Drosophila raft lipid microdomains. J Biol Chem 274(17):12049–12054
21. Clark A, Gellman W (1985) A rapid spectrophotometric assay of triglycerides in Drosophila. Drosophila Inf Serv 61:190
22. Grönke S, Beller M, Fellert S, Ramakrishnan H, Jäckle H, Kühnlein RP (2003) Control of fat storage by a Drosophila PAT domain protein. Curr Biol 13(7):603–606
23. Touchstone JC (1995) Thin-layer chromatographic procedures for lipid separation. J Chromatogr B Biomed Sci Appl 671(1–2):169–195
24. Van Veldhoven PP, Swinnen JV, Esquenet M, Verhoeven G (1997) Lipase-based quantitation of triacylglycerols in cellular lipid extracts: requirement for presence of detergent and prior separation by thin-layer chromatography. Lipids 32(12):1297–1300
25. Bradford MM (1976) A rapid and sensitive method for the quantitation of microgram quantities of protein utilizing the principle of -protein-dye binding. Anal Biochem 72(1–2):248–254
26. Stoscheck CM (1990) [6] Quantitation of protein. In: Methods in enzymology, vol 182. Elsevier, pp 50–68
27. He F (2011) Bradford protein assay. Bio-protocol 1(6):e45. https://doi.org/10.21769/BioProtoc.45
28. Smith Pe KRI, Hermanson G, Mallia A, Gartner F, Provenzano M, Fujimoto E, Goeke N, Olson B, Klenk D (1985) Measurement of protein using bicinchoninic acid. Anal Biochem 150(1):76–85
29. Bainor A, Chang L, McQuade TJ, Webb B, Gestwicki JE (2011) Bicinchoninic acid (BCA) assay in low volume. Anal Biochem 410(2):310–312

Chapter 13

Biochemical Assays to Detect the Antioxidant Level in *Drosophila melanogaster*

Janmejaya Bag and Monalisa Mishra

Abstract

Oxygen is a key molecule for the survivability of the aerobic animals, which is converted into toxic byproduct known as reactive oxygen species (ROS) under stress. ROS can induce damage to the cells. To protect the organism from ROS-induced damage, living organisms develop antioxidant defence system. The antioxidative enzymes include superoxide dismutase (SOD), catalase (Cat), glutathione peroxidase (GPx) and 1,1-diphenyl-2-picrylhydrazyl (DPPH). These antioxidative enzymes are also present in *Drosophila* to invade the ROS generated within the body. As a result, the antioxidant enzymes prevent disease development and increase the life span of the organism. Thus, indirectly the determination of antioxidant enzymes is a pointer of the health of the animal. The protocol describes various antioxidant assays in *Drosophila melanogaster*.

Key words Reactive oxygen species, Superoxide dismutase, Catalase (Cat), Glutathione peroxidase (GPx), Lipid peroxidation (LPP), 1,1-Diphenyl-2-picrylhydrazyl (DPPH)

1 Introduction

Every organism possesses a defence mechanism to protect itself from oxidative insult. Oxidative stress causes damage to DNA, protein oxidation, lipid peroxidation and enzyme inactivation. Thus, living cells possess many enzymes, chemical and biomolecular substances as the defence system to fight against the foreign and unwanted toxic free radicals [1]. Free radicals are generated within the body during metabolism, as an intermediate product due to the utilization of oxygen, a byproduct of the metabolic processes of the biomolecules [2–4]. Free radicals include superoxide anions, oxygen radicals, hydroxyl radicals and hydrogen peroxide which are commonly termed as reactive oxygen species (ROS) [5]. To combat the ROS generated within the body, an organism possesses a well-developed defence mechanism to reduce the level of ROS [2, 6].

Monalisa Mishra (ed.), *Fundamental Approaches to Screen Abnormalities in Drosophila*, Springer Protocols Handbooks, https://doi.org/10.1007/978-1-4939-9756-5_13,

Drosophila develops a well-developed antioxidant system to protect itself from oxidative stress [7]. This antioxidant pathway shares common homology with mammals, and the level of antioxidant enzyme varies with the disease. For example, the altered antioxidant pathway is observed in the fly model for Parkinson's disease as well as for human beings [8]. Like mammals, increased ROS level mediates the differentiation of haematopoietic stem cells of *Drosophila* [9, 10]. Antioxidant properties of nanomaterials were investigated using fly [11]. Like vertebrates, *Drosophila* antioxidant pathway is composed of superoxide dismutase (SOD), thioredoxin reductase (TxR) and catalase (Cat) [7]. H_2O_2 is produced from superoxides by SOD, which is degraded by catalase to form H_2O and O_2. TrxR and GR help in the formation of Trx (SH)2 and GSH by transferring reduced equivalents from NADPH to thioredoxin and glutathione disulphide (GSSG). These molecules behave as antioxidants [12].

H_2O_2 is primarily produced from the oxidation of the NADPH oxidase within the mitochondria [13, 14]. The superoxide anion is formed inside the cell by the loss of one electron of oxygen molecules which rapidly converted by the superoxide dismutase (SOD1 and SOD2) into H_2O_2. SOD1 is located between the intermembrane of mitochondria and the cytosol. SOD2 is present within the mitochondrial matrix. SOD prevents the accumulation of the superoxide (SO) ions by scavenging, damaging or inactivating the proteins containing metallic clusters [15]. Thus, SO does not damage the proteins; rather a set of specific proteins which are sensitive to inactivation by the signalling cascade gets activated by the superoxide [16]. Therefore, the increased amount of ROS is responsible for the hyperactivation of the immune cells. ROS signalling is required for the maintenance of the tissue and ageing [17]. ROS is essential for the differentiation of the stem cells, muscles and adipocytes [18]. Likewise, an increase in ROS generation in *Drosophila* triggers differentiation of haematopoietic progenitors, and a decrease in ROS impairs differentiation [18]. Therefore, ROS acts like a morphogen where at low level it helps in the construction of the cells and at high-level destruction of the cells.

1.1 Protein Quantification

The amount of protein present in our body has a role in maintaining metabolic activity. Thus, the quantification is necessary to understand the total protein content in that sample or in a specimen. Accurate protein quantification is necessary to conduct the other critical assays. Proteins are the building block of enzymes which shows catalytic properties [19, 20]. Therefore, any change in enzymes resulted in alteration of the protein level. Enzymes are the biocatalyst, which enhances the rate of biological reaction. Thus, before performing any enzymatic assays or the body metabolites, it is essential to determine the protein quantity of that sample. In *Drosophila*, protein quantity and enzymatic activity play a major role in the regulation of body metabolism and development [21–23].

1.2 GPx Act as Detoxification of Peroxidase

Glutathione peroxidase (GPx) helps in the detoxification of the peroxides present in the living cells. GPx catalyses the reduction of the H_2O_2 and another organic peroxide such as hydroperoxide (R-OOH) into the complementary stable alcohol (R-OH) and H_2O using the reducing reagent glutathione (GSH). This step has a significant role in cell defence from the oxidative damages to the free radicals. The lipid molecules present within the cells are prone to free radicals and cause lipid peroxidation(LPP). GPx acts as the reducing agent that prevents the formation of free radicals [1]. GPx has four 22 kDa monomers which have selenocysteine moiety in each active site which is involved in the direct transfer of peroxide and thus become oxidized. Alteration of cellular GPx is involved in various disease.

1.3 Catalase Prevents from Hydrogen Peroxide Accumulation

Catalase is an enzyme, which catalyses the H_2O_2 into H_2O molecule and O_2 and thus protects the cells from the oxygen free radicals. In *Drosophila*, the catalase gene is present on chromosome 3 of the left arm and is mapped to 75D1-76A. At the larval stage and during metamorphosis, high catalase activity is seen [24]. Overexpression of catalase increases with the life span and makes the fly more resistant towards stress [25, 26]. Inside the cell, hydrogen peroxide acts as a secondary messenger, which is produced from various cell activity during the synthesis of ROS [27]. However, the toxicity of the hydrogen peroxide varies from cell to cell. Similarly, the exposure duration also varies with cells. Therefore, increased hydrogen peroxide concentration inside cells exerts oxidative stress and causes irreversible damage to the cells. Catalase breaks down the hydrogen peroxide into water, and oxygen molecule thus protects the cells from being damaged [28].

1.4 Lipid Peroxidation Is a Marker of ROS Production

Lipid peroxidation (LPP) involves the degradation of cellular lipids resulting in aldehyde products as 4-hydroxynonenal and malondialdehyde [29]. Excess of the LPP damages the cell membrane and also haemolyses the RBCs' cell membrane. The end product of the lipid peroxidation can bind with the DNA and form adducts, which leads to a mutation in DNA [30] However, the antioxidant of the body can inhibit the oxidative damage to the targeted cells and molecules [31]. Therefore, lipid peroxidation is a marker to determine the ROS production within the cells. LPP is also used as a marker to detect the metabolic byproduct of various diseases in *Drosophila* [32, 33]. The byproduct further acts as a mutagen for *Drosophila* [34]. Similarly, the treatment of various compounds also alters the LPP level in *Drosophila* [35].

1.5 DPPH Determine the Strength of Antioxidant

DPPH (2,2-diphenyl-1-picrylhydrazyl) is found in free radical form, which is stable at the room temperature (25°C) and produces purple colour with methanol. With the disappearance of the colour, ROS pairs with the nitrogen free radicals. So, by determining the DPPH level, the amount of the antioxidant and the radical scavenging activity can be observed [36]. Various experiments measured the DPPH level in *Drosophila* [37–39].

The current protocol summarizes various methods to detect the level of antioxidant status of the fly from the hemolymph and whole body homogenate using various pathological and treated conditions.

2 Materials and Method

2.1 Hemolymph Collection

1. Petri plate
2. 1X PBS
3. 0.5 and 1.5 mL centrifuge tube
4. Fine needle/pins
5. 70% ethanol
6. Ice-box
7. Centrifuge
8. −20 °C freezer

2.2 Larvae/Adult Whole Body Protein Isolation

1. Third instar larvae/adult fly
2. 1X PBS
3. 1.5 mL centrifuge tube
4. Micropestle
5. 0.1% Triton X-100 (HiMedia-MB031)
6. Ice-box
7. 20 mM Tris-acetate buffer (pH 7.8, SRL-77-86-1)
8. 1 mM phenylmethylsulfonyl fluoride (PMSF) (ThermoScientific-87786).
9. Centrifuge (ThermoFisher, Heraeus Megafuge 16R centrifuge)
10. −20 °C freezer
11. Microplate reader (BIOBASE-EL10A ELISA reader)

2.3 Protein Estimation by Lowry Method

1. 2% alkaline sodium carbonate solution (HiMedia-GRM254)
2. 0.1 N NaOH (HiMedia-GRM1183)
3. 0.5% copper sulphate solution (HiMedia-GRM6391)
4. 1% sodium potassium tartrate (HiMedia-GRM598)
5. Folin and Ciocalteu's phenol reagent (HiMedia-RM10822)
6. BSA (1 mg/mL) (HiMedia-MB038)

7. The unknown concentration of hemolymph/tissue homogenate
8. Micropipette
9. 96-well plate
10. Microplate reader (BIOBASE-EL10A ELISA reader)

2.4 SOD

1. 50 mM phosphate buffer
2. 50 μM of EDTA (HiMedia-TC115)
3. L-methionine (HiMedia-GRM200)
4. 50 μM riboflavin (HiMedia-CMS181)
5. 1% sulphanilamide (w/v) (HiMedia-GRM1558)
6. 0.1% N-(1-Naphthyl) ethylenediamine dihydrochloric acid (NED-HCl) (HiMedia-RM1073)
7. 10 mM hydroxylamine hydrochloride (HiMedia-GRM1028)
8. 2.5% orthophosphoric acid (v/v) (HiMedia-AS010)
9. 1% of Triton X-100 (HiMedia-MB031)
10. Third instar larvae hemolymph sample
11. Eppendorf tube
12. 96-well plate
13. Micropipette (10 μL, 200 μL and 1000 μL)
14. A 20 + 20 watt electric bulb containing a dark box
15. Microplate reader

2.5 Catalase Assay

1. 50 mM phosphate buffer
2. 17 mM hydrogen peroxide
3. Pipette
4. Test tube
5. Test tube stand
6. Protein homogenate
7. UV spectrophotometer
8. Protein sample

2.6 Glutathione Peroxidase

1. 0.3 M sodium phosphate buffer, pH 7.0
2. 4 mM reduced glutathione (HiMedia-MB116)
3. 10 mM sodium azide (HiMedia-MB075)
4. 2.5 mM hydrogen peroxide
5. 0.3 M disodium hydrogen phosphate (HiMedia-MB024)
6. 0.8 mM Ethylene diamine tetra acetic acid (EDTA) (HiMedia-TC115)
7. 4% 5,5′-Dithiobis(2-nitrobenzoic acid) DTNB (HiMedia-GRM 1677)
8. 100 mM Trichloro acetic acid (TCA) (HiMedia-GRM627)

2.7 Lipid Peroxidation Assay (LPP)

1. Thiobarbituric acid (TBA) (HiMedia-RM1594)
2. TCA (HiMedia-GRM627)
3. 0.1 M KCl (HiMedia-MB043)
4. 0.24 N HCl (HiMedia-AS003)
5. Malondialdehyde (MDA)
6. Hemolymph sample

2.8 DPPH Assay

1. 0.2 mM DPPH (2,2-diphenyl-1-picrylhydrazyl) (HiMedia-RM5169)
2. Methanol (HiMedia-AS058)
3. Aluminium foil
4. Ascorbic acid (HiMedia-RM393)
5. Hemolymph sample

3 Methods

3.1 Hemolymph Collection

1. Take out 20–25 third instar larvae from the rearing media, and rinse with 1X PBS.
2. Dry it with tissue paper and transfer to a 0.5 mL tube.
3. Make a small cut in the tube at the bottom with a fine razor blade.
4. Put the 0.5 mL tube into one 1.5 mL tube and transfer it into ice-box.
5. Prick the larvae with the fine needle carefully. Before pricking clean the needle with 70% ethanol.
6. Centrifuge with 10,000 rpm for 5 min at 4 °C.
7. Collect the hemolymph at the bottom of 1.5 mL tube.
8. Store the collected hemolymph in −20 °C freezer for further experimental uses (Fig. 1).

3.2 Larval/Adult Tissue Homogenate Preparation

3.2.1 Chemical Preparation

1. For the preparation of 10X PBS solution: Take 40 g of NaCl, 1 g of KCl, 7.7 g of Na_2HPO_4 and 1.2 gm of KH_2PO_4.
2. Mix all the chemicals on a glass bottle containing 400 mL of milliQ water.
3. Adjust the pH to 7.4 with HCl.
4. Finally, adjust the volume to 500 mL.
5. 1X PBS: Dilute 1 ml of 10X PBS with 9 ml of milliQ water.
6. Lysis buffer: For 10 mL of buffer, take 10 mL of 20 mM Tris-acetate buffer (pH 7.8), add 100 μL of Triton X-100 (0.1%), and add 10 μL of 100X PMSF (0.1X).

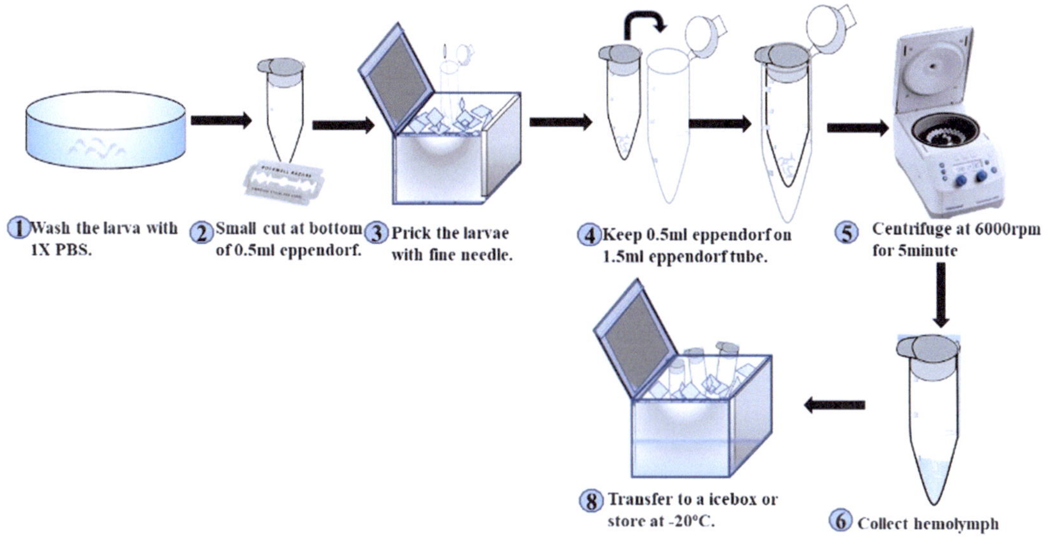

Fig. 1 Hemolymph collection procedure from third instar larvae [40]

3.2.2 Procedure

1. Take 25 third instar larvae or 25 adult flies from the food vial. Wash the larvae in a petri plate with 1X PBS to remove the food from the larval body.
2. Take the larva/fly in a 1.5 mL tube, and homogenate it in 100 μL lysis buffer on ice by using a micropestle.
3. Centrifuge at 10,000 rpm for 10 min at 4 °C.
4. Transfer the supernatant to another set of Eppendorf tube, and keep in ice-box. The samples can be stored at −20 °C freezer or in −80 °C.

3.3 Protein Estimation

Lowry Method

3.3.1 Chemical Preparation

1. Reagent A: 2% alkaline sodium carbonate solution—take 2 g of sodium carbonate in 0.1 N NaOH.
2. Reagent B: 0.5% copper sulphate solution—take 0.5 g of copper sulphate in 100 mL of milliQ water.
3. Reagent C: 1% sodium potassium tartrate—take 1 g of sodium potassium tartrate in 100 mL of milliQ water.
4. Lowry reagent: Add above all the reagents in A:B:C::100:2:2 proportions.
5. Folin and Ciocalteu's phenol reagent: Dilute the reagent by 1:2 part of milliQ water.

Table 1
Standardization table for using BSA as a protein to plot the standard curve

Sl no.	BSA Standard (µL)	milliQ water (µL)	Conc. (inµg/mL)	Lowry reagent (µL)	Incubation	Folin reagent (µL)	Incubation	Absorbance (at 660nm)
0	Blank	100	-----	25	**10 minutes**	10	**30 minutes**	
1	0	100	0	25		10		
2	10	90	20	25		10		
3	20	80	40	25		10		
4	30	70	60	25		10		
5	40	60	80	25		10		
6	50	50	100	25		10		
Unknown	20	80	-----	25		10		

6. BSA standard: Take 1 mg/mL conc. of BSA. Dilute it five times (1 ml BSA + 4 ml milliQ water). Prepare the standard solutions as mentioned in Table 1. Vortex well till dissolved completely and avoid foaming. Keep the BSA in ice-box.
7. Unknown sample (stock): Dilute the stock protein sample five times with 1X PBS (working protein sample).

3.3.2 Procedure

1. Take fresh 96-well plates and keep in the ice-box.
2. Pipette BSA solution in wells, by taking 0, 10, 20, 30, 40 and 50 µL in triplicate to make 0, 20, 40, 60, 80 and 100 µg/mL concentrations, respectively.
3. Add milliQ water to individual well as mentioned in Table 1.
4. Take 100 µL of milliQ water as a blank (Table 1).
5. Pipette 20 µL of unknown protein sample from stock, and add 80 µL of milliQ water.
6. Add 25 µL of Lowry reagent and incubate for 10 min at 25 °C.
7. Add 10 µL of Folin and Ciocalteu's phenol reagent.
8. Incubate the plate 30 min at 25 °C in the dark.
9. Take the absorbance at 660 nm in a microplate reader (Fig. 2).

3.3.3 Calculation

Plot the standard graph and note down the straight-line equation, $Y = mX + C$,

where Y = absorbance at 660 nm

m = slope,

X = conc. of unknown sample

C = constant

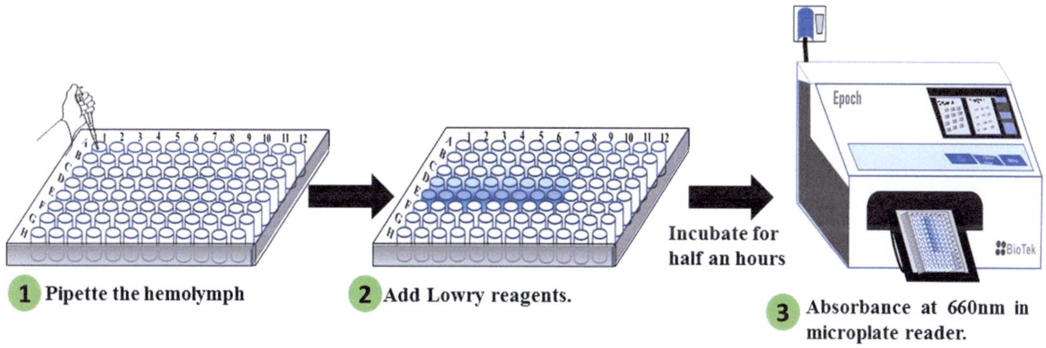

Fig. 2 Protein estimation of hemolymph by Lowry method [41, 42]

Thus, $X = Y - C/m$

So the protein concentration will be $(X) = Y - C/m * DF$

where DF = dilution factor

3.4 SOD

3.4.1 Chemical Preparation

A. 50 mM phosphate buffer (PB):

1. Take 800 mL of milliQ water in a glass bottle.
2. Add 10.105 g of Na_2HPO_4 to it.
3. Add 1.697 g of NaH_2PO_4 to the solution.
4. Adjust the solution to final pH 7.4 using HCl or NaOH.
5. Add milliQ water to make 1 L of the final solution.

B. 20 mM L-methionine:

1. Take 0.0298 g of L-methionine in 10 mL of milliQ water.
2. Mix well and vortex.
3. Always prepare it freshly before the experiment.

C. 1% Triton X-100:

1. Add 1 mL of Triton X-100 in 99 mL of milliQ water.
2. Avoid foam formation by carefully adding water.

D. 10 mM hydroxylamine hydrochloride(Hxl-HCl):

1. Add 0.695 g of Hxl-HCl in 10 mL of milliQ water to make a stock concentration of 100 mM.
2. Make 10 mM as the final working concentration.

E. 50 μM EDTA:

1. Take 2.923 g of EDTA in 80 mL of milliQ water; adjust the pH to 8.0 with 1 N NaOH or 1 N HCl to get 100 mM EDTA.
2. Make the final volume to 100 mL.
3. Dilute the solution with milliQ water to get a 50 μM concentration of EDTA.

F. 50 μM riboflavin:
   1. Take 3.7 mg of riboflavin in 1 mL of methanol to have 10 mM stock of riboflavin.
   2. Vortex well and avoid light exposure.
   3. Make a series of serial dilution to have a 50 μM of concentration.

G. 2.5% orthophosphoric acid (OPA) (V/V):
   1. Take 2.94 mL of 85% OPA and dilute it with 97.06 mL of milliQ water.

H. 1% sulphanilamide (W/V): Take 1 g of sulphanilamide in 100 mL of 2.5% OPA.

I. 0.1% naphthylethylenediamine dihydrochloride (NED-HCl): Take 0.1 g of NED-HCl in 100 mL of milliQ water.

J. Griess reagents: Mix 1% sulphanilamide and 0.1% NED-HCl in 1:1 ratio immediately before the experiment.

K. Standard ascorbic acid: Make a stock of 1 mg/mL ascorbic acid with milliQ water, and dilute it with phosphate buffer to make a different concentration of it.

L. Sample: Take different concentrations of protein to study the scavenging activity of SOD.

3.4.2 Procedure

1. Make the cocktail (see Table 2).
2. Add different concentrations of ascorbic acid and samples. Add the samples (20–40 μg) on the basis of the concentration of protein measured by Lowry method.
3. Add 8 μL of 50 mM riboflavin (see Table 3).
4. Expose the microplate to two 40 W white light for 15 min.
5. After 15 min immediately add 100 μL of Griess reagent.
6. Then take the absorbance at 543 nm on microplate reader spectrophotometer (Fig. 3).

Unknown Sample: *After knowing 50% scavenging activity of ascorbic acid take that much amount of (μg/mL) protein sample. This much amount of sample will be used in place of ascorbic acid, and all other values will remain as it is. Repeat the same procedure as the standard. This will give the value of 'S' which is the unknown value of our sample (tissue homogenate). Next calculate the % inhibition using the below formula:*

Calculation: *% of inhibition* $= C - S/C * 100$

where C=Control

S=Sample

Table 2
Preparation of cocktail for SOD assay

Sl no.	Components	Concentration	Cocktail per well(in μL)
1	Phosphate buffer (pH 7.4)	50 mM	111
2	L-methionine	20 mM	7.5
3	Triton X-100	1%	4
4	Hydroxylamine hydrochloride	10 Mm	7.5
5	EDTA	50 μM	10
		Total	140

Table 3
Table for standardization of SOD activity by using ascorbic acid as a standard. Note the value which will give you 50% scavenging of ascorbic acid

Well name	Cocktail (μL)	Ascorbic acid standard or sample (s) (μL)	PB (μL)	Conc. (In μg/ mL)	Riboflavin (μL)	Incubation	Griess reagent In μL	Absorbance at 543 nm
Blank	140	0	18	0	0	15	100	
Control	140	0	10	0	8	min.	100	
A	140	1	9	1	8	light	100	
B	140	2.5	7.5	2.5	8	expose	100	
C	140	5	5	5	8		100	
D	140	7.5	2.5	7.5	8		100	
E	140	10	0	10	8		100	

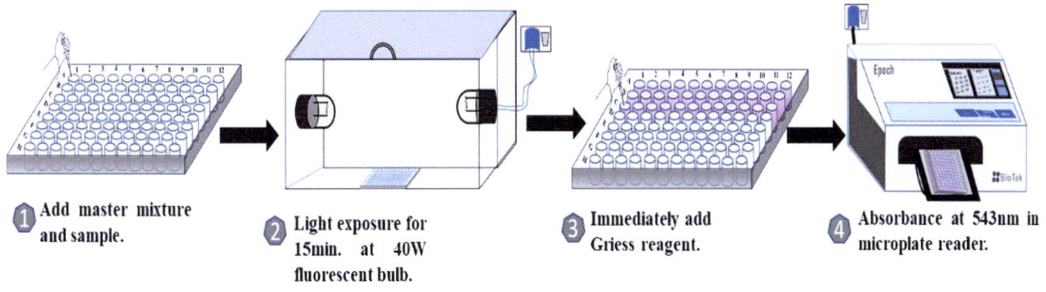

Fig. 3 SOD assay of hemolymph [2–4]

3.5 Catalase Assay

3.5.1 Chemical Preparation

A. 50 mM phosphate buffer (PB):

1. Take 800 mL of milliQ water in a glass bottle.
2. Add 10.105 g of Na_2HPO_4 (50 mM) to it.
3. Add 6.80 g of KH_2PO_4 (50 mM) to the solution.
4. Adjust the final pH to 7.4 using HCl or NaOH.
5. Add milliQ water to have 1 L of the final solution.

B. Add about 0.75 mL of 30% H_2O_2 to 100 mL phosphate buffer. Prepare freshly before measurement and store at 4 °C.

3.5.2 Procedure

1. Pipette 2.0 mL of sample into a cuvette (path length: 1 cm).
2. Equilibrate at 25 °C for about 5 min.
3. Add 1.0 mL of H_2O_2 solution.
4. Take two blanks. In one, take only phosphate buffer, and in other take phosphate buffer with H_2O_2.
5. Take absorbance at 240 nm in UV visible spectroscope (Fig. 4).

3.5.3 Calculation

Volume activity (U/mL) = $\Delta A \times 34.4 \times DF$

Weight activity (U/ms) = (U/mL) × 1/C

0.0436 = Millimolar extinction coefficient of H_2O_2 at 240 nm

DF = Dilution factor

C = Content of catalase preparation in sample (mg/mL)

3.6 Glutathione Peroxidase

3.6.1 Chemical Preparation

1. Prepare 0.3 M sodium phosphate buffer with pH 7.0: Add 2.04 gm of disodium hydrogen phosphate and 2.56 gm of sodium dihydrogen phosphate in 80 mL of milliQ water. Adjust the pH 7.0 using HCl or NaOH. After that add the milliQ water until volume is 100 mL.

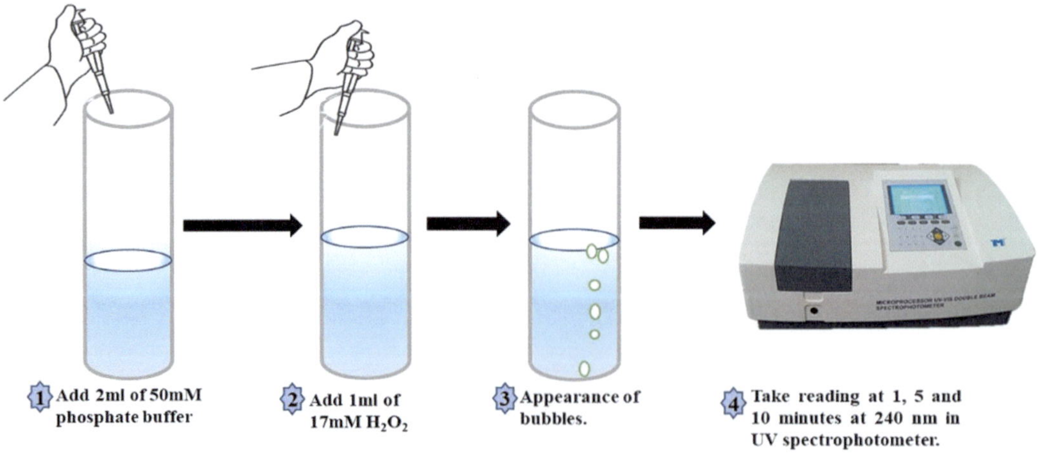

Fig. 4 Catalase assay of hemolymph [1]

2. 10 mM sodium azide: Take 6.5 mg of sodium azide in 10 mL milliQ water.
3. 4 mM reduced glutathione: Take 0.012 g of reduced glutathione in 10 mL of milliQ water.
4. 2.5 mM of H_2O_2: Add 5 μL of stock H_2O_2 in 9995 μL of sodium phosphate buffer.
5. 100 mM TCA: Take 1.6339 g of TCA in 100 mL of milliQ water.
6. 0.3 M disodium hydrogen phosphate solution: Take 4.258 g of disodium hydrogen phosphate in 100 mL of milliQ water.
7. 0.4% DTNB: Take 40 mg of DTNB in 100 mL of 1% sodium phosphate buffer having 0.1 mM of EDTA.
8. 0.8 mM EDTA: Take 23.3 mg of EDTA in 100 mL of milliQ water.

3.6.2 Procedure

1. Tissue homogenizing buffer recipe: Take 0.4 mL of phosphate buffer, 0.2 mL sodium azide, 0.2 mL EDTA, 0.2 mL hydrogen peroxide and 0.2 mL of reduced glutathione with 0.5 ml of protein sample and 0.3 ml of milliQ water.
2. Divide the mixture into two tubes each having 1 ml of the sample.
3. Incubate one sample for 90 min and the other for 180 min at the 37 °C incubator.
4. After 90 min, add 500 μL of TCA as a precipitating agent to one tube.
5. Centrifuge at 10000 rpm for 10 min at 4 °C.
6. Transfer the supernatant and store at 4 °C.
7. Repeat the step number 4–6 for another tube incubated for 180 min.
8. For microplate assay: take 20 μL of the supernatant from both the tubes.
9. Add 120 μL of 0.3 M disodium hydrogen phosphate.
10. And 20 μL of DTNB.
11. Take blank disodium hydrogen phosphate with DTNB.
12. Take reading at 412 nm on microplate reader (Fig. 5).

3.7 Lipid Peroxidation Assay (LPP)

3.7.1 Chemical Preparation

1. TBA reagent: Add 0.04 g of TBA, 15% of TCA and 0.24 N HCl per 10 mL.
2. 0.1 M KCl.
3. Different concentrations of MDA standard.

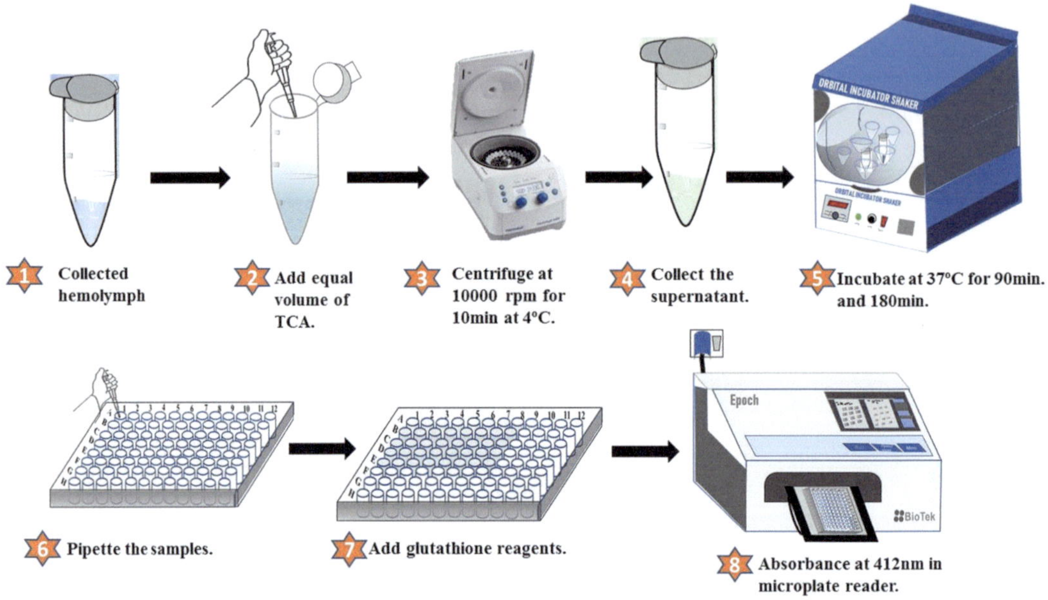

Fig. 5 Glutathione peroxidase assay of hemolymph [1]

3.7.2 Procedure

1. Take 200 μl of the working sample.
2. Add 200 μL of TBA reagent.
3. Centrifuge at 10,000 rpm for 10 min at 4 °C.
4. Transfer the supernatant.
5. Heat the sample for 20 min at 65 °C (Estimate the protein before using for enzyme activity).
6. Measure the absorbance at 532 nm on a microplate reader (Fig. 6).
7. Convert the calculated absorbance into standard MDA plot.

3.7.3 Calculation

$$\text{MDA}(\text{mgMDA/kg}) = \text{Y} = \text{mx} + \text{C}$$

$\text{x} = (\text{Y}-\text{C}/\text{m}) \times \text{DF}$

Y = absorbance of sample

m = slop of straight line

x = concentration of sample

DF = dilution factor

extinction coefficient $= 1.56 \times 10^5\ \text{M}^{-1}$

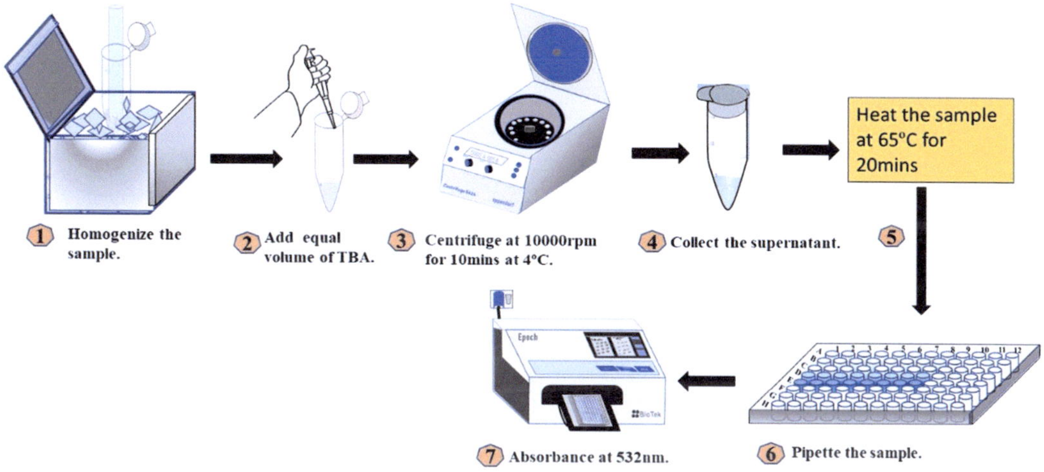

Fig. 6 Lipid peroxidation (LPP) assay of whole body protein extraction from third instar larvae [43]

Table 4
Table indicating addition of various chemicals to detect the scavenging activity by DPPH

Sl no	Sample (in μL)	Conc. (in μg/mL)	DW (in μL)	DPPH (in μL)	Methanol (in μL)	Absorbance (ABS) at 517 nm
Solvent	0	0		0	250	
1	2.5	10	97.5	125	50	
2	5	5	95	125	50	
3	10	2.5	90	125	50	
4	20	1.25	80	125	50	
$DPPH_{control}$	125	0.1 mM			125	

3.8 DPPH Assay

3.8.1 Chemical Preparation

1. 0.2 mM DPPH: Weigh 7.9 mg of DPPH, mix with 100 mL of methanol, and vortex well. Store at 4 °C covering with aluminium foil.

3.8.2 Procedure

1. Take 2.5, 5, 10 and 20 μL of homogenate in a microplate in triplicates
2. Add 125 μL of DPPH to it.
3. Add 50 μL of absolute methanol to the sample well (Table 4).
4. Incubate for 20 min at room temperature in the dark.
5. Read the absorbance at 517 nm in a microplate spectrophotometer (Fig. 7).

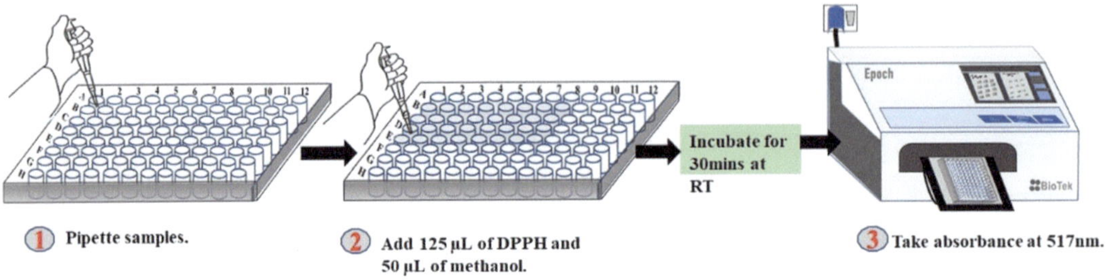

Fig. 7 DPPH assay of hemolymph [1]

3.8.3 Calculation

$$\textbf{Percentage of DPPH scavenging} = \left(\text{ABS}_{\text{control}} - \text{ABS}_{\text{sample}}/\text{ABS}_{\text{control}}\right) \times 100$$

4 Notes

1. **Protein Estimation.**
    1. Wear the gloves and perform the experiment in the ice-box.
    2. Avoid direct light exposure after addition of Folin reagent.
    3. Avoid the foam formation since it can alter the experimental reading.
2. **SOD.**
    1. The assay must be done in dark condition.
    2. Methionine, ascorbic acid and Griess reagents are prepared just before the experiment.
    3. Exposure of light must be uniform.
3. **GPx.**
    1. Avoid the foam formation, since it can alter the experimental values.
    2. Sodium azide is carcinogenic and can cause suffocation or irritation to the skin and thus needs careful handling.
4. **Catalase Assay.**
    1. Degradation of H_2O_2 occurs in a time-dependent manner. Thus the reading should be taken carefully with respect to time.
5. **LPP Assay.**
    1. Must wear the gloves while handling of HCl, TBA, etc.
    2. Avoid chemical contact on skin and eyes.
6. **DPPH.**
    1. Perform all the tests in triplicates, and calculate the average mean values.

Acknowledgements

JB is thankful to BT/PR21857/NNT/28/1238/2017 for financial support. MM lab is supported by Grant No. BT/PR21857/NNT/28/1238/2017, EMR/2017/003054 and Odisha DBT 3325/ST(BIO)-02/2017.

References

1. Soni NO (2014) Antioxidant assay in vivo and vitro. Int J Phytopharmacol 5(1)
2. Paital B, Chainy G (2010) Antioxidant defenses and oxidative stress parameters in tissues of mud crab (Scylla serrata) with reference to changing salinity. Comp Biochem Physiol Part C: Toxicol Pharmacol 151(1):142–151
3. Flohe L (1984) [10] Superoxide dismutase assays. In: Methods in enzymology, vol 105. Elsevier, pp 93–104
4. Nandi A, Chatterjee I (1987) Scavenging of superoxide radical by ascorbic acid. J Biosci 11(1–4):435–441
5. Cross CE, Halliwell B, Borish ET, Pryor WA, Ames BN, Saul RL, McCORD JM, Harman D (1987) Oxygen radicals and human disease. Ann Intern Med 107(4):526–545
6. Halliwell B, Gutteridge JM (2015) Free radicals in biology and medicine. Oxford University Press, Oxford
7. Missirlis F, Phillips JP, Jäckle H (2001) Cooperative action of antioxidant defense systems in Drosophila. Curr Biol 11(16):1272–1277
8. Faust K, Gehrke S, Yang Y, Yang L, Beal MF, Lu B (2009) Neuroprotective effects of compounds with antioxidant and anti-inflammatory properties in a Drosophila model of Parkinson's disease. BMC Neurosci 10(1):109
9. Owusu-Ansah E, Banerjee U (2009) Reactive oxygen species prime Drosophila haematopoietic progenitors for differentiation. Nature 461(7263):537
10. Juntilla MM, Patil VD, Calamito M, Joshi RP, Birnbaum MJ, Koretzky GA (2010) AKT1 and AKT2 maintain hematopoietic stem cell function by regulating reactive oxygen species. Blood 115(20):4030–4038
11. Alaraby M, Hernández A, Annangi B, Demir E, Bach J, Rubio L, Creus A, Marcos R (2015) Antioxidant and antigenotoxic properties of CeO2 NPs and cerium sulphate: Studies with Drosophila melanogaster as a promising in vivo model. Nanotoxicology 9(6):749–759
12. Carmel-Harel O, Storz G (2000) Roles of the -glutathione-and thioredoxin-dependent reduction systems in the Escherichia coli and Saccharomyces cerevisiae responses to oxidative stress. Ann Rev Microbiol 54 (1):439–461
13. Lambeth JD (2004) NOX enzymes and the biology of reactive oxygen. Nat Rev Immunol 4(3):181
14. Schieber M, Chandel NS (2014) ROS function in redox signaling and oxidative stress. Curr Biol 24(10):R453–R462
15. Fridovich I (1997) Superoxide anion radical (O· 2), superoxide dismutases, and related matters. J Biol Chem 272(30):18515–18517
16. Chen Y, Azad M, Gibson S (2009) Superoxide is the major reactive oxygen species regulating autophagy. Cell Death Differ 16(7):1040
17. Ristow M, Schmeisser S (2011) Extending life span by increasing oxidative stress. Free Radic Biol Med 51(2):327–336
18. Tormos KV, Anso E, Hamanaka RB, Eisenbart J, Joseph J, Kalyanaraman B, Chandel NS (2011) Mitochondrial complex III ROS regulate adipocyte differentiation. Cell Metab 14(4):537–544
19. Branden C-I, Tooze J (1999) Introduction to protein structure. Taylor & Francis
20. Illanes A (2008) Enzyme biocatalysis. Principles and applications editorial. Springer, New York
21. Parker CG, Fessler LI, Nelson RE, Fessler JH (1995) Drosophila UDP-glucose: glycoprotein glucosyltransferase: sequence and characterization of an enzyme that distinguishes between denatured and native proteins. EMBO J 14(7):1294–1303
22. Nelson RE, Fessler L, Takagi Y, Blumberg B, Keene D, Olson P, Parker C, Fessler J (1994) Peroxidasin: a novel enzyme-matrix protein of Drosophila development. EMBO J 13(15):3438–3447
23. Birley A, Barnes B (1973) Genetical variation for enzyme activity in a population of Drosophila melanogaster. Heredity 31(3):413
24. Nahmias JA, Bewley GC (1984) Characterization of catalase purified from Drosophila melanogaster by hydrophobic interaction

chromatography. Comp Biochem Physiol Part B Comp Biochem 77(2):355–364
25. Orr WC, Sohal RS (1992) The effects of catalase gene overexpression on life span and resistance to oxidative stress in transgenic Drosophila melanogaster. Arch Biochem Biophys 297(1):35–41
26. Orr WC, Sohal RS (1994) Extension of life-span by overexpression of superoxide dismutase and catalase in Drosophila melanogaster. Science 263(5150):1128–1130
27. Bienert GP, Schjoerring JK, Jahn TP (2006) Membrane transport of hydrogen peroxide. Biochim Biophys Acta (BBA)-Biomembr 1758(8):994–1003
28. Subedi RP, Vartak RR, Kale PG (2017) Management of stress exerted by hydrogen peroxide in Drosophila melanogaster using Abhrak bhasma. J Appl Pharm Sci 7(12):065–071
29. Marnett LJ (1999) Lipid peroxidation—DNA damage by malondialdehyde. Mutat Res Fundam Mol Mech Mutagen 424(1):83–95
30. Ostrea JREM, Cepeda EE, Fleury CA, Balun JE (1985) Red cell membrane lipid peroxidation and hemolysis secondary to phototherapy. Acta Paediatr 74(3):378–381
31. Jacob RA (1995) The integrated antioxidant system. Nutr Res 15(5):755–766
32. Miquel J, Fleming J, Economos AC (1982) Antioxidants, metabolic rate and aging in Drosophila. Arch Gerontol Geriatr 1(2):159–165
33. Le Bourg É (2001) Oxidative stress, aging and longevity in Drosophila melanogaster. FEBS Lett 498(2–3):183–186
34. Demir E, Turna F, Kaya B, Creus A, Marcos R (2013) Mutagenic/recombinogenic effects of four lipid peroxidation products in Drosophila. Food Chem Toxicol 53:221–227
35. Bonilla E, Contreras R, Medina-Leendertz S, Mora M, Villalobos V, Bravo Y (2012) Minocycline increases the life span and motor activity and decreases lipid peroxidation in manganese treated Drosophila melanogaster. Toxicology 294(1):50–53
36. Sortibrán A, Téllez M, Arnaiz R (2015) Assessment of the genotoxic and antioxidant activities of several vegetables, spices and herbs in combination with cyclophosphamide and 4-nitroquinoline n-oxide in Drosophila melanogaster. Adv Tech Biol Med 1:2379–1764
37. Macedo GE, Gomes KK, Rodrigues NR, Martins IK, da Luz Wallau G, de Carvalho NR, da Cruz LC, da Costa Silva DG, Boligon AA, Franco JL (2017) Senecio brasiliensis impairs eclosion rate and induces apoptotic cell death in larvae of Drosophila melanogaster. Comp Biochem Physiol Part C Toxicol Pharmacol 198:45–57
38. Mora M, Medina-Leendertz SJ, Bonilla E, Terán RE, Paz MC, Arcaya JL (2013) Minocycline, but not ascorbic acid, increases motor activity and extends the life span of Drosophila melanogaster. Investig Clin 54(2):161–170
39. Girish C (2012) Propensity of Selaginella delicatula aqueous extract to offset rotenone-induced oxidative dysfunctions and neurotoxicity in Drosophila melanogaster: Implications for Parkinson's disease. Neurotoxicology 33(3):444–456
40. Handke B, Poernbacher I, Goetze S, Ahrens CH, Omasits U, Marty F, Simigdala N, Meyer I, Wollscheid B, Brunner E (2013) The hemolymph proteome of fed and starved Drosophila larvae. PLoS One 8(6):e67208
41. Lowry OH, Rosebrough NJ, Farr AL, Randall RJ (1951) Protein measurement with the Folin phenol reagent. J Biol Chem 193:265–275
42. Waterborg JH, Matthews HR (1994) The Lowry method for protein quantitation. In: Basic protein and peptide protocols. Springer, pp 1–4
43. Sharma OP, Bhat TK (2009) DPPH antioxidant assay revisited. Food Chem 113(4):1202–1205

Chapter 14

Molecular Approaches for Analysis of *Drosophila* MicroRNAs

Manish Pandey, Sakshi Bansal, and Geetanjali Chawla

Abstract

MicroRNAs (miRNAs) belong to a class of small non-coding endogenous RNAs that regulate gene expression at the post-transcriptional level. These small RNAs recognize sequences within 3′ untranslated regions of target mRNAs in complexes referred to as a miRNA-induced silencing complex (miRISC). *Drosophila melanogaster* has served as an indispensable model system for defining the diverse biological roles of miRNAs, their mechanism of action as well as the role of miRNA biogenesis factors. In this chapter we describe some of the assays used for molecular analysis of *Drosophila* miRNAs.

Key words MiRNAs, Luciferase sensors, Post-transcriptional, *Drosophila*, Northern analysis, TaqMan assays

1 Introduction

Drosophila melanogaster has served as an important experimental organism for understanding the role of miRNAs in diverse cellular processes. Currently there are 258 *Drosophila melanogaster* miRNAs listed in the miRNA database (miRbase 22 release) that presumably regulate several hundreds of mRNA targets, thereby modulating a significant proportion of cellular proteins. In this section, we summarize the miRNA biogenesis pathways and functional roles of miRNAs that have been studied in *Drosophila melanogaster*.

1.1 Biogenesis of miRNAs

Mature miRNAs arise from longer precursor transcripts that are characterized by the presence of hairpins and are processed by two RNAse III cleavage reactions [1]. The first step in the generation of a processed miRNA is a transcription of its primary miRNA transcript (pri-miRNA) by RNA polymerase II [2]. These primary transcripts are cleaved by a microprocessor complex that contains the nuclear RNAse III enzyme, Drosha, to result in 50–70-nucleotide-long stem-loop precursor (pre) miRNAs [3, 4]. The

Monalisa Mishra (ed.), *Fundamental Approaches to Screen Abnormalities in Drosophila*, Springer Protocols Handbooks, https://doi.org/10.1007/978-1-4939-9756-5_14,

pre-miRNAs undergo a second round of cleavage in the cytoplasm by a second RNAse III enzyme, Dicer-1, to generate 21–23 nucleotide miRNA duplexes [5–7]. One strand of this miRNA duplex preferentially associates with a miRNA-induced silencing complex (miRISC) and directs the effector complex to complementary sites in target mRNA(s) [8–12]. Analogous to other RNA polymerase II transcripts, the expression of mature miRNAs is regulated at both transcriptional and post-transcriptional levels. Northern, in situ and transcriptome analysis of pri-miRNAs in *Drosophila* have indicated that transcription of miRNAs is spatially and temporally regulated [13–16]. Studies have also uncovered post-transcriptional mechanisms that can regulate miRNA processing in a gene-specific or global manner. RNA binding proteins such as Lin-28, hnRNP A1 and KSRP can bind to pri- and pre-miRNAs and influence their processing in a positive or negative manner [17–20]. Changes in expression level and post-translational modifications of core miRNA biogenesis factors such as Drosha and DGCR8 have been shown to regulate miRNA processing activity [21–23]. Additionally, adenosine-to-inosine (A-to-I) RNA editing of miRNA sequences by the *a*denosine *d*eaminase *a*cting on *R*NA (ADAR) enzyme can alter stability, biogenesis and target recognition of miRNAs [24, 25]. Thus, precise regulation at multiple levels is required for achieving the dynamics of miRNA production and degradation [26]. In higher eukaryotes, defects in miRNA processing have been linked to several diseased states, including cancer. Thus, understanding conserved molecular mechanisms that control miRNA production in model organisms like *Drosophila melanogaster* is important to identify novel regulatory molecules that could serve as potential targets for therapeutic interventions.

1.2 Biological Functions of miRNAs

Drosophila melanogaster has contributed immensely towards an understanding of the diverse functional roles of miRNAs and the alterations in regulatory pathways that are caused by these non-coding RNAs [27, 28]. These small non-coding RNAs are involved in either fine-tuning the levels of their target mRNAs or are involved in silencing of noisy target genes [12, 29–33]. Analysis of miRNA mutants has revealed a role for *Drosophila* miRNAs in cell fate specification and differentiation, ageing, neurodegeneration, metabolism, circadian rhythm, stem cell maintenance and proliferation, behaviour, tissue growth and robustness [30, 31, 34–50]. Some of the currently available resources for analysis of *Drosophila* miRNAs include resources for overexpression of 165 miRNAs and a large collection of miRNA knockout mutants that encompass 130 individual miRNAs [51–54]. One of the key challenges for the future is to identify mRNA targets that are most responsive to each miRNA and to understand the roles of miRNA in large biochemical networks [27, 55]. In this chapter we have described detailed protocols of four molecular techniques that have been used extensively by fly researchers for functional characterization of miRNAs.

2 Materials

2.1 RNA Preparation

1. Polypropylene pestles (Sigma Z359947).
2. Disposable gloves (Kimberly Clark).
3. Pipettes.
4. Barrier tips.
5. Microcentrifuge tube stand.
6. TRIzol Reagent (ThermoFisher Scientific 15596026).
7. Chloroform (Sigma 650471).
8. 2-Propanol (Sigma I9516).
9. 75% ethanol (prepared by diluting absolute ethanol in DEPC water).
10. GlycoBlue coprecipitant (15 mg/ml) (ThermoFisher Scientific AM9515).
11. 1.5 ml microcentrifuge tubes (RNase-DNase free).
12. DNAse I (RNAse-free) (NEB M303S).
13. Diethyl pyrocarbonate (DEPC) (Sigma D5758).
14. DEPC water (prepared by adding 0.1% DEPC to milliQ water and stirring overnight at RT followed by autoclaving).
15. Phenol: Chloroform (Sigma P1944).
16. 3M sodium acetate pH 5.3: Dissolve 246.1 g of sodium acetate in 500 mL of DEPC treated water. Adjust the pH to 5.3 with glacial acetic acid. Allow the solution to cool overnight. Adjust the pH once more to 5.3 with glacial acetic acid. Adjust the final volume to 1 L with DEPC-treated water and filter sterilize.
17. Vortex Genie 2 (Scientific Industries SI-0236).
18. Microcentrifuge.
19. −80 °C freezer (Eppendorf CryoCube −86 °C).
20. Eppendorf mixer (Eppendorf 2231000574).
21. Autoclave.

2.2 Small RNA Northern Blot

1. TEMED (Sigma T9281).
2. Heat block (95 °C).
3. UV Stratalinker (Stratagene model UV Stratalinker 1800).
4. 10% ammonium persulfate: Prepare by dissolving 1gm of ammonium persulfate (Sigma A3678) in 10 ml DEPC-treated milliQ water, and store the solution at 4 °C.
5. Electrophoresis and blotting apparatus (BioRad Mini-PROTEAN tetra cell and blotting module).

6. Amersham Hybond-N+ membrane (GE Healthcare Life Sciences RPN203B).
7. Phosphorimager (Amersham Typhoon 5 Biomolecular Imager 29187191).
8. Phosphorimager screen and cassettes (GE Healthcare Life Sciences).
9. [α-^{32}P] dATP (6000 Ci/mmol, 10 mCi/mL).
10. Exo- Klenow DNA polymerase (NEB M0212S).
11. Hybridization oven.
12. Hybridization bottles.
13. UV lamp hand-held.
14. Starfire oligo (Integrated DNA Technologies; designed based on miRNA to be detected) and Starfire Universal template (Integrated DNA Technologies).
15. Sephadex TM G-25 (Cat. No. 17-0033-01).
16. 5X TBE: 1 litre is prepared by mixing 54 g Tris base, 27.5 g boric acid and 20 ml of EDTA (pH 8.0).
17. Ethidium bromide 10 mg/ml (Sigma E1510).
18. 20 X SSPE: 1 l is prepared by mixing 175.3 g of sodium chloride and 27.6 g of sodium phosphate in monobasic in 800 ml water and then adjusting pH to 7.4 with sodium hydroxide and adjusting the volume to 1 litre followed by autoclaving for 20 min.
19. 2X gel loading dye is prepared by mixing 95% v/v deionized formamide, 20 mM EDTA (pH 8.0), 0.025% bromophenol blue, 0.025% xylene cyanol and 0.025% SDS.
20. Blotting paper.
21. Acrylamide: bisacrylamide 19:1, 40% w/v (Fisher BP1406-1).
22. Urea (Sigma U6504).
23. 20% sodium dodecyl sulphate (SDS) (Sigma-Aldrich L3771) is prepared by dissolving 20 gm of SDS in 90 ml milliQ water. Heat to 60 °C and stir with a magnetic stirrer to dissolve SDS. Adjust the volume to 100 ml. Filter sterilize and store at room temperature.
24. Sodium phosphate dibasic heptahydrate (Sigma-Aldrich S9390).
25. TES buffer is prepared by mixing Tris-Cl (10 mM, pH 7.6), EDTA (1 mM, pH 8.0) and sodium dodecyl sulphate (0.5%, w/v).
26. Micro Bio-spin chromatography columns (Bio-Rad, 7326204).
27. TE buffer is prepared by mixing Tris-Cl (10 mM, pH 8.0) and EDTA (1 mM, pH 8.0).

2.3 TaqMan miRNA Assays

1. Superscript III (ThermoFisher Scientific 18080051).
2. SUPERase-In (ThermoFisher Scientific AM2696).
3. 25 mM dNTPs (Prepared by mixing equal volumes of 100 mM dATP, dCTP, dGTP, dTTP) (Fermentas Life Sciences R0181).
4. RNAse-free water (Prepared by treating milliQ water with DEPC).
5. Vortex Genie2 (Scientific Industries SI-0236).
6. PCR machine (Bio-Rad Laboratories C1000 Touch Thermal Cycler).
7. QuantStudio 6 Real-Time PCR system (ThermoFisher Scientific 4485697).
8. Microcentrifuge (Eppendorf 5424R).
9. MicroAmp Clear Adhesive Film (ThermoFisher Scientific 4306311).
10. MicroAmp Fast Optical 96-well Reaction Plate, 0.1 ml (ThermoFisher Scientific 4346907).
11. TaqMan microRNA assays comprise of an RT primer and a TaqMan probe (ThermoFisher Scientific catalog no. determined by the miRNA to be detected).
12. TaqMan Fast Advanced mix (ThermoFisher Scientific 4444557).
13. Barrier Pipette tips.

2.4 Luciferase Sensor Assays

1. psiCHECK2 vector (Promega C8021).
2. Dual luciferase reporter assay system (Promega E1910).
3. Turner Model TD-20/20 luminometer or GloMax® 20/20 Luminometer (Promega E5311).
4. Polypropylene tubes 8 × 50 mm or 96-well plate of luminometer tubes.
5. Haemocytometer.
6. Effectene (Qiagen 301425).
7. Kc167/S2 cell line (DGRC).
8. CCM3 medium with glutamine (Hyclone SH30065.02).
9. 48-well tissue culture plates (Corning Costar TC-treated multiple well plates, CLS 3548).
10. 10× phosphate-buffered saline (PBS) is prepared by dissolving 80 g of sodium chloride, 2.0 g of potassium chloride, 14.4 g of disodium hydrogen phosphate and 2.4 g of potassium dihydrogen phosphate in 800 ml of milliQ water. Adjust the volume to 1 l and filter sterilize. Dilute to 1X for working solution.

3 Methods

3.1 Total RNA Preparation

TRIzol method of RNA extraction has been effectively used for isolating small RNAs such as miRNAs from cells, tissues and whole flies. The basic set-up for RNA preparation includes gloves, microfuge tube stand, pipettes, barrier tips, DEPC water, chloroform, 2-propanol and polypropylene pestle (Fig. 1). The name TRIzol is derived from the fact that this reagent facilitates purification of RNA, DNA and protein from a single sample.

1. Collect samples (5–10 whole larvae or adult fruit flies) in a 1.5 ml microfuge tube. Add 200 μl of TRIzol reagent to sample and homogenize with polypropylene pestles (autoclaved).
2. Add 800 μl of TRIzol reagent and shake vigorously by vortexing at room temperature for 10 min.
3. Add 200 μl of chloroform and vortex the sample briefly. Spin the sample for 15 min at 12,000 rpm at 4 °C.
4. Transfer the supernatant by pipetting to a new microcentrifuge tube and add 500 μl of 2-propanol. Add 1.5 μl GlycoBlue coprecipitant and mix by inverting the tube. Incubate the sample at room temperature for 10 min.

   Note: Using GlycoBlue coprecipitant aids in the recovery of RNA from limited tissue samples by precipitating with the RNA. In addition, the presence of the blue dye increases the visibility of the pellet.
5. Spin the sample for 15 min at maximum speed (12,000 rpm) at 4 °C.
6. Remove the supernatant and wash the RNA pellet with 500 μl of 75% ethanol. (Optional: The sample can be stored at −80 °C for overnight at this step.) Spin the sample at maximum speed for 5 min at 4 °C. Remove the 75% ethanol by pipetting and air-dry the RNA pellet.
7. Resuspend the dried pellet in 90 μl of DEPC-treated water, and vortex for 10 min to dissolve the RNA. Set up DNAse I

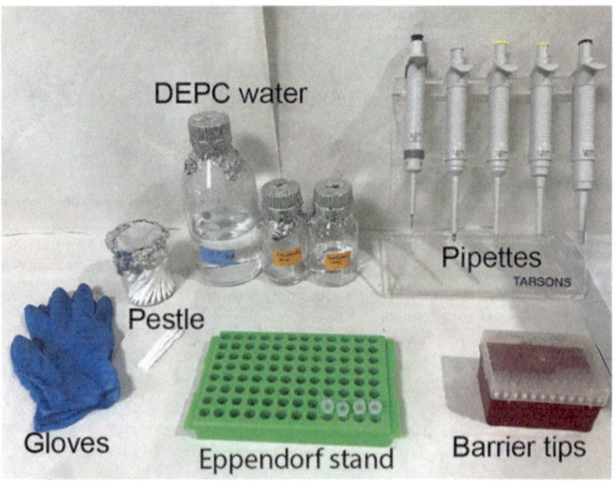

Fig. 1 The basic set-up for RNA preparation

digestion by adding 10 μl of 10X DNAse I buffer and 1 μl DNAse I enzyme. Incubate the sample at 37 °C for 30 min.

8. Perform Phenol: Chloroform extraction as follows. Add 170 μl of DEPC treated water and 30 μl of 3 M sodium acetate (pH 5.3) and 250 μl of phenol: chloroform (pH 4.3–4.7). Vortex and spin for 5 min at RT at max speed. Remove the top aqueous layer into a new tube (autoclaved) and 750 μl (3 volumes) of 100% ethanol. Incubate at −80 °C for 2 h or overnight.
9. Centrifuge at 4 °C for 15 min at 12,000 rpm. Remove supernatant and add 500 μl of 75% ethanol (removes salt from preparation). Spin at 12,000 rpm for 5 min.
10. Remove ethanol and air-dry pellet.
11. Resuspend in 15–50 μl (depending on the amount of starting material) of DEPC water.
12. Quantitate RNA.

3.2 Northern Blots for miRNAs

This protocol has been used for detection of miRNAs (19–24 nucleotides) and precursor miRNAs (50–70 nucleotides) by hybridization of membrane blots [56, 57]. The technique involves separation of RNAs on a denaturing polyacrylamide gel, followed by its transfer to membrane. The RNA is fixed onto the membrane by UV cross-linking and incubated with antisense radioactive probes (Fig. 2). Several variations of this technique, including LNA probes and chemical cross-linking of RNA, have been utilized to increase its sensitivity and allow detection of less abundant miRNAs [58, 59]. We describe here a protocol that has been used for detection of miRNAs from *Drosophila melanogaster* whole animals, dissected tissues from different stages and cell lines [14, 25, 31].

1. Prepare 1 l 15% PAGE; 7.5 M Urea gel by as follows:

5X TBE	200 ml
Acrylamide: bisacrylamide (40%)	375 ml
Urea	450 g
MilliQ water	To 1 l

Stir the mixture until the urea dissolves (may need to heat to 45 °C), and filter and store in a dark-coloured glass bottle at 4 °C.

2. Assemble the gel casting unit, and pour gel into the glass mini gel set-up after mixing 15 ml 15% PAGE; 7.5 M Urea mix with 15 μl TEMED and 150 μl of 10% ammonium persulfate. Use a 10–12-well comb based on the volume of RNA to be loaded. Allow the gel to polymerize for 20–30 min.

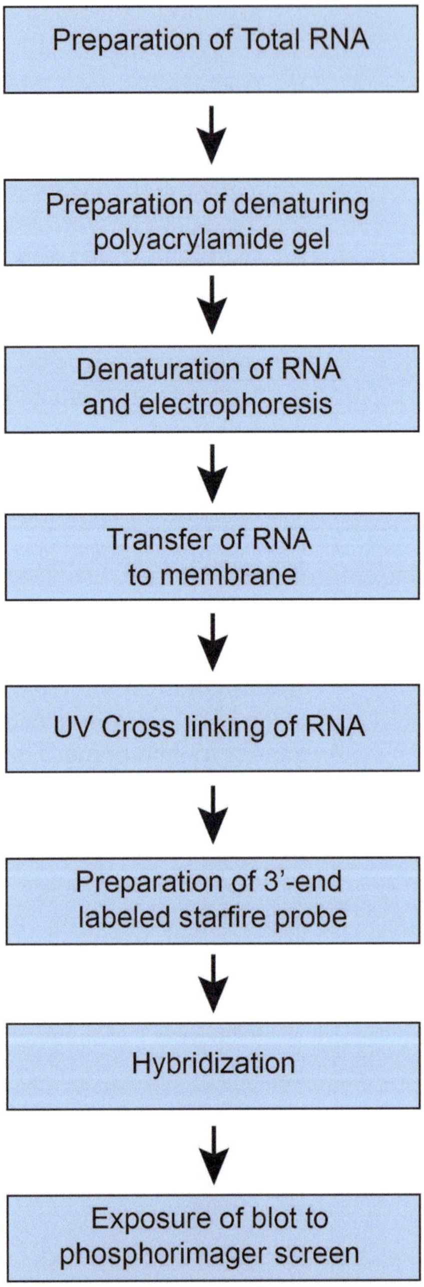

Fig. 2 Workflow for small RNA northern blot analysis. The key steps in the workflow are the preparation of total RNA, preparation of denaturing polyacrylamide gel, denaturation of RNA and electrophoresis, transfer of RNA to the membrane, UV cross-linking of the membrane, preparation of probe, hybridization, exposure of blot to phosphorimager screen and analysis

3. Mix 15–20 μg of total RNA with an equal volume of 2X formamide dye, and heat the samples at 90 °C for 4–5 min. The samples can be stored in ice until ready to load.
4. Remove the gels from casting unit and assemble the gel-running set-up. Fill the tanks with 1X TBE. Wash the wells prior to loading the gel and load the denatured RNA sample in the lane.
5. Run the gel with constant stirring (by loading gel box on a magnetic stirrer) in 1X TBE at 125 V (constant V) for 1.5 h (until the bromophenol blue reaches the bottom of the gel).
6. Cut the Hybond N+ membrane and two blotting papers to the dimension of the small glass plate, and soak the membrane in 0.5X TBE until the gel run is complete.
7. After the gel run, disassemble the glass plates and lift the gel on a blotting paper after wetting it with 0.5X TBE. Wear gloves while loading and setting up transfer to prevent RNAse contamination.
8. Stain the gel with ethidium bromide to confirm uniform loading in all lanes.
9. Prepare the set-up required for transfer as follows:
   - (i) Place a sponge pad that has been wetted with 0.5X TBE on one of the sides of the cassette.
   - (ii) Place the filter paper with gel on the fibre/sponge pad.
   - (iii) Place the prewetted membrane on top of the gel using forceps. Remove air bubbles with a roller.
   - (iv) Place another wet blotting paper on top of the membrane.
   - (v) Place the prewetted fibre/sponge pad on top of the blotting paper.
   - (vi) Close the cassette carefully, trying to avoid moving the gel or membrane, and lock the cassette by sliding the white latch.
   - (vii) Place the cassette in the transfer module, and assemble the second cassette if two gels were run.
   - (viii) Add the frozen blue cooling unit and fill the tank with chilled 0.5X TBE.
   - (ix) Add a magnetic stir bar and place the tank on a magnetic stirrer. The stir bar aids in maintaining uniform buffer temperature and ion distribution in the tank during transfer.
   - (x) Place the lid and plug the cables and transfer at 80 V (constant V) for 1 h. Buffer from running gel can be diluted to 0.5X and reused for transfer.

10. After the transfer, UV cross-link membrane with the auto-cross-link setting of Stratalinker 1800 (twice). The membrane should be dry prior to cross-linking.
11. Prepare hybridization buffer as follows:

DEPC-treated MilliQ water	25 ml
0.5 M Na_2HPO_4 pH 7.0	40 ml
20% SDS	35 ml
Total volume	100 ml

12. For prehybridization, place the cross-linked membrane into a hybridization tube with the RNA side facing inwards. For each blot to be probed, add 10 ml of prehybridization solution. Prehybridize the blot at 33–35 °C for 15 min to 1 h with rotation.
13. Prepare the probe by assembling the labelling reaction in a 0.5 ml tube as follows:
    (i) 0.5 μl of Starfire primer
    (ii) 0.5 μl of Starfire template
    (iii) 0.5 μl of NEB exo-Klenow 10X reaction buffer
    (iv) Vortex and incubate/denature in a heat block set at 94 °C for 1 min. Spin the tube in a centrifuge and let it cool at room temperature for 5 min. Then add 3 μl of [α-^{32}P] dATP (6000 μCi/ml) and 0.5 μL of NEB exo-Klenow.
14. Incubate the tube at room temperature for at least 1 h.
15. Stop the reaction by adding 45 μl of TES buffer.
16. Remove unincorporated [α-^{32}P] dATP with a Sephadex G-25 spin column as follows:
    (i) Put the micro bio-spin column in an empty 1.5 ml microfuge tube.
    (ii) Pipette 500 μl of a 50% suspension of completely hydrated Sephadex G-25 into the column and spin at 3000 rpm in a microcentrifuge for 2 min. Remove the eluted water and wash two times with 50 μl of TE buffer. Discard the flow through and tube.
    (iii) Place the micro bio-spin column in a new 1.5 ml microfuge tube, and apply 50 μl of the labelled reaction from Step 15.
    (iv) Centrifuge the column for 3 min at 3000 rpm.
    (v) Add an additional 50 μL of TES buffer and re-elute. Discard the column in radioactive waste.
    (vi) The eluted sample can be directly used in hybridization.
17. After prehybridization, add the entire probe to the prehybridization buffer. Hybridize at 33–35 °C for ~20 h.

18. After hybridization, the hybridization solution can be transferred to a 50 ml conical tube and stored at −20 °C.
19. Wash the blots in several volumes excess of 2xSSPE, 0.1% SDS for a total of three times 20 min or more at 33 °C.
20. Wrap the washed blots in a clear plastic wrap and expose to phosphorimager screen overnight.
21. Read the hybridization signals by scanning the screen in a phosphorimager, and analyse with the appropriate software.
22. For analysis of relative abundance, the blot can be stripped and reprobed with an internal control such as U6 snRNA. The hybridization signal from the internal control can be used for normalization.
23. Northern blots can be reused after stripping the probe by boiling the membrane in 0.1% SDS and then cooling it to room temperature. The blot can be exposed to determine whether the probe is completely stripped. The blot can then be probed for another abundantly expressed miRNA or a normalization control.

3.3 TaqMan miRNA Assays

The TaqMan miRNA assays are extensively utilized for detection and quantification of mature miRNAs. Some of the advantages of these assays are that they are relatively easy and fast to perform. In addition, these assays provide high specificity and sensitivity and allow accurate quantitation and detection of miRNAs from a very limited amount of RNA [60]. The assay is performed using a two-step RT-PCR (Fig. 3). The first step is the reverse transcription, where single-stranded cDNA is reverse transcribed from total RNA using a specific primer for the miRNA.

1. 25 ng of total RNA is used per 10 μl reverse transcription (RT) reaction.
2. The RT reaction is prepared as follows:
   (i) Thaw the reagents on ice, and add the following reagents in a 1.5 ml microcentrifuge tube:

Reagent	Volume in a 10 μl reaction
25 mM dNTPs	0.1 μl
10X RT Buffer	1 μl
25 mM $MgCl_2$	2 μl
0.1 M DTT	1 μl
Superscript III	0.1 μl
Superase-In	0.1 μl
Total	4.3 μl per reaction

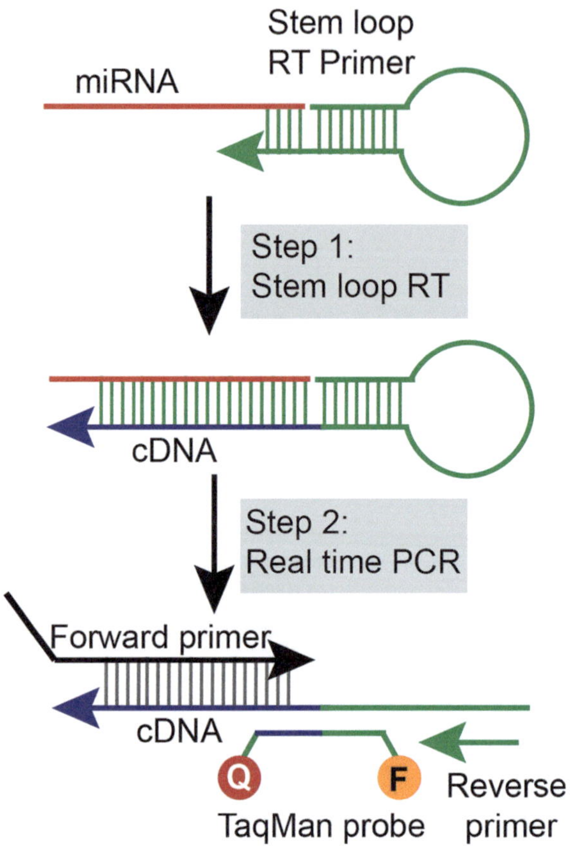

Fig. 3 TaqMan quantitative real-time PCR. TaqMan-based miRNA assays are performed in two steps. The first step is the stem-loop RT reaction in which stem-loop RT primers bind to the 3′end of the miRNA and are reverse transcribed to single-stranded cDNA. In the second step, the RT product is quantitated using TaqMan assays that include a tailed miRNA-specific forward primer, reverse primer and a TaqMan probe. The tailed primer is designed at the 5′ end to increase the melting temperature of the primer

(ii) To each 10 μl reaction, dispense 4.3 μl of the RT reaction mix and 1 μl of RNA sample up to a total of 3 μl RT primer (0.75 μl each of up to 4 miRNA RT primers can be added to each reaction) and RNAse-free water to a total volume 5.7 μl in a 0.2 ml polypropylene tube.

3. Mix the contents in the tube by tapping, and centrifuge briefly to bring the components to the bottom of the tube.
4. Use the following conditions to perform the reverse transcription in the thermal cycler.

Temperature (°C)	Time (min)
16	30
42	30
85	5
4	∝

5. Set the reaction volume to 10 μl and place your tubes in the thermal cycler.
6. Start the reverse transcription program run.
7. The second step is the PCR reaction, where the cDNA is amplified using the TaqMan miRNA assay and PCR master mix.
8. Thaw the miRNA assays on ice and centrifuge briefly to bring spin down. Prepare 1:5 dilution of the cDNA template.
9. The PCR reaction should be performed in two technical replicates for each RT reaction. The PCR reactions are set in a 96-well plate with each PCR reaction containing the following reagents:

Component	Volume (μl)/10 μl reaction[a]
TaqMan probe	0.5
TaqMan mix	5
cDNA	1
RNAse-free water	3.5
Total	10

[a]Scale the volumes to 2.25 reactions for two technical replicates and 12.5% extra to account for pipetting losses

10. Mix the contents and centrifuge briefly to bring the solution to the bottom of the tube.
11. Transfer 10 μl of the PCR reaction for each cDNA to each well of the PCR reaction plate. Seal the reaction plate with an adhesive cover after transferring all the PCR reactions.
12. Centrifuge the reaction plate at 1200 rpm for 1 min to spin down the contents.
13. Load the reaction plate in the PCR machine and set the following cycling conditions:

Temperature	Time	Cycles
95 °C	20 s	1
95 °C 60 °C	1 s 20 s	40

14. Set the appropriate reaction volume for the reaction plate and start the run.
15. In order to use the comparative C_t method, endogenous controls such as 2S rRNA can be used for normalization of expression levels of the target miRNAs.

Note: TaqMan assays are extremely sensitive, and care should be taken to avoid any fluorescent contamination on the outside of the plate or gloves as those will be detected as a signal by the PCR machine block. Hence, extreme caution should be exercised while setting up the reaction, so that the block is not contaminated with the reaction cocktail or RT product. One recommended way to monitor contamination is to employ "no template controls" (NTC) in each experiment.

3.4 Luciferase Reporter Assays

A critical step in functional analysis of a miRNA is the identification of its functionally relevant endogenous target(s), and a number of approaches are being utilized by researchers to reliably identify and validate targets that have been predicted using computational tools [12, 61–64]. One widely used experimental technique for validation of miRNA targets and quantitation of the effects of specific miRNA-target interactions in the *Drosophila* cell culture system is the 3′UTR-luciferase reporter assay [25, 31, 51, 65–67]. In this assay, the 3′UTR from a potential miRNA target is fused to the end of a luciferase reporter gene. If a 3′UTR harbours a miRNA binding site, the reporter activity will change with variation in the concentration of the functional miRNA. This reporter system can also be used to assess the relevance of a particular miRNA binding site by engineering mutations to disrupt binding of a miRNA. Other controls that can be used for the luciferase miRNA sensor assay include the empty vector control (no 3′UTR) or control with the 3′UTR of *Drosophila melanogaster white* gene lacking any miRNA binding sites. This serves as a control for transfection efficiency and will provide a high luciferase signal. In addition, a synthetic 3′UTR harbouring the perfect miRNA binding sites may be utilized as a control for miRNA expression or miRNA mimic activity. The key steps in the luciferase assay include designing the constructs, transfection of reporter constructs and pri-miRNA reporters, luciferase reporter assays and analysis of data (Fig. 4a).

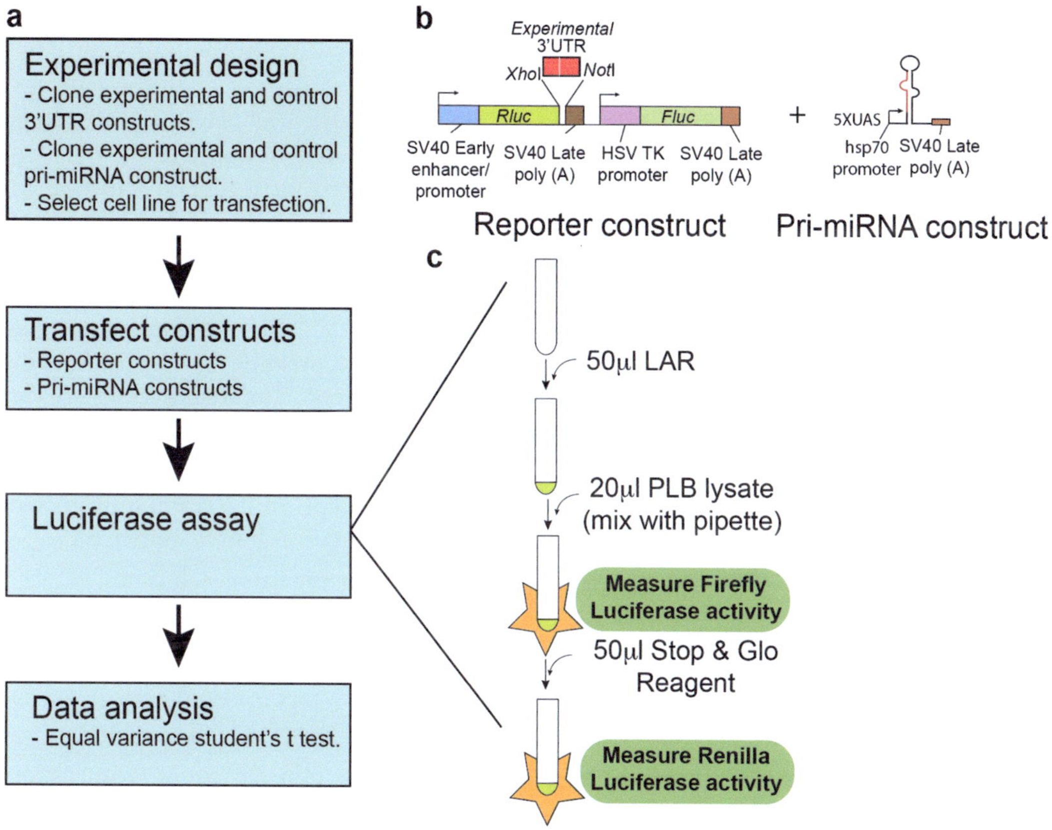

Fig. 4 Workflow for luciferase miRNA sensor assays. (**a**) The key steps in this workflow are experimental design, transfection of the reporter and pri-miRNA constructs, luciferase assay and data analysis. (**b**) Schematic of the psiCHECK 2 vector used for cloning the experimental and control 3′UTR sequences (left) and the pri-miRNA construct cloned in pUAST attB vector (right). (**c**) Format of the dual luciferase assay using a manual luminometer

1. Clone the 3′ UTR of the potential miRNA target into siCHECK-2 series of vectors. The primary reporter gene in this vector is a synthetic version of Renilla luciferase, *hRluc*. To facilitate fusion of the target gene sequences or synthetic miRNA binding sites to Renilla luciferase gene, a multiple cloning region has been engineered downstream of its stop codon. The restriction sites in the multiple cloning site can be used to create genetic fusions of the Renilla reporter with any gene of interest (Fig. 4b).
2. These vectors also include the firefly luciferase gene as a second reporter gene that can be utilized for normalization of the Renilla luciferase signal (Fig. 4b).
3. In addition to cloning the 3′UTR of the target gene, the pri-miRNA gene of interest also needs to be cloned into a vector that allows high expression (e.g. pUASTattB)

(Fig. 4b). Alternatively, miRNA mimics (synthetic mature miRNAs from Dharmafect) can be transfected along with luciferase reporters.

4. Seed 200 μl cells (Kc167/S2) in 48-well plates (at a density of 1.0–3.5 × 10^6 cells/ml), on the day of transfection.

   Note: The cell line to be used for luciferase sensor assays should be highly transfectable and should preferably have low endogenous levels of the miRNA being analysed. If the endogenous level of the miRNA is high, it will be hard to detect the changes in luciferase activity.

5. Transfect reporter constructs and pri-miRNA constructs using the Effectene Transfection Reagent. The transfections are carried out in triplicates
6. Dilute 300 ng of the Gal4 plasmid (to drive expression of the primary miRNA transcript), UAS pri-miRNA and miRNA sensor in 50 μl of the DNA-condensation buffer, Buffer EC.
7. Add 6 μl of enhancer and mix by pipetting 3–5 times.
8. Incubate at room temperature (15–25 °C) for 5 min.
9. Add 12 μl Effectene Transfection Reagent to the DNA-Enhancer mixture. Mix by pipetting up and down five times.
10. Incubate the samples for 10 min at room temperature (15–25 °C) to allow transfection complex formation.
11. Add 70 μl growth medium (can contain serum and antibiotics) to the tube containing the transfection complexes. Mix by pipetting up and down twice, and immediately add 40 μL of the transfection complexes dropwise onto the cells in each well (three replicates). Gently swirl the dish to ensure uniform distribution of the transfection complexes.
12. Seventy-two hours post-transfection, remove the 48-well tissue culture plates with the Kc167/S2 cell line from the incubator.
13. Transfer the cells to 1.5 ml microcentrifuge tubes with a 1 ml pipette.
14. Spin down the cells in a microcentrifuge at 3000 rpm for 3 min, and aspirate the cell culture medium.
15. Wash the cells with 500 μl of phosphate-buffered saline (PBS). Spin down the cells at 3000 rpm for 3 min and aspirate the PBS.
16. Apply 50 μl of 1X passive lysis buffer (PLB) to the cell pellet, and resuspend with a P200 pipette to obtain a homogenous suspension.
17. Incubate the tubes at RT for at least 10 min.
18. Clear the lysate of cellular debris by spinning at 1200 rpm for 2 min. Use the supernatant for luciferase assay.

19. Prepare Luciferase Assay Reagent II (LAR II) by resuspending the provided lyophilized Luciferase Assay Substrate in 10 ml of the supplied Luciferase Assay Buffer II (Fig. 4c).

    It is recommended that once the luciferase assay reagent is reconstituted, working aliquots be made and stored at −70 °C for the longer term.
20. Prepare an adequate volume of Stop & Glo to perform the required number of assays (50 μl reagent per assay). Stop & Glo substrate is supplied at a 50X concentration. Add 1 volume of 50X Stop & Glo substrate to 50 volumes of Stop & Glo buffer in a polypropylene tube. The Stop & Glo Reagent should always be prepared immediately before use as the reconstituted reagent is not very stable at room temperature or at 4 °C.
21. Luciferase assays are carried out with a Dual-Glo luciferase assay kit in a Turner Model TD-20/20. Predispense 50 μl of LAR II into the required number of luminometer tubes and program the luminometer. For single-sample luminometers such as Turner Model TD-20/20, the equipment should be configured to measure light emission over a defined period. The company recommends programming the luminometer to provide a 2 s pre-read delay, followed by a 10 s measurement period.
22. Transfer 15–20 μl of cell lysate into the luminometer tube containing the LAR II and mix by pipetting. Take care to avoid bubbles. Vortexing is not recommended when mixing the lysate with LAR II as it may lead to the coating of the sides of the tube with a luminescent solution that can escape mixing with the subsequently added Stop & Glo solution reagent.
23. Place the tube in the luminometer and note the reading for firefly luciferase activity (Fig. 4c).
24. Dispense 50 μl of Stop & Glo reagent in the tube containing the sample lysate and LAR II, and mix by pipetting.
25. Replace the tube in the luminometer and record reading of Renilla luciferase activity (Fig. 4c).
26. Discard the tube and proceed to the next sample.
27. Fold repression by a miRNA is calculated by comparison with assays in which an empty vector (not expressing the miRNA) or a non-targeting control miRNA is used.
28. The assays are usually performed in three replicates, and the data are analysed by equal variance student's t test.

    Note: As an extension of the luciferase assays, researchers can also test the significance of individual miRNA binding sites by mutagenizing individual miRNA binding sites in these reporters, in a scenario where there are multiple miRNA binding sites or multiple miRNAs regulating the same target mRNA.

Acknowledgements

We would like to acknowledge the efforts of all researchers involved in establishing and refining the techniques and assays used to analyse miRNA function. This work was supported by the Wellcome Trust/DBT India Alliance Fellowship [grant number IA/I (S)/17/1/503085] awarded to GC.

References

1. Ha M, Kim VN (2014) Regulation of microRNA biogenesis. Nat Rev Mol Cell Biol 15(8):509–524
2. Kim VN, Han J, Siomi MC (2009) Biogenesis of small RNAs in animals. Nat Rev Mol Cell Biol 10(2):126–139
3. Lee Y et al (2003) The nuclear RNase III Drosha initiates microRNA processing. Nature 425(6956):415–419
4. Denli AM et al (2004) Processing of primary microRNAs by the microprocessor complex. Nature 432(7014):231–235
5. Saito K et al (2005) Processing of pre-microRNAs by the Dicer-1-loquacious complex in Drosophila cells. PLoS Biol 3(7): e235
6. Jiang F et al (2005) Dicer-1 and R3D1-L catalyze microRNA maturation in Drosophila. Genes Dev 19(14):1674–1679
7. Fukunaga R et al (2012) Dicer partner proteins tune the length of mature miRNAs in flies and mammals. Cell 151(3):533–546
8. Forstemann K et al (2007) Drosophila microRNAs are sorted into functionally distinct argonaute complexes after production by dicer-1. Cell 130(2):287–297
9. Tomari Y, Du T, Zamore PD (2007) Sorting of Drosophila small silencing RNAs. Cell 130(2):299–308
10. Lai EC (2002) Micro RNAs are complementary to 3' UTR sequence motifs that mediate negative post-transcriptional regulation. Nat Genet 30(4):363–364
11. Brennecke J et al (2005) Principles of microRNA-target recognition. PLoS Biol 3(3):e85
12. Bartel DP (2009) MicroRNAs: target recognition and regulatory functions. Cell 136(2):215–233
13. Sokol NS, Ambros V (2005) Mesodermally expressed Drosophila microRNA-1 is regulated by twist and is required in muscles during larval growth. Genes Dev 19(19):2343–2354
14. Chawla G, Sokol NS (2012) Hormonal activation of let-7-C microRNAs via EcR is required for adult Drosophila melanogaster morphology and function. Development 139(10):1788–1797
15. Liu Q et al (2017) Identification of active miRNA promoters from nuclear run-on RNA sequencing. Nucleic Acids Res 45(13):e121
16. Aboobaker AA et al (2005) Drosophila microRNAs exhibit diverse spatial expression patterns during embryonic development. Proc Natl Acad Sci U S A 102(50):18017–18022
17. Newman MA, Thomson JM, Hammond SM (2008) Lin-28 interaction with the Let-7 precursor loop mediates regulated microRNA processing. RNA 14(8):1539–1549
18. Guil S, Caceres JF (2007) The multifunctional RNA-binding protein hnRNP A1 is required for processing of miR-18a. Nat Struct Mol Biol 14(7):591–596
19. Trabucchi M et al (2009) The RNA-binding protein KSRP promotes the biogenesis of a subset of microRNAs. Nature 459(7249):1010–1014
20. Heo I et al (2008) Lin28 mediates the terminal uridylation of let-7 precursor MicroRNA. Mol Cell 32(2):276–284
21. Tang X et al (2010) Phosphorylation of the RNase III enzyme Drosha at Serine300 or Serine302 is required for its nuclear localization. Nucleic Acids Res 38(19):6610–6619
22. Tang X et al (2013) Acetylation of drosha on the N-terminus inhibits its degradation by ubiquitination. PLoS One 8(8):e72503
23. Herbert KM et al (2013) Phosphorylation of DGCR8 increases its intracellular stability and induces a progrowth miRNA profile. Cell Rep 5(4):1070–1081
24. Berezikov E et al (2011) Deep annotation of Drosophila melanogaster microRNAs yields insights into their processing, modification, and emergence. Genome Res 21(2):203–215

25. Chawla G, Sokol NS (2014) ADAR mediates differential expression of polycistronic microRNAs. Nucleic Acids Res 42(8):5245–5255
26. Zhou L et al (2018) Importance of miRNA stability and alternative primary miRNA isoforms in gene regulation during Drosophila development. Elife:7
27. Carthew RW, Agbu P, Giri R (2017) MicroRNA function in Drosophila melanogaster. Semin Cell Dev Biol 65:29–37
28. Pal AS, Kasinski AL (2017) Animal models to study MicroRNA function. Adv Cancer Res 135:53–118
29. Flynt AS, Lai EC (2008) Biological principles of microRNA-mediated regulation: shared themes amid diversity. Nat Rev Genet 9(11):831–842
30. Karres JS et al (2007) The conserved microRNA miR-8 tunes atrophin levels to prevent neurodegeneration in Drosophila. Cell 131(1):136–145
31. Chawla G et al (2016) A let-7-to-miR-125 MicroRNA switch regulates neuronal integrity and lifespan in Drosophila. PLoS Genet 12(8): e1006247
32. Bushati N et al (2008) Temporal reciprocity of miRNAs and their targets during the maternal-to-zygotic transition in Drosophila. Curr Biol 18(7):501–506
33. Giraldez AJ et al (2006) Zebrafish MiR-430 promotes deadenylation and clearance of maternal mRNAs. Science 312(5770):75–79
34. Brennecke J et al (2003) Bantam encodes a developmentally regulated microRNA that controls cell proliferation and regulates the proapoptotic gene hid in Drosophila. Cell 113(1):25–36
35. Thompson BJ, Cohen SM (2006) The Hippo pathway regulates the bantam microRNA to control cell proliferation and apoptosis in Drosophila. Cell 126(4):767–774
36. Xu P et al (2003) The Drosophila microRNA Mir-14 suppresses cell death and is required for normal fat metabolism. Curr Biol 13(9):790–795
37. Zhang W, Cohen SM (2013) The Hippo pathway acts via p53 and microRNAs to control proliferation and proapoptotic gene expression during tissue growth. Biol Open 2(8):822–828
38. Li Y et al (2006) MicroRNA-9a ensures the precise specification of sensory organ precursors in Drosophila. Genes Dev 20(20):2793–2805
39. Li X et al (2009) A microRNA imparts robustness against environmental fluctuation during development. Cell 137(2):273–282
40. Wu YC et al (2012) Let-7-complex microRNAs regulate the temporal identity of Drosophila mushroom body neurons via chinmo. Dev Cell 23(1):202–209
41. Liu N et al (2012) The microRNA miR-34 modulates ageing and neurodegeneration in Drosophila. Nature 482(7386):519–523
42. Hyun S et al (2009) Conserved MicroRNA miR-8/miR-200 and its target USH/FOG2 control growth by regulating PI3K. Cell 139(6):1096–1108
43. Varghese J, Lim SF, Cohen SM (2010) Drosophila miR-14 regulates insulin production and metabolism through its target, sugarbabe. Genes Dev 24(24):2748–2753
44. Weng R, Cohen SM (2012) Drosophila miR-124 regulates neuroblast proliferation through its target anachronism. Development 139(8):1427–1434
45. Nahvi A, Shoemaker CJ, Green R (2009) An expanded seed sequence definition accounts for full regulation of the hid 3′ UTR by bantam miRNA. RNA 15(5):814–822
46. Vodala S et al (2012) The oscillating miRNA 959-964 cluster impacts Drosophila feeding time and other circadian outputs. Cell Metab 16(5):601–612
47. Weng R et al (2013) miR-124 controls male reproductive success in Drosophila. elife 2: e00640
48. Yu JY et al (2009) Dicer-1-dependent Dacapo suppression acts downstream of insulin receptor in regulating cell division of Drosophila germline stem cells. Development 136(9):1497–1507
49. Hilgers V, Bushati N, Cohen SM (2010) Drosophila microRNAs 263a/b confer robustness during development by protecting nascent sense organs from apoptosis. PLoS Biol 8(6): e1000396
50. Chen W et al (2014) Regulation of Drosophila circadian rhythms by miRNA let-7 is mediated by a regulatory cycle. Nat Commun 5:5549
51. Bejarano F et al (2012) A genome-wide transgenic resource for conditional expression of Drosophila microRNAs. Development 139(15):2821–2831
52. Schertel C et al (2012) Functional characterization of Drosophila microRNAs by a novel in vivo library. Genetics 192(4):1543–1552
53. Szuplewski S et al (2012) MicroRNA transgene overexpression complements deficiency-based modifier screens in Drosophila. Genetics 190(2):617–626
54. Chen YW et al (2014) Systematic study of Drosophila microRNA functions using a collection of targeted knockout mutations. Dev Cell 31(6):784–800

55. Agarwal V et al (2018) Predicting microRNA targeting efficacy in Drosophila. Genome Biol 19(1):152
56. Rio DC (2014) Northern blots for small RNAs and microRNAs. Cold Spring Harb Protoc 2014(7):793–797
57. Li C, Zamore PD (2018) Analysis of small RNAs by Northern hybridization. Cold Spring Harb Protoc 2018(8):pdb prot097493
58. Laneve P, Giangrande A (2014) Enhanced Northern Blot detection of small RNA species in Drosophila melanogaster. J Vis Exp (90)
59. Varallyay E, Burgyan J, Havelda Z (2007) Detection of microRNAs by Northern blot analyses using LNA probes. Methods 43(2):140–145
60. Chen C et al (2005) Real-time quantification of microRNAs by stem-loop RT-PCR. Nucleic Acids Res 33(20):e179
61. Griffiths-Jones S et al (2006) miRBase: microRNA sequences, targets and gene nomenclature. Nucleic Acids Res 34(Database issue): D140–D144
62. Ruby JG et al (2007) Evolution, biogenesis, expression, and target predictions of a substantially expanded set of Drosophila microRNAs. Genome Res 17(12):1850–1864
63. Grun D et al (2005) microRNA target predictions across seven Drosophila species and comparison to mammalian targets. PLoS Comput Biol 1(1):e13
64. Schnall-Levin M et al (2010) Conserved microRNA targeting in Drosophila is as widespread in coding regions as in 3'UTRs. Proc Natl Acad Sci U S A 107(36):15751–15756
65. Horwich MD, Zamore PD (2008) Design and delivery of antisense oligonucleotides to block microRNA function in cultured Drosophila and human cells. Nat Protoc 3(10):1537–1549
66. Yang JS et al (2014) Intertwined pathways for -Argonaute-mediated microRNA biogenesis in Drosophila. Nucleic Acids Res 42 (3):1987–2002
67. Okamura K et al (2007) The mirtron pathway generates microRNA-class regulatory RNAs in Drosophila. Cell 130(1):89–100

Chapter 15

CRISPR Technique for Gene Deletions in *Drosophila*

Abhinandan Patnaik and Monalisa Mishra

Abstract

Clustered regularly interspaced short palindromic repeat is popularly known worldwide by its acronym CRISPR. It is a modern technique that has revolutionized molecular studies by its unique coherence in genome engineering. The robust nature and specificity of this novel approach in molecular biology have enabled researchers worldwide to adapt this method in various developmental studies. *Drosophila melanogaster*, the model organism which paved the path for groundbreaking discoveries in the field of developmental biology, is also one among many other such model systems in which the CRISPR-Cas9 system has been integrated. Despite there being other practices of mutagenesis available, the CRISPR-Cas9 system provides an edge over other such methods in terms of accuracy and efficiency, which ensures an expeditious screening process thereafter. Here we make an attempt to introduce the detailed protocol and essential steps therein involved in generating knockout allele mutants in *Drosophila*.

Key words CRISPR-Cas9, Guide RNA, PAM (protospacer adjacent motifs), NHEJ (non-homologous end joining), Gene deletions

1 Introduction

CRISPR-Cas9 is yet another heterologous system adapted in fruit flies just like the UAS-Gal4 system which transformed genetic experiments involving *Drosophila*. The clustered regularly interspaced short palindromic repeats were first discovered in the bacterial genome in the 1980s, although researchers then could not settle on a definite function of these repeats in the genetic material. Until 2007 the function of these repeats remained as a mystery to the scientific community. Barrangou and group determined the role of these repeats in bacterial defence mechanism and showed how bacteria integrate fragments of foreign genetic materials into their CRISPR locus and eventually acquire resistance against the infectious virus [1]. CRISPR and its associated Cas endonuclease genes together constitute an inherent defence system present in selected bacteria and archaeal biology, providing immunity and protection against several foreign genetic materials from invasive bacteriophages. So far, three distinct pathways are known for the

Monalisa Mishra (ed.), *Fundamental Approaches to Screen Abnormalities in Drosophila*, Springer Protocols Handbooks, https://doi.org/10.1007/978-1-4939-9756-5_15,

CRISPR system [2], and all of them have different molecular mechanisms. All the existing mechanisms vary from one another based on their integration and processing of systems which depend on the availability of various endonuclease systems.

The system adopted for flies is a simplified version of the type II CRISPR system, and it comprises of two components: (1) a chimeric RNA sequence (henceforth will be mentioned as guide RNA or gRNA) and (2) an endonuclease Cas-9 expressed under tissue-specific promoters. Once the transgenic template for CRISPR gRNA is transcribed, along with Cas-9, it forms a ribonucleoprotein (RNP) complex [3]. The active complex thus formed is guided to the appropriate gene region for which the guide RNA was designed. The endonuclease Cas-9 recognizes three nucleotide long motifs of the form NGG to the three prime ends of the target sequence. These motifs are otherwise called protospacer adjacent motifs or PAM sequence. Different domains of Cas-9 are responsible for its different interactions. While the Rec1 domain of Cas-9 binds the guide RNA, the PAM-interacting domain confers specificity for PAM sequences. Moreover, these interactions bring about conformational changes in Cas-9, which is believed to be essential for its activation and functioning [4, 5]. Followed by binding to the appropriate sites, double-stranded breaks (DSBs) are generated. Damage to the genome triggers and activates the repair machinery which stiches back the breaks either through homology-directed repair (HDR) or through the more common non-homologous end joining (NHEJ) [6]. NHEJ is a highly erroneous repair pathway and generates indels (insertions and deletions). Using this, the overall integrity of any gene can be easily compromised. The protocol mentioned herewith describes a simple approach to generate null mutant alleles of a gene of interest. The method includes deletion of an entire gene locus by using two guide RNAs specific for 5 prime and 3 prime ends of the gene. We introduce the reader to the available tools and reagents to accomplish the task and also describe a general strategy to acquire loss of function alleles in *Drosophila*.

2 Integrating the CRISPR Technique in Flies

Since its inception, CRISPR-Cas9 has been adapted to several cell culture experiments as well as in many model organisms. *Drosophila*, an extremely robust and easy-to-handle model organism, was not to be left behind. The amalgamation of the CRISPR-Cas9 system into *Drosophila* has reshaped fly genetics and studies involving various essential pathways. From generating complete deletion alleles to manipulating certain genes using HDR (homology-directed repair), CRISPR has come up as a promising tool which has empowered researchers worldwide to reengineer the fly genome to one's own requirement.

Both guide RNAs and the enzyme Cas9 can be integrated into the genome by directly injecting DNA or RNA into the germline pole cells of *Drosophila* at the syncytial blastoderm stage. Transgenic fly lines with guide RNA inserts are generated using expression vectors. The vector pCFD4 is commonly used to integrate into the fly genome using the phiC31 integrase system for site-specific insertion. In expression vectors like pCFD4, the guide RNAs are cloned under strong U6 promoters which are RNA polymerase type III promoters [6]. Cas9 is brought into play by using transgenic fly lines expressing the endonuclease under tissue-specific promoters. In nos-Cas9 fly lines, the ORF is cloned into a plasmid containing the *nanos* promoter which gives germline-specific expression of the gene. Once both the transgenic fly lines are available, a simple mating scheme has to be designed to bring together the guide RNA source and the endonuclease Cas9 source. When brought together both the components are expressed and combined to form the gRNA-Cas9 RNP complex, which then recognizes the target regions (Fig. 1) and generates double-stranded breaks.

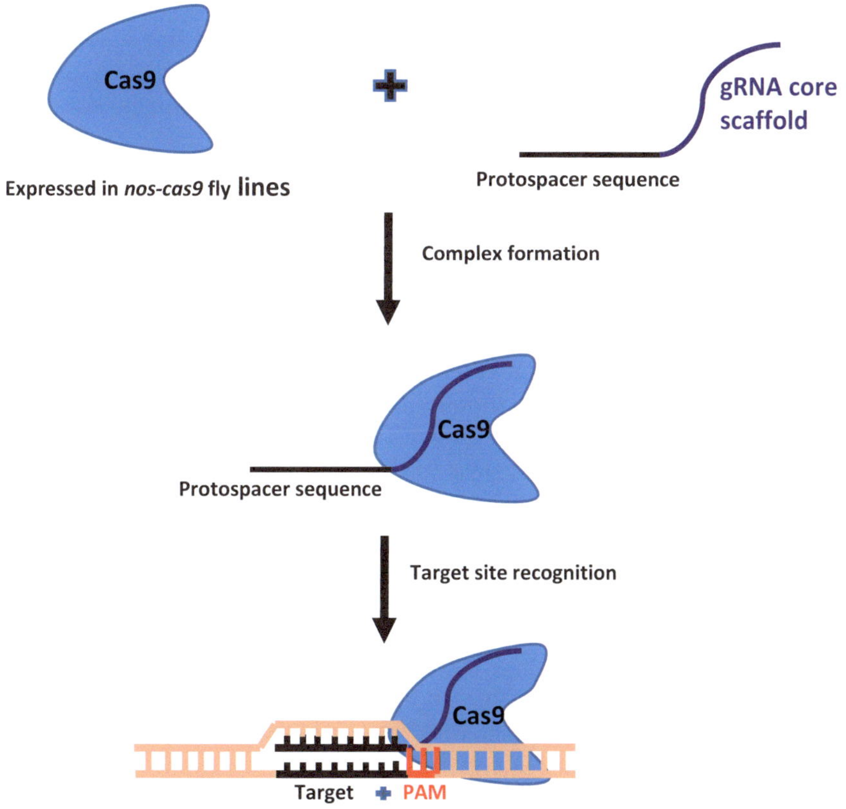

Fig. 1 Schematic diagram depicting the expression and formation of Cas9-gRNA complex and subsequent recognition of the target sequence

3 CRISPR over Other Methods of Mutagenesis

In order to analyse the roles of a gene in a certain essential pathway, it is first necessary to interpret its interaction with other genes with known functions. The level of interactions and indispensability of a molecule can only be determined when either the expression of the gene is supressed or a case where the gene is knocked out or deleted. From time to time with the advancement in the understanding of molecules, many tools have been introduced into studies in life sciences to manipulate a gene at expression levels.

Early attempts at the generation of mutants in forwarding genetics involved usage of DNA alkylating agents like EMS (ethyl methanesulfonate). EMS generally instigates transition mutations in bases [7]. The induced mutations then have to be mapped to determine the gene locus where the mutation occurred. Mostly, mutants generated by EMS are point mutations which create hypomorphic mutants [8] with relatively subtle phenotypes. Hence the functional analysis of the gene would be highly unreliable.

Another fairly reliable method applied in *Drosophila* genetics is P-element mobilization. P elements are class-II transposons which mobilize via a cut and paste mechanism throughout the genome, landing in and disrupting an actively transcribing gene [9]. Additionally, it may also generate deletions after subsequent translocation events. Excisions produced after mobilization cannot often be predicted before the screening initiates, which is another limiting factor considering the fact that the process may take days to months. The truncated gene products may still have the essential functional domains compensating for the deletions.

Compared to the above-mentioned methods, the CRISPR-Cas9 system has now gained popularity in the era of genome editing and engineering. CRISPR-Cas9 uses the inherent repair mechanism of *Drosophila* to generate null or truncated mutant genes. In contrast to the uncontrolled and undetermined alterations that occur in EMS or P-element mutants, CRISPR is very specific and advanced. The system can be designed as per the need and the results are more accurate and predictable. Increased specificity of the technique facilitates the avoidance of any of the target effects. The technique also allows the user to choose any target of their choice, and hence the method can be modified to delete specific regions of a gene to determine the roles of the functional domains of its product. Moreover, the screening process is hustle free and gives optimum results. The current protocol describes the CRISPR-Cas9 techniques to create the mutation in *Drosophila*.

4 Materials Required

4.1 Designing Target Sequence for Guide RNAs

1. FlyBase repository (http://flybase.org/)
2. E-CRISP online software platform (http://www.e-crisp.org/E-CRISP/)

4.2 Cloning of Guide RNA in Expression Vectors

1. pCFD4 guide RNA expression vectors available with Addgene [Catalog no:49411] (https://www.addgene.org/49411/)
2. Primers for cloning protospacer sequences
3. BbsI nuclease (NEB #R3539)
4. Gibson Assembly Master Mix (NEB Catalog no: E2611S)
5. Standard plasmid extraction kit (Qiagen QIAprep Spin Miniprep Kit; Catalog no. 27106)

4.3 Mutant Generation and Screening

1. Transgenic flies with guide RNA inserts
2. nos-Cas9 fly lines [Bloomington stock number:54591]
3. Oligo primers for PCR screening (Custom DNA oligos from Sigma-Aldrich)
4. Conventional *Drosophila* genomic DNA isolation protocol [10]

5 Method

5.1 Designing Target Sequence for Guide RNAs

The protospacer sequences to be incorporated into the vector should be 19–20 nucleotide long depending on whether the sequence starts from N or G. Select these fragments carefully in order to avoid any off-target deletions, which might complicate the situation at a later stage. Several online tools and platforms like E-CRISP and flyCRISPR Design, to name a few, are available to design optimal guide RNAs suitable for the users.

1. Acquire the gene region sequence data from FlyBase repository.
2. Download the FASTA file format or type in the gene symbol in E-CRISP, and then click on Start sgRNA search (*see* **Note 1**).
3. The software would display optimal sites for designing the primers for the target sequence (Fig. 2a) (*see* **Note 2**).
4. The sequences with minimum or zero homologies with any other regions in the genome must be selected for designing the protospacer sequence in primers.

5.2 Cloning Guide RNA in Expression Vectors

1. Set up a PCR reaction with high fidelity polymerase using the CRISPR primers (Fig. 2b) and pCFD4 as the template. Optimal annealing temperatures for the primers could vary

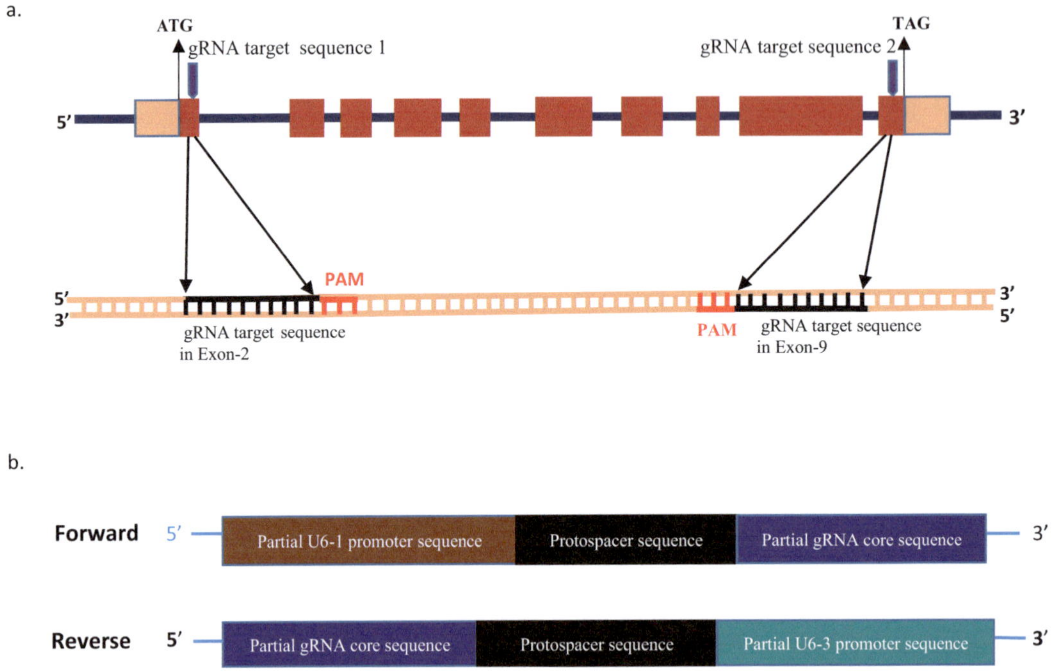

Fig. 2 Determining protospacer sequence from gene region. (**a**) A hypothetical gene region depicting exons, introns and UTR regions along with the highlighted target regions. (**b**) Basic components of the primers to be designed for cloning

depending on target sequence integrated into the primers. Hence, troubleshooting might be required.

2. Run the 600 bp amplicon from the PCR reaction alongside a DNA ladder to confirm the amplicon length (Fig. 3b). Subsequently, purify the amplicon with PCR purification kit or gel extraction kit (*see* **Note 3**).
3. Digest pCFD4 with BbsI. The digestion product would release a 600 bp fragment which has to be confirmed on 1% agarose gel (Fig. 3a).
4. Purify the gel to get the 6.4 kb backbone of the vector.
5. Assemble the pCFD4 vector backbone and the 600 bp amplicon using Gibson assembly (NEB). Follow the instruction manual provided by NEB carefully for details related to the concentration (*see* **Note 4**).
6. Transform the assembly reaction into 50 μl DH5alpha competent cells and plate on Ampicillin plates.
7. Select 4–5 colonies from the transformed competent cells, and verify the guide RNA inserts by sequencing using the pCFD4 sequencing primer (Fig. 3c) (*see* **Note 5**).

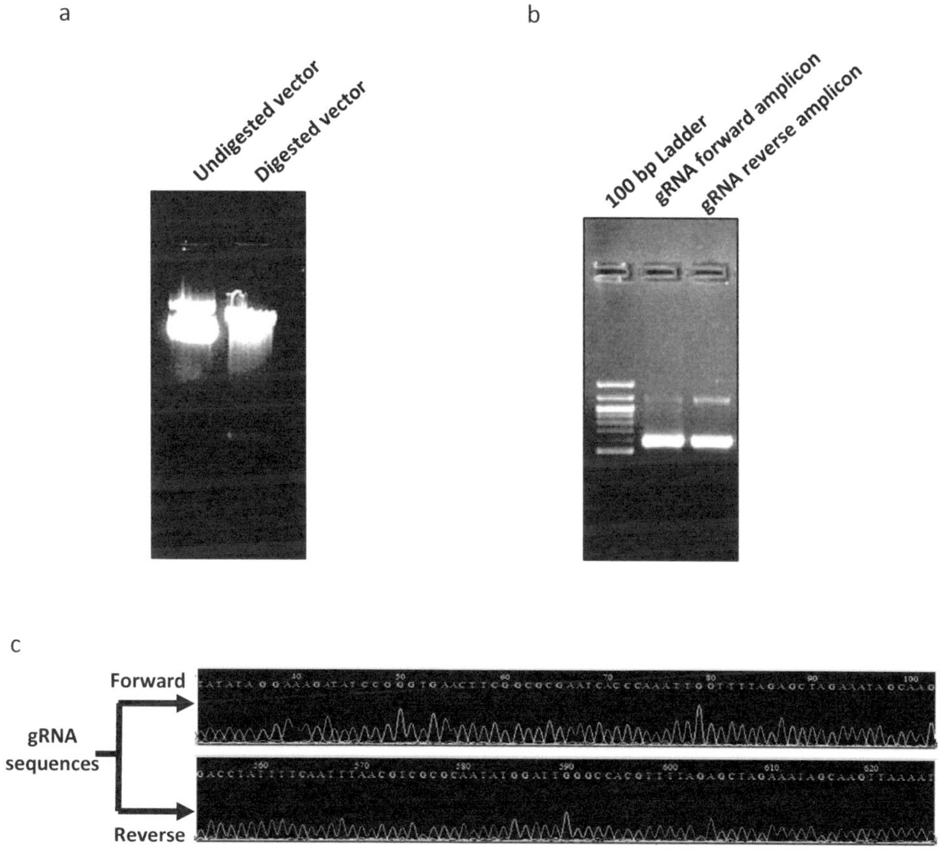

Fig. 3 Gel images of molecular work. (**a**) Digestion of pCFD4 vector with BbsI releasing a 600 bp fragment. (**b**) The gRNA insert amplicons of 600 bp size. (**c**) Sequencing result verifying the gRNA inserts in the vector

5.3 Mutant Generation and Screening

5.3.1 Injection of Cloned Expression Vectors

Transgenic flies with guide RNA inserts are generated by microinjection of positive constructs in fly embryos followed by a screening of the transformants (Fig. 4). Site-specific integration of gRNAs into the fly genome is achieved by phiC31-based transgenesis. The injections can be done in any other mutant background embryos as well, as per the need. The process of generating integrated gRNA flies is often time consuming but gives the advantage of increased potency; hence outsourcing the injections is a common approach.

Alternately, gRNAs can also be injected directly in the form of DNA or RNA in embryos with Cas9 as a transgenic source for quick experiments. The same can also be done for endonuclease Cas-9 in which Cas9 mRNA is injected in fly embryos. This method saves time and increases the pace of further experiments. However, generating transgenic fly lines with gRNA inserts is reliable and more effective.

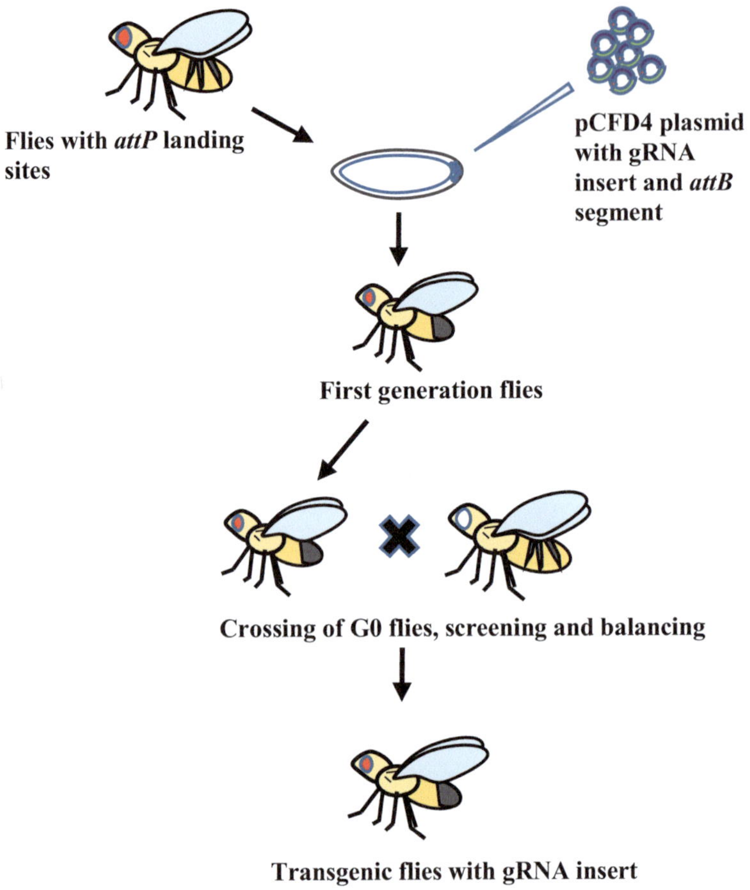

Fig. 4 A schematic diagram depicting the workflow of generating transgenic fly lines with gRNA insert

5.3.2 Mating Scheme for Mutant Generation

1. The transgenic flies with guide RNA inserts over a balancer are mated with nos-Cas9 flies in a mass cross (*see* **Note 6**).
2. Collect males from the first mass cross, and mate them with double balancer virgin females. The progeny of this cross is now devoid of the nos-Cas9 endonuclease activity.
3. Set up single line crosses of the males collected from a second mass cross with double balancer females.
4. Progeny from the third step can now be self-crossed and further screened using PCR-based screening (*see* **Note 7**) (Fig. 5).

6 Notes

1. If the gRNA is to be designed for specific region/regions in the gene, then selected portions of the gene region can be used to check for target sequences in place of the entire sequence.

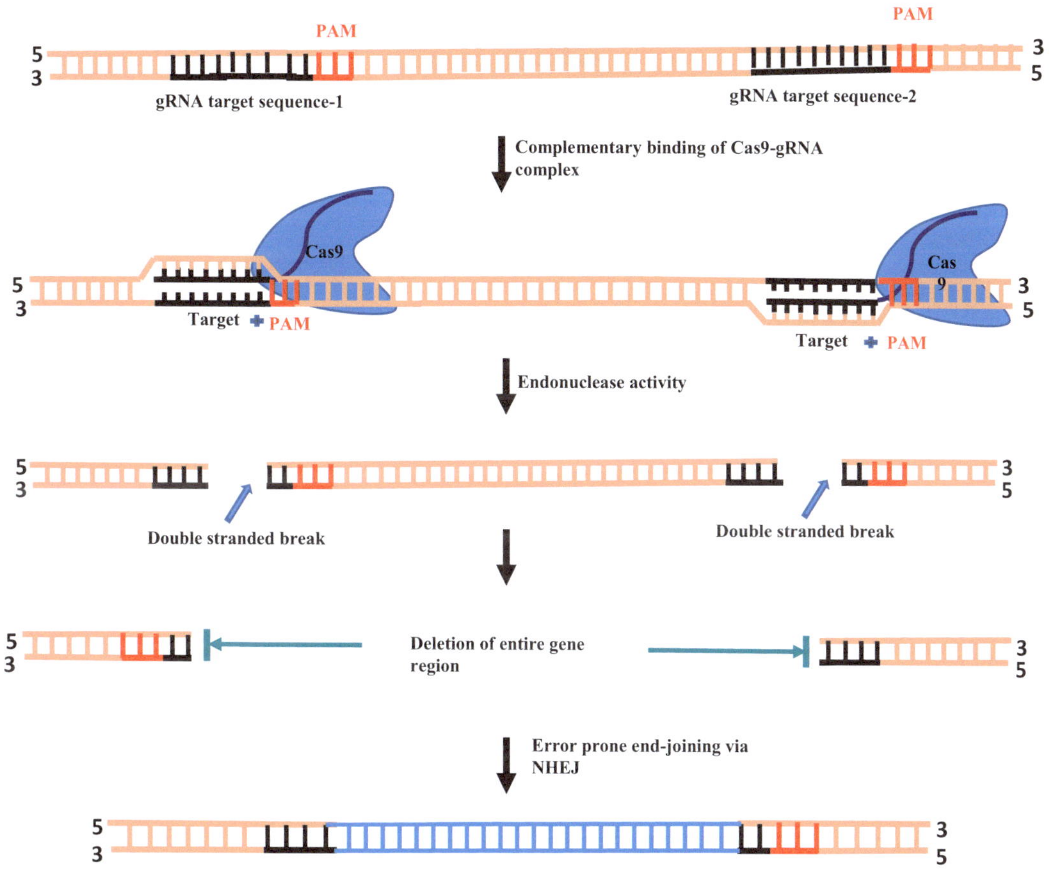

Fig. 5 The figure shows the process of generation of double-stranded breaks and deletion of the entire gene region with CRISPR-Cas9-mediated gene deletion. Subsequently, the deletion can be screened using primers designed at least 100 base pairs upstream of designed guide RNA targets

2. In order to delete the entire gene region, it is essential to select target sequences which would include the start and stop codons of the gene. Retaining the coding sequence of the genes might result in truncated translational products due to frameshift mutations.
3. Gel extraction kit must be used only when the amplicons obtained are non-specific. The gel extraction process may result in decreased DNA concentration after the final elution step; hence large volumes of PCR reactions must be set up for this process. Generally, a 200 microliter volume reaction is sufficient.
4. The Gibson Assembly Master Mix is highly efficient; therefore small volume reactions (5 μl) can be set up.
5. As Gibson assembly provides high cloning efficiency, 4–5 of the obtained positive colonies can be directly sent for sequencing.

6. The male to female ratio in a mass cross is generally 1:2.
7. The primers designed for PCR screening must be 100–200 base pairs upstream of the designed target sequences so that the exact number of bases deleted can be verified by sequencing.

References

1. Barrangou R, Fremaux C, Deveau H, Richards M, Boyaval P, Moineau S, Romero DA, Horvath P (2007) CRISPR provides acquired resistance against viruses in prokaryotes. Science 315(5819):1709–1712
2. Carte J, Christopher RT, Smith JT, Olson S, Barrangou R, Moineau S, Glover CV III, Graveley BR, Terns RM, Terns MP (2014) The three major types of CRISPR-Cas systems function independently in CRISPR RNA biogenesis in Streptococcus thermophilus. Mol Microbiol 93(1):98–112
3. Makarova KS, Haft DH, Barrangou R, Brouns SJ, Charpentier E, Horvath P, Moineau S, Mojica FJ, Wolf YI, Yakunin AF, Van Der Oost J (2011) Evolution and classification of the CRISPR–Cas systems. Nat Rev Microbiol 9 (6):467
4. Jinek M, Jiang F, Taylor DW, Sternberg SH, Kaya E, Ma E, Anders C, Hauer M, Zhou K, Lin S, Kaplan M (2014) Structures of Cas9 endonucleases reveal RNA-mediated conformational activation. Science 343(6176):1247997
5. Nishimasu H, Ran FA, Hsu PD, Konermann S, Shehata SI, Dohmae N, Ishitani R, Zhang F, Nureki O (2014) Crystal structure of Cas9 in complex with guide RNA and target DNA. Cell 156(5):935–949
6. Port F, Chen HM, Lee T, Bullock SL (2014) Optimized CRISPR/Cas tools for efficient germline and somatic genome engineering in Drosophila. Proc Natl Acad Sci 111(29): E2967–E2976
7. Pastink A, Heemskerk E, Nivard MJ, van Vliet CJ, Vogel EW (1991) Mutational specificity of ethyl methanesulfonate in excision-repair-proficient and-deficient strains of Drosophila melanogaster. Mol Gen Genet MGG 229 (2):213–218
8. Lin SC, Chang YY, Chan CC (2014) Strategies for gene disruption in Drosophila. Cell Biosci 4 (1):63
9. Liao GC, Rehm EJ, Rubin GM (2000) Insertion site preferences of the P transposable element in Drosophila melanogaster. Proc Natl Acad Sci 97(7):3347–3351
10. Huang AM, Jay Rehm E, Rubin GM (2009) Quick preparation of genomic DNA from Drosophila. Cold Spring Harb Protoc 2009 (4):pdb–prot5198

Chapter 16

Analysis of Various Body Parts of *Drosophila* Under a Scanning Electron Microscope

Janmejaya Bag and Monalisa Mishra

Abstract

Scanning electron microscope (SEM) helps to analyse the surface of any cell or organ by magnifying it several times. SEM analysis provides an indispensable tool for people working with a small model organism like *Drosophila*. Various morphological abnormalities originated in different body parts during development can be visualized after SEM analysis. The images generated from SEM make it easy to characterize the morphological defect. SEM is recently attached with Energy-Dispersive X-ray Spectroscopy (EDX/EDS). This allows detecting all the metals present within the organ. The current protocol describes the fixation of *Drosophila* head, haltere, hemolymph and gut for SEM analysis. We are also describing the metal analysis of the gut using EDS or EDX analysis.

Key words Head, Eye, Haltere, EDS analysis, Gut, Wing, Hemolymph

1 Introduction

1.1 Principle Behind Image Formation

Scanning electron microscope (SEM) is a powerful and versatile tool to observe the outer surface of the specimens by magnifying it to a great extent. The conventional high-vacuum SEM requires proper sample preparation to keep the sample at its own shape. Sample preparation includes proper fixation and dehydration to retain its original structure. An electron microscope uses an electron beam [1]. Thus, the resolution of the image produced by SEM is very high. The electron source is placed within a vacuum to protect itself from contamination and vibration. Furthermore, in the absence of the vacuum, other atoms present inside the chamber interfere with the electron resulting low-quality image. The electrons are generated by heating the cathode. Once the electrons are produced, it is guided by electromagnetic lenses. The electron beam passes through the condenser lens and then the objective lens and then to the sample. Once it hits the sample, either it passes through the sample or reflects back. The interaction of the electron beam and sample is of two types: (1) elastic and (2) non-elastic.

Monalisa Mishra (ed.), *Fundamental Approaches to Screen Abnormalities in Drosophila*, Springer Protocols Handbooks, https://doi.org/10.1007/978-1-4939-9756-5_16,

The electrons after hitting the sample generate backscattered electrons (BSE). BSE is produced due to elastic interaction, while secondary electrons (SE) are due to non-elastic interaction. BSE is generated from the deepest part, while SE is generated from the surface. All the electrons are detected by backscatter and secondary electron detectors [2, 3]. Usually, SE describes the topographic feature of the sample and BSE shows the elemental contrast. Secondary electron imaging (SEI) is mediated through the secondary electrons. When high energy electrons strike through the sample, they can be detected by Energy-Dispersive X-ray Spectroscopy (EDS) [4, 5].

1.2 Types of SEM

Based on the sample preparation, SEM can be (1) environmental scanning electron microscope (ESEM) and (2) field emission scanning electron microscope (FESEM) [6, 7]. In ESEM very less time is needed for sample preparation. Thus the chance of producing an artefact in a biological sample is very less. However, both ESEM and FESEM produce high-resolution, black and white digital images [8]. Various alterations of structure and metal in a biological sample can be detected using scanning electron microscope. However, *Drosophila* tissues need certain modification to get better contrast of the images. One of such modification is the sample coating.

1.3 Coating for Biological Sample

Drosophila samples are non-conductive. To make it conductive after proper dehydration, samples are coated with metals. For coating, metals like gold, tungsten, platinum, chromium, osmium, etc. are used. Due to the conductive layer around the sample, the chance of getting charged while examining under an electron microscope is less [3, 9]. If the samples are charged, then it interferes with the contrast of imaging and boosts the signal for secondary electrons for 3D structures. For coating of metal, a sputter coater is used, and the time of coating depends on the type of sample [1]. In the next section, we are describing the requirement of analysis of various *Drosophila* tissues under SEM.

1.4 Phenotypic Analysis of Various Parts Under SEM: Head

The head is the most important organ of the body, which possesses the eye, antenna, mouthparts and peripheral bristles. During development, a defective gene may alter the structures of these organs. Thus, based on the phenotypic defect, the alteration of the gene can be chased. For example, the eye is formed of 750–800 functional units called ommatidia. Any fusion of ommatidia [10–12], alteration of the arrangement of bristles [13, 14] and the arrangement of ommatidia cannot be seen properly under a light microscope [15, 16]. However, these defects can be observed properly under a scanning electron microscope. Similarly, *Drosophila* antennae possess three different segments and many bristles. The functionality of the antennae further depends on a structure called aristae which is present in the A3 segment. Any alteration in the arista causes a faulty antenna [17]. Thus, to observe various morphological defects present in the antennae, we need to examine the head under a scanning electron microscope. Similarly, *Drosophila* mouthparts are formed of many segments and offer it as a nice model to study genetic interaction [18]. *Drosophila* mouthparts are formed of the clypeus, maxillary palp, labrum, labium and labellum. Due to the small size, any defect cannot be observed properly under a light microscope. In such a case, an SEM image is indispensable to identify phenotypic defect if there is any, in the mouthparts. Thus, proper fixation of the head is needed to analyse it under SEM.

1.4.1 Wings

Drosophila wing is an integrated structure which may alter due to genetic, developmental and various environmental factors. It has a distinct shape, venation pattern and proper arrangement of bristles. The arrangement of bristle can be seen under the scanning electron microscope. Similarly, there is any alteration of change in the size of the bristle that can be seen under SEM [19]. At the base of the wing, the haltere is present which is essential for flying. It has three different segments, and it includes (1) scabellus, (2) pedicellus (3) and capitellum. The morphology of the pedicellus can be seen under SEM [20, 21].

1.4.2 Hemocytes

Hemocytes, the major component of hemolymph, play a crucial role in the defence mechanism of *Drosophila*. The number and type vary with physiological condition and infection. Thus, by analysing the hemoctyes, the physiological condition of the animal can be known. *Drosophila* possesses crystallocyte, lamellocyte and plasmatocyte cells in the hemolymph [22, 23]. These three cells have different function and structure and can be visualized under SEM.

1.4.3 Larva or Adult Gut

The metal composition of the gut alters during infection, metabolic disorder and with alteration of food. Fly needs a certain amount of metals in the body for physiological activity. Key metals like Ca^{+2} and phosphorous have an important role in the metabolism of the fly [24]. This can be detected within the gut using SEM with EDS analysis. The number of metals obtained from EDS analysis of the treated fly can be compared with the control flies.

The current protocol describes the sample preparation of the head, wing, haltere, hemocyte and gut for scanning electron microscopy.

2 Material

2.1 SEM Image Analysis

1. Glass slide (ABDOS, G10101 25 × 75 mm)
2. Diamond pencil
3. Adhesive black tape (High Strength Self Adhesive Bond Tape, 12 mm × 5 M)

4. Forceps (SCHOFIC, stainless steel tweezer)
5. Fine razor blade (Gillette, WILKINSON®)
6. Third instar larva
7. Adult fly
8. Sputter coater (Quorum, 150TES plus)
9. Scanning electron microscope (Nova NanoSEM 450)
10. For the dehydration process: 30%, 50%, 70% and 90% absolute ethanol (analytical CS reagent, 1170)
11. Propylene oxide (ACS reagent, 20401)

3 Methods

3.1 SEM Image Analysis

3.1.1 Sample Fixation of Solid Tissue (Eye, Head, Halter)

1. Take the adult fly or larva from the rearing media on a vial.
2. Anaesthetize it with the diethyl ether for 1 min.
3. Transfer the fly or larva to a glass slide and dissect under the stereomicroscope.
4. Dissect the section of interest (head, wing, haltere, gut) from adult or larva.
5. Keep the tissue of interest in 4% PFA at room temperature for 2 h or overnight at 4 °C in a 1.5 ml Eppendorf tube (Fig. 1).

3.1.2 Sample Dehydration

1. On the next day, wash the sample with PBS, and dehydrate with a graded series of ethanol starting from 30% ethanol.
2. In each step keep for 10 min.
3. The ethanol series includes 30%, 50%, 70% and 90% and absolute ethanol two times (*see* **Note 1**).
4. Next dehydrate with 100% propylene oxide (Fig. 1) (*see* **Note 2**).

3.2 Mounting Procedure

1. Divide the glass slide into eight parts. Carefully cut out small squares of the glass slide with the diamond knife.

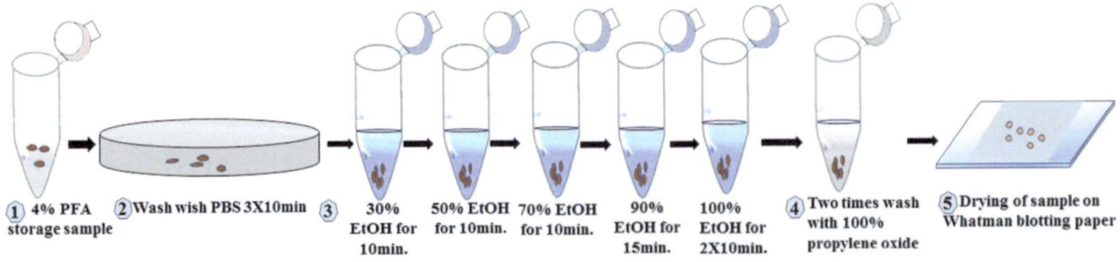

Fig. 1 Dehydration and mounting of the sample over carbon-coated glass slide

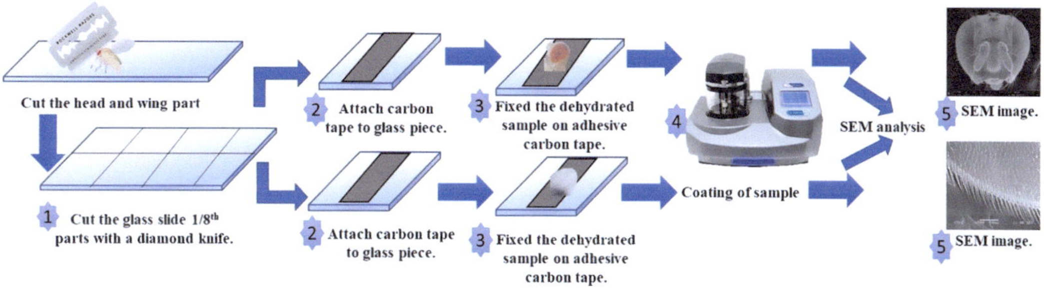

Fig. 2 Mounting of the dried sample over carbon tape for coating

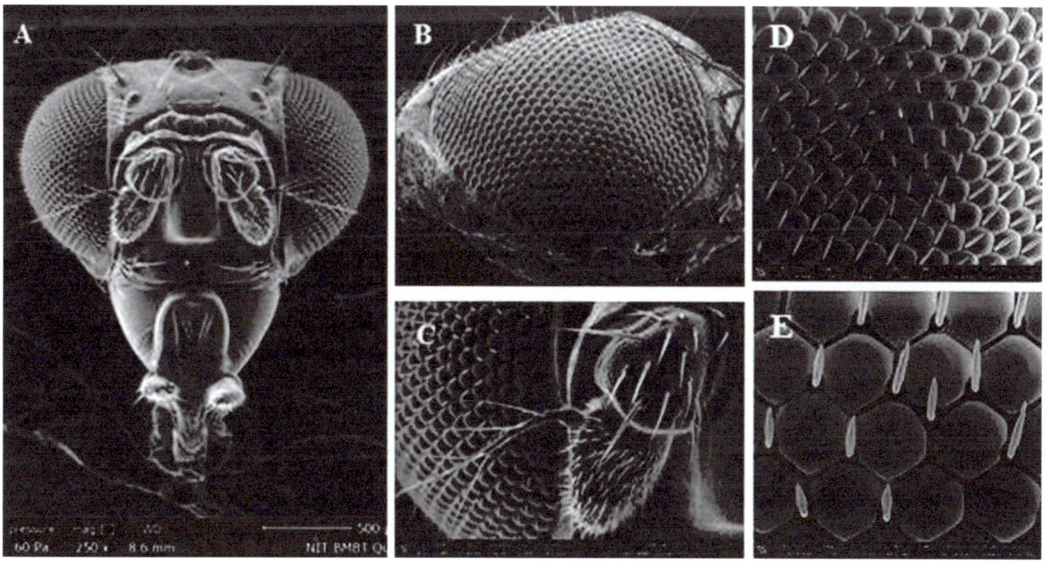

Fig. 3 SEM image of fly head. (**a**) Head. (**b**) Eye. (**c**) Antennae. (**d**) Ommatidia. (**e**) Hexagonal shape of ommatidia

2. Take one part of the small square size glass piece, and fix the adhesive black tape over it (Fig. 2).
3. Then fix the dry sample (obtained from Sect. 3.1.2) over the black tape under the stereomicroscope.
4. Align the sample position with the help of a forcep (Fig. 2).
5. Coat the samples with gold for 1 min by following the manual of the coating instrument. This is done at a 100 K temperature in a vacuum medium.
6. Once the coating is done, transfer the slide for imaging under SEM, FESEM or ESEM (Fig. 2).
7. Various tissues like the head, eye, antennae, haltere and wing are shown in Figs. 3 and 4.

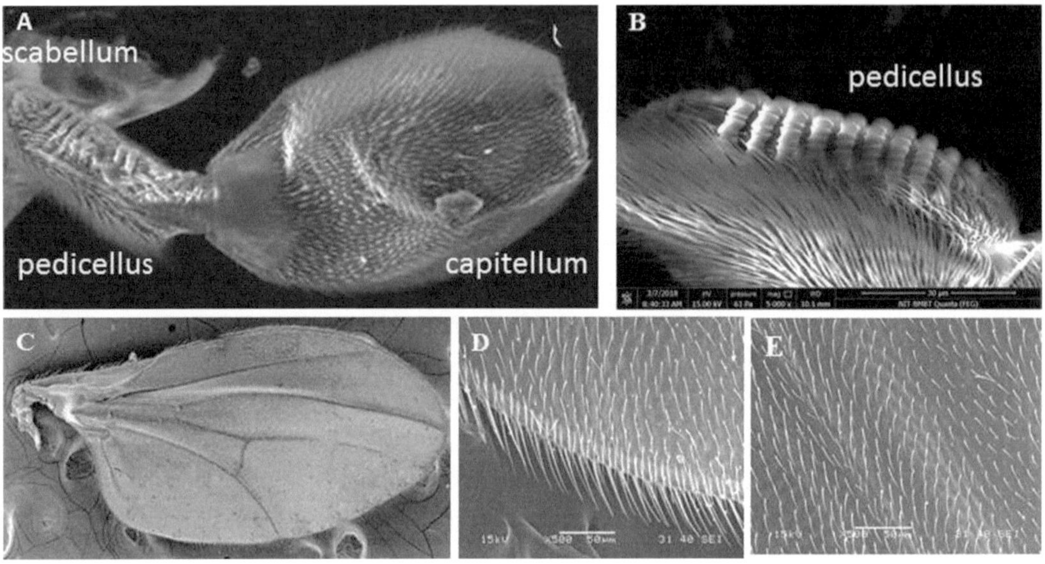

Fig. 4 SEM image of haltere and wing. (**a**) Various segments of haltere (**b**). Pedicellus of the haltere. (**c**) Wing. (**d**) Marginal wing blades. (**e**) Arrangement of hairs over the wing

3.3 SEM Image Preparation for Hemolymph

1. Take out three larvae from the rearing media.
2. Wash it with 1X PBS.
3. Transfer the larvae to the one-eighth part cut slide.
4. Put the larvae on the slide, and prick the hemolymph directly with the help of a fine pin, by puncturing the cuticle.
5. Make a smear of the hemolymph (*see* **Note 3**).
6. Allow the slides to dry. It takes 1 h to dry the sample.
7. Then fix with 70% EtOH
8. Keep it for 2 days for air-drying or in an incubator at 37 °C for 2 days.
9. Then, mount the slides on a stub and coat it with gold.
10. Examine the samples under SEM, ESEM or in FESEM (Fig. 5).
11. The shape of crystallocyte, lamellocyte and plasmatocyte under SEM is shown in Fig. 6.

3.4 Gut Sample Preparation

1. Take out the larvae from the rearing media.
2. Wash it with 1X PBS, to remove the food particles.
3. Keep the larva on a slide for dissection under a stereomicroscope.
4. Take out the larval gut, and transfer it to the 4% PFA, and store it in the 4°C freezer.

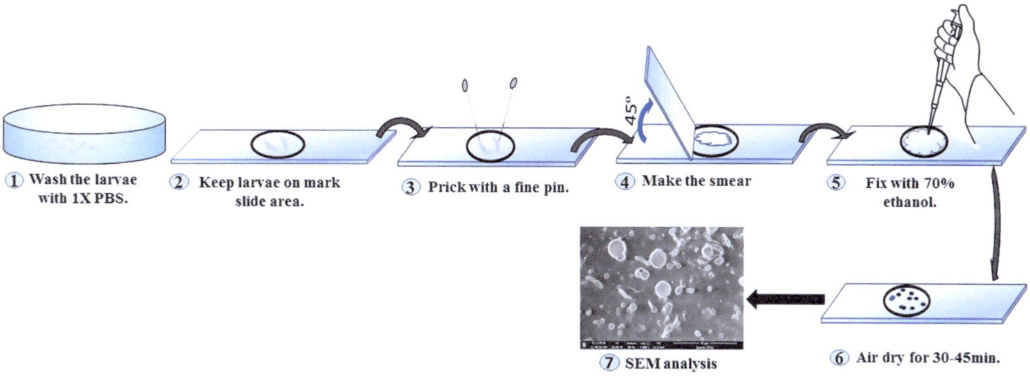

Fig. 5 SEM analysis of larval isolated hemolymph [25]

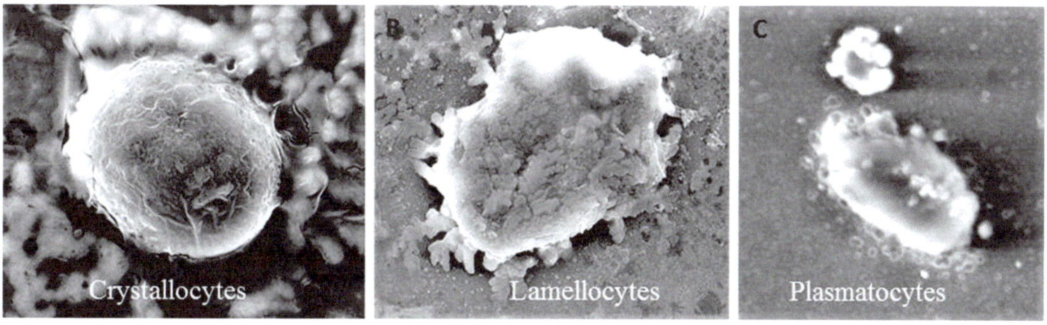

Fig. 6 Types of hemocytes under SEM

3.5 Dehydration Procedure

1. Take the sample from the fixative and wash once with 1X PBS.
2. First, keep the sample in 30% EtOH for 10 min, and followed by 50%, 70%, 90% and 100% of EtOH 10 min each.
3. Take out the sample, and dehydrate it with Whatman 40 blotting paper, so that the sample will be dry completely.
4. Examine the samples either under SEM, ESEM or in FESEM (Fig. 7).

3.6 EDS Analysis of Larval Gut

1. For the EDS analysis, puncture the gut with the help of a fine needle at mid gut regions. This will allow the elements of the gut to be exposed to the electron beam which ultimately creates a map of the element present on the gut (*see* **Note 4**).
2. Compare the elements of interest from the control data to treated data to observe variation in elemental content of the gut (Fig. 8).

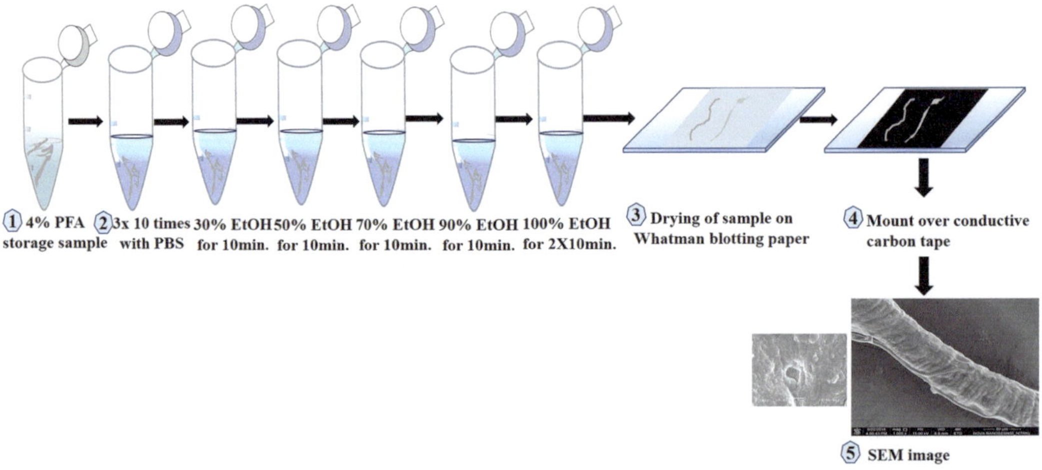

Fig. 7 SEM analysis of larval Gut

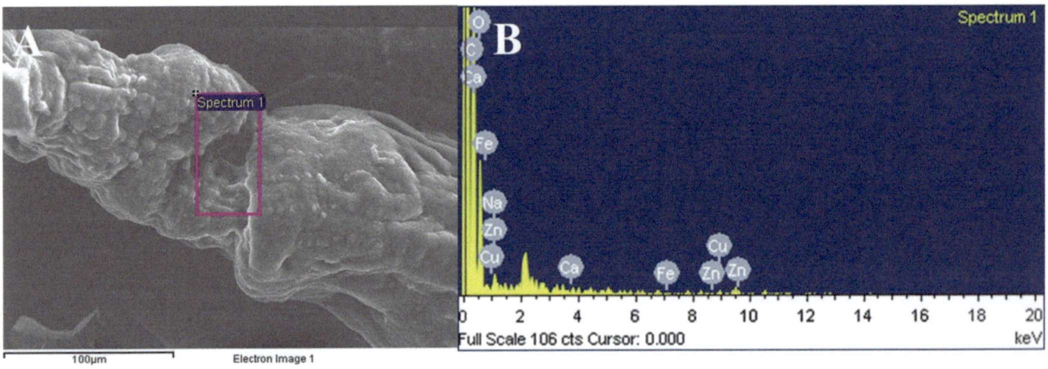

Fig. 8 EDS analysis of larval gut. (**a**) Area of gut under EDS analysis. (**b**) Elemental map of EDS analysis

4 Notes

1. In the dehydration process, ethanol concentration should be maintained.
2. Avoid exposing of propylene oxide to the eye, because propylene oxide causes severe eye irritation and is toxic to the cornea of the eye.
3. Make the smear uniform and thin.
4. While mounting the gut, mount it under the stereomicroscope so that the hole of the gut will be upward.

Acknowledgements

JB is thankful to BT/PR21857/NNT/28/1238/2017 for financial support. MM lab is supported by Grant No. BT/PR21857/

NNT/28/1238/2017, EMR/2017/003054, Odisha DBT 3325/ST(BIO)-02/2017.

References

1. Bozzola JJ, Russell LD (1999) Electron microscopy: principles and techniques for biologists. Jones & Bartlett Learning, Sudbury
2. Egerton RF (2011) Electron energy-loss spectroscopy in the electron microscope. Springer Science & Business Media, New York
3. Williams DB, Carter CB (1996) The transmission electron microscope. In: Transmission electron microscopy. Springer, pp 3–17
4. Wollman DA, Irwin KD, Hilton GC, Dulcie L, Newbury DE, Martinis JM (1997) High-resolution, energy-dispersive microcalorimeter spectrometer for X-ray microanalysis. J Microsc 188(3):196–223
5. Kutchko BG, Kim AG (2006) Fly ash characterization by SEM–EDS. Fuel 85 (17–18):2537–2544
6. Goldstein JI, Newbury DE, Michael JR, Ritchie NW, Scott JHJ, Joy DC (2017) Scanning electron microscopy and X-ray microanalysis. Springer, New York
7. Choudhary OP, Malik P (2017) Scanning electron microscope: advantages and disadvantages in imaging components. Int J Curr Microbiol Appl Sci 6:1877. https://doi.org/10.20546/ijcmas.2017.605.207
8. McMullan D (1995) Scanning electron microscopy 1928–1965. Scanning 17(3):175–185
9. Reed SJB (2005) Electron microprobe analysis and scanning electron microscopy in geology. Cambridge university press, Cambridge
10. Henchcliffe C, García-Alonso L, Tang J, Goodman CS (1993) Genetic analysis of laminin A reveals diverse functions during morphogenesis in Drosophila. Development 118(2):325–337
11. Alone DP, Tiwari AK, Mandal L, Li M, Mechler BM, Roy JK (2004) Rab11 is required during Drosophila eye development. Int J Dev Biol 49(7):873–879
12. Geisbrecht ER, Haralalka S, Swanson SK, Florens L, Washburn MP, Abmayr SM (2008) Drosophila ELMO/CED-12 interacts with Myoblast city to direct myoblast fusion and ommatidial organization. Dev Biol 314 (1):137–149
13. Tokunaga C, Gerhart JC (1976) The effect of growth and joint formation on bristle pattern in D. melanogaster. J Exp Zool 198(1):79–95
14. Lees AD, Waddington CH (1942) The development of the bristles in normal and some mutant types of Drosophila melanogaster. Proc R Soc Lond Ser B-Biol Sci 131 (862):87–110
15. Yasuzumi G, Deguchi N (1958) Submicroscopic structure of the compound eye as revealed by electron microscopy. J Ultrastruct Res 1(3):259–270
16. Mishra M, Knust E (2012) Analysis of the Drosophila compound eye with light and electron microscopy. In: Retinal degeneration. Springer, pp 161–182
17. Cribbs D, Benassayag C, Randazzo F, Kaufman T (1995) Levels of homeotic protein function can determine developmental identity: evidence from low level expression of the Drosophila homeotic gene proboscipedia under Hsp70 control. EMBO J 14(4):767–778
18. Benassayag C, Plaza S, Callaerts P, Clements J, Romeo Y, Gehring WJ, Cribbs DL (2003) Evidence for a direct functional antagonism of the selector genes proboscipedia and eyeless in Drosophila head development. Development 130(3):575–586
19. Szabad J, Bellen HJ, Venken KJ (2012) An assay to detect in vivo Y chromosome loss in Drosophila wing disc cells. G3 2(9):1095–1102
20. Johnson SA, Milner MJ (1987) The final stages of wing development in Drosophila melanogaster. Tissue Cell 19(4):505–513
21. Foelix R, Stocker R, Steinbrecht R (1989) Fine structure of a sensory organ in the arista of Drosophila melanogaster and some other dipterans. Cell Tissue Res 258(2):277–287
22. Lanot R, Zachary D, Holder F, Meister M (2001) Postembryonic hematopoiesis in Drosophila. Dev Biol 230(2):243–257
23. Stofanko M, Kwon SY, Badenhorst P (2010) Lineage tracing of lamellocytes demonstrates Drosophila macrophage plasticity. PLoS One 5(11):e14051
24. Pappus SA, Ekka B, Sahu S, Sabat D, Dash P, Mishra M (2017) A toxicity assessment of hydroxyapatite nanoparticles on development and behaviour of Drosophila melanogaster. J Nanopart Res 19(4):136
25. Silva J, Boleli I, Simões Z (2002) Hemocyte types and total and differential counts in unparasitized and parasitized Anastrepha obliqua (Diptera, Tephritidae) larvae. Braz J Biol 62 (4A):689–699

Chapter 17

Fourier Transform Infrared Spectroscopy: A Novel Approach for Biomolecular Characterization of *Drosophila* Hemolymph

Sanjeev Paikra and Monalisa Mishra

Abstract

Fourier transform infrared (FTIR) spectroscopy is the technique which allows the analysis of a biological sample or biofluid to characterize the functional group. It reveals the molecular information without disturbing the chemical nature of the biological sample. It doesn't require any labeling. We have used vibrational spectroscopic techniques to analyse the functional group of hemolymph. We used various treatments in flies to see the alteration of functional group in the hemolymph. Flies were fed with bacteria-contaminated food and nanoparticle (alumina NP and lignin NP) mixed food to check the alteration of the biochemical composition of hemolymph. Parameters like the composition of lipid, protein, nucleic acid and DNA undergo changes with respect to chemicals. The current protocol describes the analysis of carbohydrate, lipid, protein, nucleic acid and DNA damage from the hemolymph using FTIR.

Key words *Drosophila*, FTIR, Hemolymph, Nanoparticles, Bacteria infection

1 Introduction

Different organic as well as inorganic elements have an essential role in the regulation of the physiological process of various animals including *Drosophila*. Biomolecules are responsible for physiological process (cell division or development) also composed of organic elements like carbon, hydrogen, oxygen, nitrogen and phosphorus. These elements form a characteristic functional group in the biofluid. *Drosophila melanogaster* is widely accepted as a model organism to study different diseases like Alzheimer's disease [1], Parkinson's disease [2], Huntington's disease [3], heart disease [4], etc. Alteration of various biochemical parameters in the different diseased state is due to abnormality or aggregation of different functional proteins or biomolecules present within the organism. With the advancement of spectroscopic technique, Fourier

Monalisa Mishra (ed.), *Fundamental Approaches to Screen Abnormalities in Drosophila*, Springer Protocols Handbooks, https://doi.org/10.1007/978-1-4939-9756-5_17,

transform infrared spectroscopy (FTIR) is used to analyse the abnormalities or physiological changes in cell or tissue [5]. In this technique, vibrational changes in the bond within the different functional groups of the molecules can be determined. In FTIR, the chemical bond undergoes various vibrational changes such as twisting, stretching and rotation. This is an easy method than biochemical estimation since it doesn't need any reagent and the equipment is easy to handle [6]. Each cell and tissue has its own molecular fingerprint, which can be distinguished in FTIR spectra [7]. Thus, the control state can be distinguished from the diseased state using the FTIR technique. Large molecules in the cells have very complex spectra which is difficult to interpret. Common bonds like C≡N, N-H, P=O, S-H, C=O and C-H bond are the characteristic features of carbohydrates, nucleic acid, proteins and lipid [8, 9]. For spectral analysis of the biological sample, fingerprint region (600–1450 cm^{-1}) which corresponds to the overlapping contribution of a number of biomolecules present in the sample, spectral range 3050–2800 cm^{-1}, is mainly attributed to lipid/hydrocarbon chain, and amide I/II region (1500–1700 cm^{-1}) can be identified [10]. Stretching vibration in O-H, S-H, C-H and N-H bond can be characterized at region 2550–3500 cm^{-1} [10]. The categorization of this band position will be helpful in disease diagnosis and other clinical investigations [11]. Information obtained from FTIR makes it unique and complementary to other tools to analyse the hemolymph including microscopy and other biochemical estimations.

Besides hemolymph FTIR was also used to analyse various states of cells and tissues. Attenuated total reflectance FTIR spectroscopy is employed to confirm protein aggregation [12]. All protein or protein aggregates have characteristic β-pleated structure with specific frequency band position. This method is used to study the secondary structure of protein and polypeptide [13]. Vibrational spectroscopy techniques such as circular dichroism and FTIR have upper hand in studying the protein dynamics. Protein aggregation is associated with the change in protein native confirmation which could be observed by the shift in peak position [12] in IR spectra. IR excite the vibrational mode of biomolecules of interest because of net dipole charge [14]. FTIR microspectroscopy technique was employed to analyse the human biopsy tissue [15]. The change in the spectra is due to the variation of biochemical components like carbohydrate, glycogen, DNA, RNA, lipid and protein amount. In this chapter, we have described the FTIR spectroscopy in attenuated total reflection, to investigate the hemolymph under various physiological conditions. To create various physiological conditions, we fed *Drosophila* with different nanoparticles and bacteria-contaminated food. The protocol describes the measurement of various biochemical parameters from *Drosophila* hemolymph using FTIR.

2 Materials

2.1 Materials Required for Food Preparation

1. Food media.
2. Sucrose (HiMedia GRM3063).
3. Corn meal (from the local market).
4. Yeast extract (subala yeast powder).
5. Type I agar (HiMedia GRM666).
6. Methyl paraben (HiMedia GRM1899): Prepare the working solution by adding 0.25 g methyl paraben to 100 ml ethanol.
7. Propionic acid (HiMedia GRM3658).
8. Cotton.
9. Autoclave bag.
10. Autoclave.
11. Weighing balance (Aczet).
12. Glass bottle (Schott).
13. Glass vials (Riviera).

2.2 Addition of Nanoparticles to the Standard Food

2.2.1 Nanoparticle Stock Solution

1. 15 ml tube (Tarson)
2. Synthesized nanoparticle of interest
3. Probe sonicator
4. MilliQ water
5. Weighing balance (Aczet)

2.3 Addition of Bacteria to the Standard Fly Food

2.3.1 Bacterial Culture

1. Luria-Bertani (LB) broth (HiMedia- M1245)
2. Test tubes (Riviera)
3. Micropipette (Tarson)
4. Tips (Tarson)
5. MilliQ water
6. Cotton
7. Shaker incubator (Remi)
8. Stock culture of bacteria

2.4 Material Required for Hemolymph Collection

1. Petri plate (90 mm diameter × 15 mm width)
2. 1X PBS
3. 0.5 and 1.5 ml Eppendorf tube
4. MilliQ water
5. Fine needle/pins (Soccerene Premium Quality pin size 1, 40 mm long × 0.35 mm diameter)
6. Seventy percent ethanol

7. Ice-box
8. Centrifuge (ThermoFisher Scientific, Heraeus Megafuge 16R)
9. −80 °C freezer

2.5 Requirement for FTIR Analysis

2.5.1 Instrument

Attenuated total reflectance—Fourier transform infrared spectroscopy (Bruker-α, Platinum-ATR).

2.5.2 Chemical Required

1. 1X PBS
2. MilliQ water

3 Method

3.1 Preparation of Standard Fly Food

1. To prepare the standard fly food (100 ml), weigh 4 g sucrose, 2.5 g yeast extract, 5 g corn meal and 0.8 g agar (type I) in a weighing balance.
2. Take 100 ml of milliQ water in a glass bottle, and add all the components to it, and stir well with a glass rod.
3. Take ten clean glass vials and put a cotton plug at the mouth of each vial.
4. Keep the vials along with the food bottle in an autoclave bag, and keep them for autoclave for 45 min.
5. Take out the vials and the food bottle after 45 min from the autoclave, and allow it to cool.
6. Take them inside a laminar airflow, and add propionic acid and nipagine to the food, and mix it thoroughly.
7. Pour 10 ml of food in each vial from the food bottle and allow it to solidify.
8. Once the food is solidified, it can be used for the fly culture.

3.2 Addition of Nanoparticles to the Standard Food

3.2.1 Preparation of Nanoparticle Stock Solution

1. To prepare the stock solution (1 mg/ml), disperse the nanoparticle in milliQ water.
2. Sonicate the solution two times for 15 min each using a probe sonicator to disperse the nanoparticles properly.

3.2.2 Preparation of Fly Food and Addition of Nanoparticles

1. Prepare the standard fly food as per the protocol described in Sect. 3.1.
2. Allow the food to solidify.

3. Vortex the nanoparticle stock solution for 10 min.
4. Add the desired volume of the nanoparticle to 10 ml of fly food to obtain the desired concentrations.
5. Allow the food to dry at room temperature and transfer the flies to it.

3.3 Addition of Bacteria to the Standard Fly Food

3.3.1 Preparation of Bacterial Culture

1. Inoculate 100 μl of any bacterial culture in 10 ml of LB broth. Keep the broth at 37 °C and shake at 120 rpm for overnight. Grow the bacterial cultures up to an exponential phase, so that the bacteria are in rapidly replicating stage.
2. Subculture the overnight grown bacterial culture (100 μl) in a fresh LB broth (10 ml).
3. Ensure the bacterial growth by determining the OD at 600 nm. Take the OD in every 30 min till the OD reaches 0.6–0.8.

3.3.2 Infection of Flies

1. Starve the flies for 2–4 h before transferring them to the food vial to ensure oral infection.
2. Prepare the standard fly food as described in Sect. 3.1.
3. Once the food is prepared, take it inside laminar airflow.
4. Allow the food to cool down for sometime, and then add propionic acid and nipagine to the food.
5. Pour the food to the vials and allow it to solidify.
6. Prepare the infection vials by adding different volumes of bacterial culture directly to the standard fly food, and allow it to absorb in the food media.
7. Once the food is dry, transfer the male and female flies to the food vial, and keep it at 25 °C for 12 h in light and dark conditions.

3.4 Hemolymph Collection

1. Take out 30–35 nos. of adult flies from the rearing media. Dry them with tissue paper and transfer them to a 0.5 mL tube.
2. Make a small cut at the bottom of the 0.5 mL tube with a fine razor blade.
3. Put 0.5 mL tube into a 1.5 mL centrifuge tube and transfer it into the ice-box.
4. Prick the adult flies thorax with a fine pin carefully. Use 70% of ethanol to clean the pins (*see* **Notes 2 and 3**).
5. After that, centrifuge at 10,000 rpm for 5 min at 4 °C.
6. Collect the hemolymph and store it in −80 °C freezer for further experimental use.
7. To prevent melanization do the hemolymph collection on ice box (*see* **Note 2**).

3.5 FTIR Spectroscopy Method to Analyse the Hemolymph

1. Use the isolated hemolymph for the analysis of the functional group of biomolecules. The sample preparation technique depends on the type of sample and chemical nature of the biological sample (*see* **Notes 4 and 5**).
2. Use the collected hemolymph (as the method discussed above Sect. 17.3.4.3) for FTIR analysis. If it was frozen at −80 °C, then take it out from the −80 °C, and thaw carefully before use.
3. After thawing remove the debris by using 1X PBS buffer.
4. Dilute the debris-free hemolymph (40 μl) with 20 μl of PBS buffer (*see* **Note 6**). This will be used for FTIR analysis.
5. To take the FTIR spectra of hemolymph, turn on the FTIR instrument (ATR-FTIR, Bruker-α, Platinum-ATR), and wait till the equipment is ready for the measurement (*see* **Note 7**).
6. Clean the ATR-IRE plate with the distilled water and dried with tissue paper.
7. Take the background spectrum before the analysis of our sample of interest.
8. Put 40 μl of diluted hemolymph on the ATR plate by using micropipette (*see* **Note 8**).
9. Keep the opus software open to analyse the spectra within the spectral range of 400–4000 cm^{-1}. Do the baseline correction and save the data for further analysis.
10. The attenuated total reflectance FTIR could be further plotted in the origin software (origin pro 2016) and analysed for different peak position to detect the functional groups associated with carbohydrate, protein, lipid and nucleic acid contents (analysis of the data was explained in Sect. 3).
11. The complete experiment from sample preparation to the analysis of data will take about 1 h.
12. Various factors can affect the quality of the spectra. Thus, care should be taken before the sample analysis.

3.5.1 Operation of Origin Software for Analysis of Data

1. Install the latest version of origin software.
2. Import the data derived from ATR-FTIR instrument (either in text format or ASC format).
3. Select the data and plot it by using the line graph pattern.
4. Do the smoothening of the curve by choosing the analysis option then signal processing which is followed by smoothening of the curve by opening the dialogue box. Set the parameter by choosing the Savitzky-Golay method to smooth and normalize the curve.

3.5.2 Precaution to Enhance the Spectral Quality

1. Adjust the signal to noise ratio during the operation to get a good quality spectrum.
2. Avoid water vapour in the instrument, since it reduces the transmission of IR light and it obscures the spectral details.
3. Subtract the vapour component from the spectrum by using the computational approach to prevent the interference of light [16].
4. After data acquisition, process the data by denoising, normalization of peak and spectral correction. For denoising use the Savitzky-Golay smoothening curve with the help of origin pro 2016 software.

4 Analysis of FTIR Spectra

4.1 Analysis of Whole Spectra

1. The result obtained from FTIR spectroscopy of hemolymph possesses various functional groups (Fig. 1).
2. The spectra show a small stretch at region 3100–2800 cm^{-1}. This region is attributed to the lipid functional group which is specifically identified as the C=C stretching at 3060 cm^{-1}, CH_3 asymmetric stretch at 2960 cm^{-1}, CH_3 symmetric stretch at 2876 cm^{-1} and contribution of ester C=O stretch at 1740 cm^{-1}.

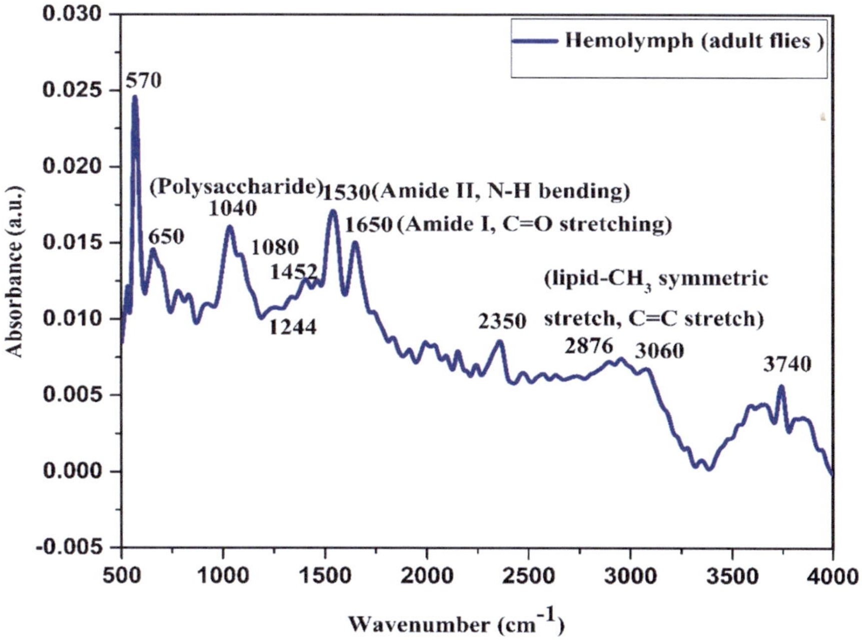

Fig. 1 The spectra of hemolymph derived from control flies representing various peaks for protein, lipid, sugar and nucleic acid

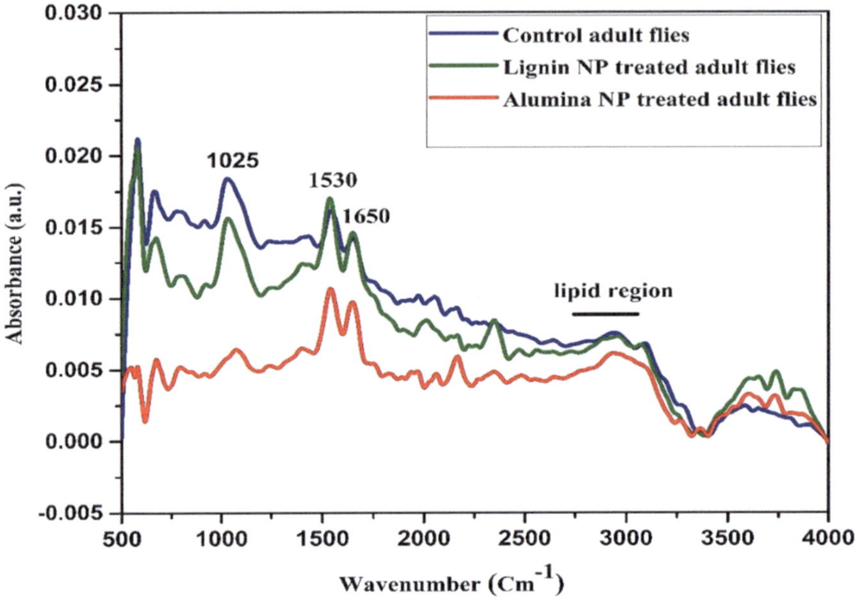

Fig. 2 Comparison between FTIR spectra of hemolymph derived from control flies and nanoparticle-treated flies

3. The band at 1452 cm^{-1} and 1244 cm^{-1} are assigned to CH_2 bending and PO_2^- asymmetric stretching, respectively.
4. Major functional group relevant to protein is observed due to the presence of amide I and amide II group. Amide I (C=O stretching) at 1650 cm^{-1} and amide II (N-H bending) at 1530 cm^{-1} are the position of the characteristic of the band corresponding to protein secondary structure [17].
5. Nucleic acids are the small biomolecules which are characterized by PO^{2-} symmetric stretching at 1080 cm^{-1}.
6. Sugar present in the hemolymph was identified by the C-O stretching of polysaccharide at 1040 cm^{-1}.
7. The spectra of NP-treated flies have shown the variation in comparison to the control (Fig. 2). The bacteria-treated flies have also shown the change in spectra of the hemolymph (Fig. 3).
8. For accuracy, the spectra were analysed separately for protein, nucleic acid and lipid components by normalizing the respective spectral region. Major components present in the hemolymph are described separately below.

4.2 Protein Structure in Hemolymph

1. The secondary structure of a protein can be identified by amide I and amide II peaks in the spectra. The vibration of a peptide group in the hemolymph is responsible for amide band in spectra. Characteristic band position for the amide groups is shown in Table 1.

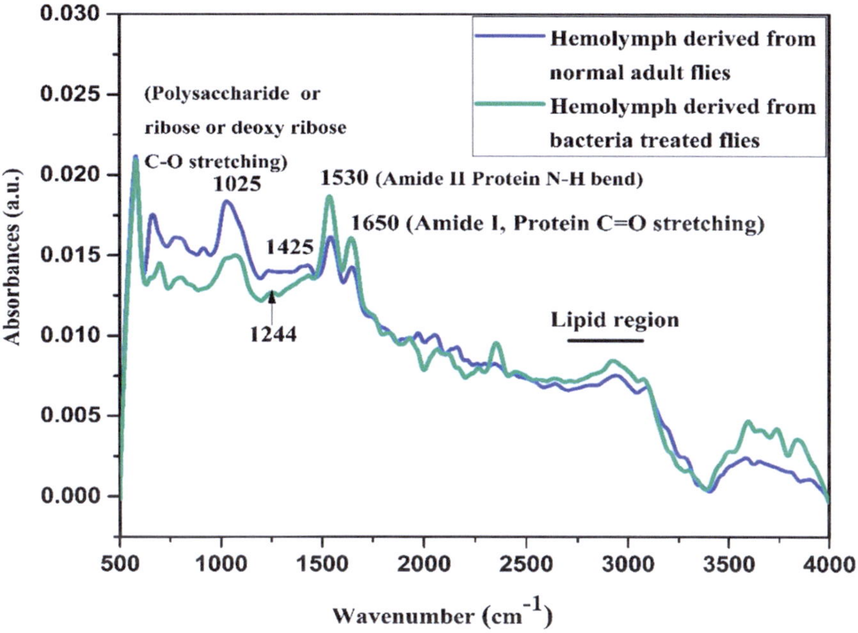

Fig. 3 Comparison between FTIR spectra of hemolymph derived from control and bacteria-treated flies

Table 1
Characteristic band position to study the protein structure [14, 17, 18]

Amide group	Approximate wavenumber (Cm^{-1})	Description
Amide I	1600–1690	C=O stretching
Amide II	1480–1575	CN stretching, NH bending
Amide III	1229–1301	CN stretching, NH bending
Amide IV	625–767	OCN bending
Amide V	640–800	Out of plane NH bending
Amide VI	537–606	Out of plane C=O bending
Amide VII	200	Skeletal torsion
Amide A	3300	NH stretching
Amide B	3100	NH stretching

2. Amide I absorption is observed in the region 1600–1700 cm^{-1}. Spectral region 1620–1640 cm^{-1} in the amide band of the protein is specifically attributed to the beta sheet. The antiparallel beta sheet in the protein could be characterized by the band at 1670–1695 cm^{-1}.
3. The spectra obtained (with the normalization of peak) depict the presence of beta sheet, antiparallel beta sheet and alpha

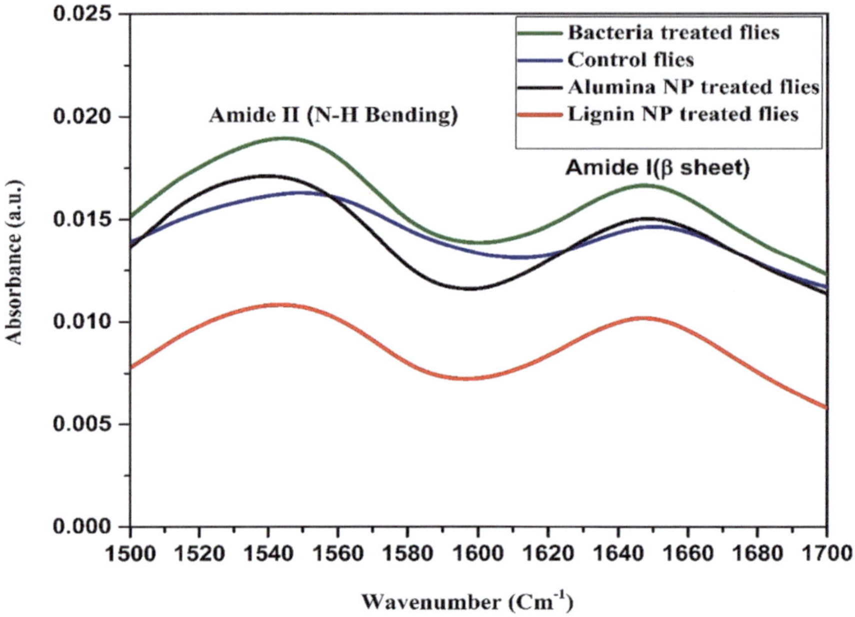

Fig. 4 FTIR spectra of hemolymph to study the secondary structure of the protein. The hemolymph was derived from control flies, bacteria-treated flies and nanoparticle-treated flies

helical structure. Second derivative spectra were obtained with the deconvolution, which confirms the secondary structure of the protein (Fig. 4).

4. The hemolymph of control flies has shown the sharp band for amide I and II, but in bacteria-treated flies, the peak intensity is higher for amide I and II.
5. The hemolymph derived from alumina oxide NP-treated flies has shown the small shift in peak position for amide I and amide II band in comparison to control flies, but in case of lignin nanoparticle-treated flies, the hemolymph has shown the sharp change in peak intensity.

4.3 Detection of DNA Damage

1. The DNA damage is associated with the change in the nucleic acid composition within the total hemolymph. FTIR spectroscopy can be helpful in analysing the DNA damage in cells [19]. Characteristic band position to study the nucleic acid is shown in table (Table 2).
2. DNA damage alters the DNA backbone structure resulting the alteration of the spectra (950–1240 cm^{-1}). The spectral change is due to molecular modification of sugar, base and redistribution of H-bonding network. This is further associated with single-stranded or double-stranded DNA break and sometime DNA-protein crosslink break. The difference in the intensity of peak may be associated with the DNA damage.

Table 2
Characteristic band position to study the nucleic acid [20–22]

Wave number (cm^{-1})	Vibration mode	Assignment
970	Ribose phosphate	DNA
1038	Ribose C-O stretching	RNA
1086	PO_2 symmetric stretching	DNA
1105	P-O-C symmetric stretching	DNA
1144	Ribose C-O stretching	RNA
1175	Sugar phosphate	DNA
1212	C-H bending	RNA
1230	O-P-O asymmetric stretching	DNA
1680	C=O stretching	RNA

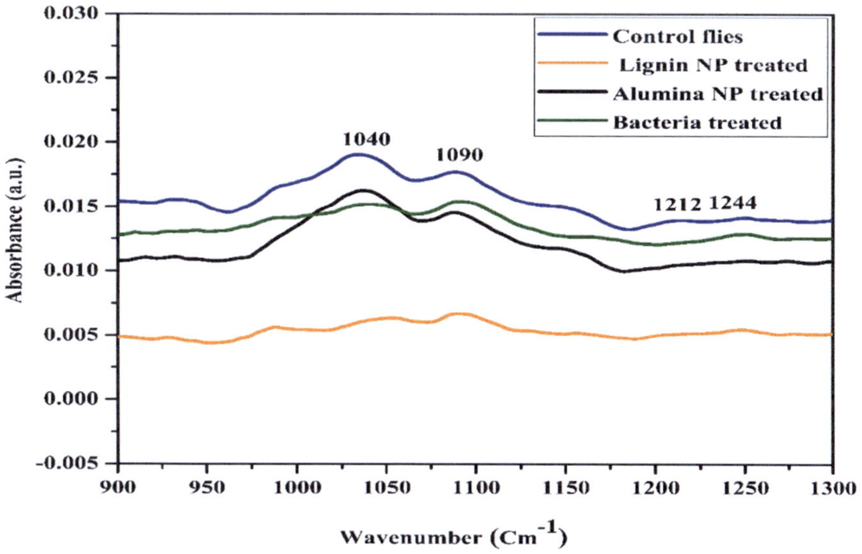

Fig. 5 FTIR spectra representing the functional group associated with nucleic acid and change in spectra due to DNA damage in the hemolymph derived from control flies, NP-treated flies and bacteria-treated flies

3. The spectra obtained from various experiments were plotted to get a significant comparison between hemolymph of control and treated flies (Fig. 5). Modification of the spectra is due to molecular changes in the vibration of C=O, N-H, C=N and C=C bond.
4. The most prominent change was observed in ribose C-O stretching of RNA at region 1040 cm^{-1}. The change in peak

intensity at 1090 cm^{-1} is observed, which is ascribed to PO_2 symmetric stretching of DNA. Some changes are also observed at the region of 1212 cm^{-1} and 1244 cm^{-1}, which is attributed to the C-H ring bending of RNA and O-P-O asymmetric stretching of DNA.

5. Overall, we can conclude that the extent of DNA damage is quite high in lignin nanoparticle-treated flies in comparison to alumina oxide NP-treated flies.

4.4 Detection of Change in Lipid Profile

1. Various band positions to analyse the lipid structure is shown in Table 3.
2. Free lipid is present in the spectra at region 2927 cm^{-1} and 3015 cm^{-1} which is attributed to the CH_2 asymmetric stretch and C=C unsaturated fatty acids. Unsaturation of the lipid component is evidenced by the signal at 3015 cm^{-1}.
3. This spectrum suggested that in case of bacteria-treated flies, degree of unsaturation is high for lipid component, but in case of NP-treated flies, degree of unsaturation is low, which could be observed in the spectra (Fig. 6).

Table 3
Characteristic band position to study the lipid structure [19, 23, 24]

Wave number (cm^{-1})	Vibration mode	Assignment
2954	$-CH_3$	Saturated fatty acid
1465–1467	$-CH_2$	Fatty acid
1378–1381	$-CH_3$	Fatty acid
1365	$-CH_2$	Fatty acid
1243	SO_3	Sulfatides
1225–1234	PO_2	Phospholipid
1174–1179	C-O	Phospholipid
1084–1090	Symmetric vibration PO2	Phospholipid
1025	C-O-P	Phospholipid
970/971	$N^+-(CH_3)_3$	Phosphatidylcholine, sphingomyelin
822/823	Asymmetric vibration P-O	Phospholipid

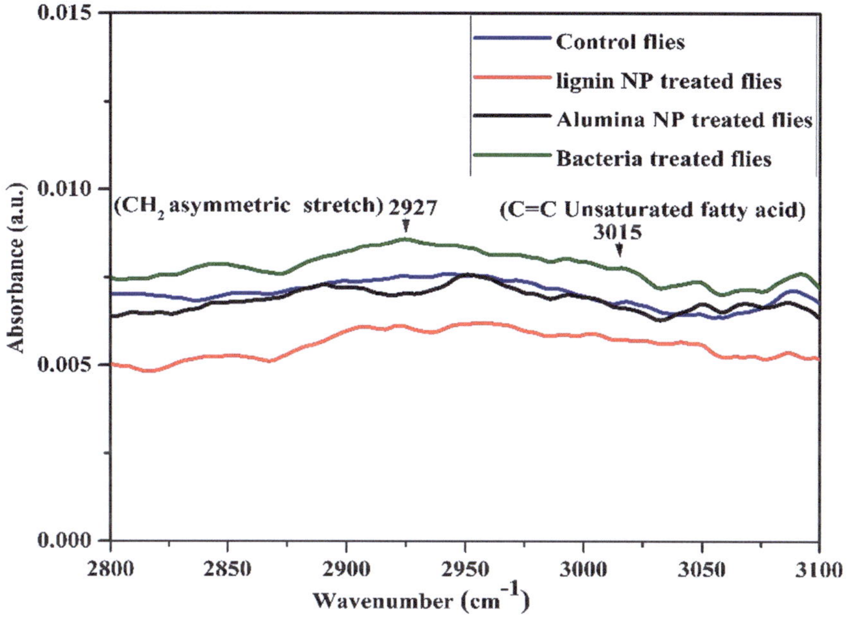

Fig. 6 FTIR spectra showing the functional groups associated with the lipid component of hemolymph derived from control flies, NP-treated flies and bacteria-treated flies

5 Notes

1. Hemolymph collection should be performed on the ice-box.
2. If you are having melanization, use CO_2 to prevent the melanization.
3. If there is an issue of aggregation, then dilute the hemolymph with anticoagulant buffer (41 mM citric acid, 98 mM NaOH, 186 mM NaCl, 1.7 mM EDTA, pH 4.5).
4. BaF_2, CaF_2-coated slides and IR reflective slides are sometimes used for the analysis of biological samples.
5. Analysis of live cells incorporates the use of synchrotron and globar light source in FTIR analysis, which enhances the spatial resolution and quality of the data.
6. Dilute the hemolymph immediately after thawing, and take the FTIR measurement within 1 or 2 min to prevent the melanization of hemolymph. Keep the instrument on before thawing your sample to avoid hurry.
7. If the instrument is switched-off, we need to make sure to check the interferogram signals for the actual location and amplitude. The system needs to be realigned, if it has been moved or components are changed.
8. Contact of biofluid with the crystal is a crucial parameter for ATR-FTIR analysis. When we are using an aperture ATR-FTIR, 1 μl of the sample has to be dried within 8 min.

Acknowledgements

SP is thankful to MHRD for the financial support. MM lab is supported by Grant No. BT/PR21857/NNT/28/1238/2017, EMR/2017/003054, Odisha DBT 3325/ST (BIO)-02/2017.

References

1. Sivanantharajah L, Mudher A, Shepherd D (2019) An evaluation of Drosophila as a model system for studying tauopathies such as Alzheimer's disease. J Neurosci Methods 319:77–88
2. Shishido T et al (2019) Synphilin-1 has neuroprotective effects on MPP+-induced Parkinson's disease model cells by inhibiting ROS production and apoptosis. Neurosci Lett 690:145–150
3. Bason M et al (2019) Astrocytic expression of the chaperone DNAJB6 results in non-cell autonomous protection in Huntington's disease. Neurobiol Dis 124:108–117
4. Saada A (2019) Sea squirt alternative oxidase bypasses fatal mitochondrial heart disease. EMBO Mol Med 11(1):e9962
5. Zelig U et al (2009) Diagnosis of cell death by means of infrared spectroscopy. Biophys J 97 (7):2107–2114
6. Diem M, Griffiths PR, Chalmers JM (2008) Vibrational spectroscopy for medical diagnosis, vol 40. Wiley, Chichester
7. Boydston-White S et al (1999) Infrared spectroscopy of human tissue. V. Infrared spectroscopic studies of myeloid leukemia (ML-1) cells at different phases of the cell cycle. Biospectroscopy 5(4):219–227
8. Le Gal J-M, Morjani H, Manfait M (1993) Ultrastructural appraisal of the multidrug resistance in K562 and LR73 cell lines from Fourier transform infrared spectroscopy. Cancer Res 53 (16):3681–3686
9. Jackson M, Mantsch HH (1996) Biomedical infrared spectroscopy. In: Infrared spectroscopy of biomolecules. Wiley, New York, pp 311–340
10. Walsh MJ et al (2008) FTIR microspectroscopy coupled with two-class discrimination segregates markers responsible for inter-and intra-category variance in exfoliative cervical cytology. Biomark Insights 3:BMI. S592
11. Bellisola G, Sorio C (2012) Infrared spectroscopy and microscopy in cancer research and diagnosis. Am J Cancer Res 2(1):1
12. Shivu B et al (2013) Distinct β-sheet structure in protein aggregates determined by ATR–FTIR spectroscopy. Biochemistry 52(31):5176–5183
13. Surewicz WK, Mantsch HH (1988) New insight into protein secondary structure from resolution-enhanced infrared spectra. Biochim Biophys Acta (BBA)-Protein Struct Mol Enzymol 952:115–130
14. Barth A (2007) Infrared spectroscopy of proteins. Biochim Biophys Acta (BBA)-Bioenerg 1767(9):1073–1101
15. Ami D et al (2016) In situ characterization of protein aggregates in human tissues affected by light chain amyloidosis: a FTIR microspectroscopy study. Sci Rep 6:29096
16. Bruun SW et al (2006) Correcting attenuated total reflection–Fourier transform infrared spectra for water vapor and carbon dioxide. Appl Spectrosc 60(9):1029–1039
17. Kong J, Yu S (2007) Fourier transform infrared spectroscopic analysis of protein secondary structures. Acta Biochim Biophys Sin 39 (8):549–559
18. Haris PI, Severcan F (1999) FTIR spectroscopic characterization of protein structure in aqueous and non-aqueous media. J Mol Catal B Enzym 7(1–4):207–221
19. Lipiec E et al (2012) FTIR microspectroscopy in studies of DNA damage induced by proton microbeam in single PC-3 cells. Acta Phys Pol A 121:506–509
20. Bassan P et al (2009) Reflection contributions to the dispersion artefact in FTIR spectra of single biological cells. Analyst 134 (6):1171–1175
21. Banyay M, Sarkar M, Gräslund A (2003) A library of IR bands of nucleic acids in solution. Biophys Chem 104(2):477–488
22. Dovbeshko GI et al (2000) FTIR spectroscopy studies of nucleic acid damage. Talanta 53 (1):233–246
23. Errico S et al (2019) A novel experimental approach for liver analysis in rats exposed to Bisphenol A by means of LC-mass spectrometry and infrared spectroscopy. J Pharm Biomed Anal 165:207–212
24. Dreissig I et al (2009) Quantification of brain lipids by FTIR spectroscopy and partial least squares regression. Spectrochim Acta A Mol Biomol Spectrosc 71(5):2069–2075

Chapter 18

Various Behavioural Assays to Detect the Neuronal Abnormality in Flies

Gyanaseni Dhar, Sumit Mukherjee, Nibedita Nayak, Swetapadma Sahu, Janmejaya Bag, Ronnaly Rout, and Monalisa Mishra

Abstract

The survivability of a motile organism depends on the ability to respond to various stimuli. *Drosophila* possesses a robust behavioural response towards the external stimuli in various developmental stages. Those behaviours are used to screen the abnormalities in larvae and adults. In the larval stage, third inster larvae are used to screen the behavioural defects. Behaviours like crawling, touch sensitivity, temperature sensitivity, light sensitivity, sound avoidance and self-right behaviours are checked in larvae. Whereas, in adults, climbing, phototaxis, temperature sensitivity, chemotaxis, sound avoidance and choice assay are commonly investigated. Each behavioural assay has a score to detect the abnormality. Accordingly, the defect in various organs at different developmental stages can be calculated. The current chapter describes various behavioural assays in larvae and adult to detect the physiological abnormality in flies.

Key words Larvae crawling, Self-right test, Temperature sensitivity, Chemotaxis, Climbing behaviour, Phototaxis, Touch sensitivity

1 Introduction

Like every living organism *Drosophila* depicts a well-developed behavioural response to the internal as well as external factors. These behavioural defects can be scored at various developmental stages. Third inster larvae [1] and adult [2] flies are used to score various behavioural defects. As a consequence, various methods are developed to score the physiological defects in larvae and adult flies using behavioural assays.

1.1 Larva Crawling Assay

Larvae crawling is a coordinated action of various neurons to detect neuronal defects [3]. It is a simple assay to understand the rhythmic behaviour and thus employed in various fields such as neurobiology, evolution and medicine [1, 4]. The larval locomotion includes a number of movements, such as turns and twists, linear crawling, burrowing and other movements [5, 6]. This is a relatively complex

Monalisa Mishra (ed.), *Fundamental Approaches to Screen Abnormalities in Drosophila*, Springer Protocols Handbooks, https://doi.org/10.1007/978-1-4939-9756-5_18,

behaviour which involves, muscle, neurons and neuromuscular junction. The crawling pattern reflects the cellular basis of the larvae [7, 8].

1.2 Touch Sensitivity Test

Touch is a coordinated action of many organs, like the central nervous system, body segments and neuromuscular junctions. The stimuli comes from the central pattern generators (CPG) present within the brain [9]. The oscillatory network produces movement even in the absence of the external sensory input. However, in the absence of a feedback loop from the peripheral nervous system (PNS), the expansion and contraction of the body segments become uncoordinated [10, 11]. The peristaltic signal from the CPG gets activated at the late embryonic stage and continues during the larval stage. The chordotonal organ, which is the primary sense organ of the PNS, gets a signal from the CPG for sense and locomotion [12]. Thus, if there is any sensory defect, then the response of larvae towards the stimuli gets hindered. From this assay, the larval behaviour can be analysed, and the neuronal defect can be scored.

1.3 Self-Right Test

Drosophila larva orients and responds towards chemical stimuli [13], light intensity and electrical fields [14]. The larval brain collects information from various segments of the body through neurons which traverse the ventral nerve cord [15]. A part of the brain possesses a premotor subesophageal zone, which contains a small set of neurons to regulate the reorientation of the larva [16]. However, recently, a biomechanical model of the larva has been explored to study the underlying mechanism of larva crawling and bending without the direct involvement of neurons [17]. Self-righting (SR) behaviour depicts the complex movement used by the larva to correct its position when it turns from dorsal to the ventral side. This movement involves the coordinated contraction of consecutive body segments of the larva and is characterized by a series of events. Those steps include (1) head turning, (2) grabbing the substrate and (3) rolling down to the normal orientation resulting in the dorsal side upward direction. Although these steps are seen in wild-type larvae, in some mutants, the slow and abnormal response in self-righting behaviour is seen. The abnormal response includes (1) delay to rollover the body to the normal dorsal position, (2) interrupted movement in between or (3) backward peristaltic waves. All these steps are under the genetic control of Hox gene *Ultrabithorax* (*Ubx*), which is regulated at the post-transcriptional level [18, 19]. The *Ubx* largely targets some specific cholinergic, sensory, interneurons and motor neurons to control the early larval locomotory behaviour [20]. More than 40% of miRNA mutants of *Drosophila* shows defective SR behaviour including miR-iab4/8 [21]. Any abnormality in the SR behaviour suggests specific alteration at the molecular and cellular level.

1.4 Thermal Avoidance and Temperature Sensitivity Test

Temperature sensitivity can be divided into two categories: the ability to detect an abnormally high temperature that might cause tissue damage (the response is known as nociception) and the recognition of innocuous temperatures [22]. *Drosophila* is an ectothermic organism which relies on behaviour to maintain body temperature within a physiologically tolerant range [23]. It possesses receptors for the detection of the hot and cold temperature. A transient receptor protein (TRP) channel and an ionotropic Ca^{2+} ion channel are responsible for the high-temperature sensing in larvae, as well as in adults [24]. Moreover, the high-temperature response is found to be dependent partially on another TRPA channel, pyrexia [25]. Besides the channel, the thermal temperature signalling is also present within the central nervous system [26]. Temperature $\geq$35 °C is painful for both the central and peripheral nervous system. Painless belongs to the TRPA subfamily of the TRP cation channels and is expressed in both central and peripheral nervous system [27]. The absence of painless increases the susceptibility of the organism at high temperature [27, 28]. In high temperature ($\geq$40 °C), painless loses its function, and the temperature regulation is controlled by pyrexia [25]. Occasionally, these two proteins overlapped at moderate temperature, i.e. 37–38 °C, and get activated between 35 and 42 °C [22, 25].

Sensing of cold is assigned to the antennae. Initial ablation studies have proposed the occurrence of both warm and cool sensors in the third antennal segment of *Drosophila* [29]. Later, it was found that the warm sensors are present in the adult brain and are controlled by dTRPA1 [30]. The gene *brv1*, a part of the TRP-polycystic channel, is responsible for sensing the cold. *brv1* is present on both the surface of the aristae and in sacculus part of the third segment of the antenna [31]. Among channel proteins, dTRPA1, painless and pyrexia are responsible for sensing cold (~20 °C). However, when the temperature is <20 °C, TRP and TRPL (collectively known as TRPCs) channel proteins are involved in the cold avoidance. These two classes directly respond to the temperature and indirectly to the cold temperature [32]. The terminal region of the neurons is selectively activated by the cold temperature [32]. Thus, if there is any defect in these genes and protein, then both larvae and adult show defective behaviour towards cold temperature.

The temperature preference of an organism can be measured within the laboratory by constructing thermal gradients to estimate the potential thermal preference [33, 34]. A gradient is created by heating or cooling opposite ends of a thermally conductive material; however, other gradient devices have also been used [35–37]. A thermal gradient is created by using a lamp as a heat source [38]. The temperature sensitivity in larvae and thermal avoidance behaviour of adult flies by minimizing the effect of light is described by Ueno et al. [39].

1.5 Chemotaxis Assay

Drosophila responds positively towards the food-related odorants [40–43] and avoids toxic [41, 43] or highly concentrated chemicals [44, 45]. The responses of *Drosophila* towards light, sound and chemicals were reported since the nineteenth century [46]. Later, the integration of visual with olfaction cues was reported [47–49]. The response of *Drosophila* larvae and adult towards the chemicals can be checked by chemotaxis assay [50]. Using chemotaxis assay response towards various chemicals was studied in *Drosophila* [51–55]. Genetics and molecular biology studies identify numerous loci or genes involved in the chemosensation [56–59]. The olfactory choice assay uses the bitter perception over the normal one in *Drosophila* larvae and adult [60]. This assay helps to screen the olfactory defects among flies [52].

1.6 Climbing Assay

Locomotion is a coordinated action of many neurons. Consequently, the locomotory defect is seen with the onset of neurodegeneration [61]. Among various locomotory behaviours, climbing assay [62] is the innate ability to move against gravity. In this assay, negative geotaxis behaviour is checked. The principle of this assay lies on keeping a certain number of flies inside a vial and tap it hard against a surface which causes the flies to fall down to the bottom. As a tendency to move upwards, the flies will try to climb up to the vial. This quantitative assay determines the number of flies climbed above a mark on the vial at a particular time [61, 63]. In a climbing assay [64, 65]. The number of flies climbed up to 2–5 cm height in 10–20 s are marked. The number of flies of the experimental group is compared with the control to find the defects [8].

1.7 Choice Assay

The detection of several chemicals within the food is assigned to the gustatory organ. The chemical cue that act as a ligand for the gustatory receptors may induce or inhibit the feeding process. [66]. In this assay, larvae are given a choice between control and the food mixed with tastant. In the control food, some food colour is added so that the food intake of the larvae can be distinguished from the tastants [67]. The chemosensory organ is able to distinguish between toxic and edible food. This sensory information is further stored as memory via signal transduction. Thus, if the animal is exposed to such kind of food in the future, the larvae will show avoidance behaviour [2]. During this signal integration, neurons play a vital role in taste detection.

Selection of the oviposition site is an important mechanism for the proper developmental growth of the progeny. Thus, in a population, female selects the right area to lay her eggs. *Drosophila* is attracted to fermented foods [68] since yeast promotes developmental process [69]. Microbiota is known to accelerate the behaviour in *Drosophila* [70] by altering the metabolism [71]. Some of the bacteria act as a commensal during the egg-laying process [72]. The fermented food possesses sucrose, which is an inducer for egg laying [73]. The odour of the food also affects the oviposition site [74]. Lactic acid bacteria present in the food accelerate the oviposition site in the food source. Bacteria having a bad smell prevent the fly to lay an egg in an unhygienic environment. Thus, the odours act as an environmental cue to activate olfactory neurons for egg-laying behaviour [45, 75]. The current study describes the choice assay for selecting the right food and the number of eggs laid by the fly in the right food area.

1.8 Phototaxis Assay

Drosophila exhibits phototaxis behaviour based on visual perception, which is controlled by a complex set of neurons. The response to light varies from larva to adults. In the larva, light sensing is assigned to Bolwig's organs, which change to an eyelet in the adult. Each eyelet possesses 12 photoreceptors (PR), out of which 8 photoreceptors express green-sensitive rhodopsin (rh6) and 4 photoreceptors express blue-sensitive rhodopsin (rh5) [76, 77]. Besides eyelet, the larval body wall possesses class IV multidendritic neurons, which can detect the light intensities [78, 79]. The light sensitivity or the phototaxis behaviour of the larva is associated with the circadian neuron. The pacemaker neurons present in the central larval brain express cryptochrome protein. This protein acts as an intrinsic blue light sensor within the brain [80]. In fly, circadian proteins vary with day and night. The L3-staged larva keeps away towards light during sunlight, and this behaviour decreases during the evening [81]. The cryptochrome pathway along with other light-sensing pathways organizes the behavioural output in a day. Acetylcholine in PRs plays a role in light avoidance behaviour of the larva. In consequence, blockage of acetylcholine from PRs to circadian pacemaker neurons decreases the light avoidance behaviour in light-dark preference assay [82]. Two symmetrical pairs of neurons switch the light preference of the third instar larva along with the time of day.

The wild-type flies are positively phototactic [83], and this behaviour is in combination with memory, learning and conditioned stimulus [84]. Flies can detect both UV and visible light; however, the preference is more towards UV light [85]. Along with compound eye (which contains photoreceptors R1–R8), the phototaxis phenomena in adult also involve ocelli [86]. Blocking the function of any of the photoreceptor alters the phototactic behaviour [87]. Other than the photoreceptors, the neural circuit histamine receptors, *Ort*, present in the interneurons of lamina and medulla of optic lobes also modulates the light response [88]. Dscam2, a cellular protein which forms boundaries in the optic lobe, regulate light response [89]. Additionally, the serotonin signalling mediated by the *white* gene, biogenic amines like dopamine and octopamine also modulates phototactic response [90, 91]. Phototactic behaviour changes during egg laying, and the female avoids UV light [92]. Defective wing [93] and aged fly show decreased light response [94]. Circadian rhythm mutants having a defect in *clock* [95], *timeless2* [96] and *cry* [97] also show altered sensitivity towards light [98]. *Drosophila* phototactic behaviour can be measured by T-maze devices [99]. The flies are placed in a T-shaped device where they were given a choice to choose between the light and dark region [100].

1.9 Sound Avoidance Assay

Sound facilitates behaviour like courtship and detection of prey and predator for the animal [18, 19]. The adult fly possesses the Johnston's organ, as the hearing organ. In *Drosophila* larvae, the chordotonal neurons mediate the hearing [20, 21]. Many transient receptor potential (TRP) channels form non-selective cation channels which include proteins like NompC, Nanchung and Inactive [22]. Any mutation in TRP can alter the behaviour of the animal [23]. Exposure of 500 Hz pure tone to larvae for 1 s in every 3 s interval makes maximum larvae burrow inside the food [22]. This behaviour is known as sound avoidance behaviour and is used to check the functionality of the hearing organ.

1.10 Hygroscopic Assay

The flies have the ability to sense the humidity level in the atmosphere, which helps to protect the flies for laying eggs in the right environment. The hygrosensory receptors are present at the bristle of the distal antenna and Nanchung, a TRPV channel is essential for sensing the moist air, whereas Waterwitch (Wtrw) is required for sensing dry air [17]. The hygrosensory cells act like a mechanosensor, and the Nan and Wtrw-expressing neurons are projected towards the mechanosensitive centres in the brain [17].

The current chapter summarizes various behavioural assays in larvae and adults to detect the neuronal abnormality in the fly.

2 Materials

2.1 Preparation of Agar Plate

1. Type I agar (HiMedia-GRM666)
2. MilliQ water
3. Spatula
4. Conical flask (100 ml)
5. Petri plates (85 × 13 mm)
6. Weighing balance (Aczet)
7. Microwave oven
8. Hand grip

2.2 Larva Crawling Assay

1. 2% agar petri plate
2. Third instar larva
3. Zero-sized paint brush

4. Phosphate-buffered saline
5. Graph paper
6. Marker pen

2.3 Touch Sensitivity Test

1. 2% agar petri plate
2. Toothpick
3. Eyelash
4. 1X PBS
5. Zero-sized paint brush
6. Third instar larvae
7. Adhesive instant glue (MIDIE, Hi Gloss PC)
8. Video camera (500X magnification USB)
9. Digital microscope portable lab video camera magnifier

2.4 Self-Right Test

1. Agar-agar type 1 (HiMedia-GRM666)
2. Petri plates (90 × 15 mm)
3. Zero-sized paint brush/rounded microneedle
4. Phosphate-buffered saline (HiMedia-RM7385)
5. Early third instar larva
6. Video camera

2.5 Temperature Sensitivity Test

2.5.1 Thermal Sensitivity Test

1. 2% agar petri plate
2. 1X PBS
3. Zero-sized paint brush
4. 1 L glass beaker
5. Hot water (35–40 °C)
6. Parafilm (BEMIS®"M")
7. Hot plate
8. Third instar larvae
9. Digital camera

2.5.2 Cold Sensitivity Test

1. 2% agar petri plate
2. 1X PBS
3. Zero-sized paint brush
4. 1 L glass beaker
5. Cold water (10–15 °C)
6. Ice cubes
7. Thermometer
8. Third instar larvae
9. Digital camera

2.6 Light Preference Test

1. 2% agar petri plate
2. Aluminium foil
3. Black marker pen/paint
4. Zero-sized paint brushes
5. Light source
6. Phosphate-buffered saline (PBS)
7. Third instar larvae

2.7 Chemotaxis Assay

1. 2% agar petri plate
2. 1.5% agar solution
3. Microcentrifuge tube caps
4. MilliQ water
5. 100 mM quinine(quinine sulphate)/sucrose solution
6. Honey
7. Benzaldehyde
8. Marker pen
9. Micropipette
10. Chemotaxis device (Y-maze)
11. Third instar larvae
12. Adult flies (2–3 days old)

2.8 Choice Assay

1. Laminar hood
2. Incubator shaker
3. Agar-agar type 1 (HiMedia-GRM666)
4. LB agar (HiMedia-M1151)
5. LB broth (HiMedia-M1245)
6. Quinine hemisulphate (SIGMA ALDRICH, CAS: 130-95-0)
7. Yeast (Subala)
8. 1X PBS
9. MilliQ water

2.9 Climbing Assay

1. Measuring cylinder (100 ml or 20 cm)
2. Cotton plug
3. Adult flies (2–3 days old)
4. Soft pad
5. Marker pen
6. Stop watch
7. Digital camera

2.10 Thermal Avoidance Behaviour

1. Glass vials
2. Connector
3. Hot plate
4. Beaker
5. Ice cubes
6. Thermometer
7. Red light source
8. Adult flies (2–3 days old)

2.11 Phototaxis Behaviour

1. Female and male *Drosophila* (*Oregon* R)
2. Glass vials (25 × 100 mm)
3. Aluminium foil
4. T-maze device
5. LED light source

2.12 Sound Avoidance Assay

1. Petri dish (70 × 15 mm)
2. Food vial (25 × 100 mm)
3. Third instar larvae
4. Spoon
5. Speaker
6. Digital camera

2.13 Hygroscopic Assay

1. T-maze apparatus
2. Moist air
3. Dry air
4. Flies
5. Traceable hygrometer

3 Method

3.1 Preparation of 2% Agar Plate

1. Weigh 2 g of agar powder in a conical flask and add 100 ml of milliQ water to it in order to make 2% agar solution.
2. Stir the solution properly and heat in a microwave oven till the suspension gets fully transparent.
3. Take out the solution and slowly pour 20 ml of the solution to one petri plate.
4. Keep the petri plate for 20 min to cool down and allow it to solidify.
5. The plate now can be used for different behavioural assays.

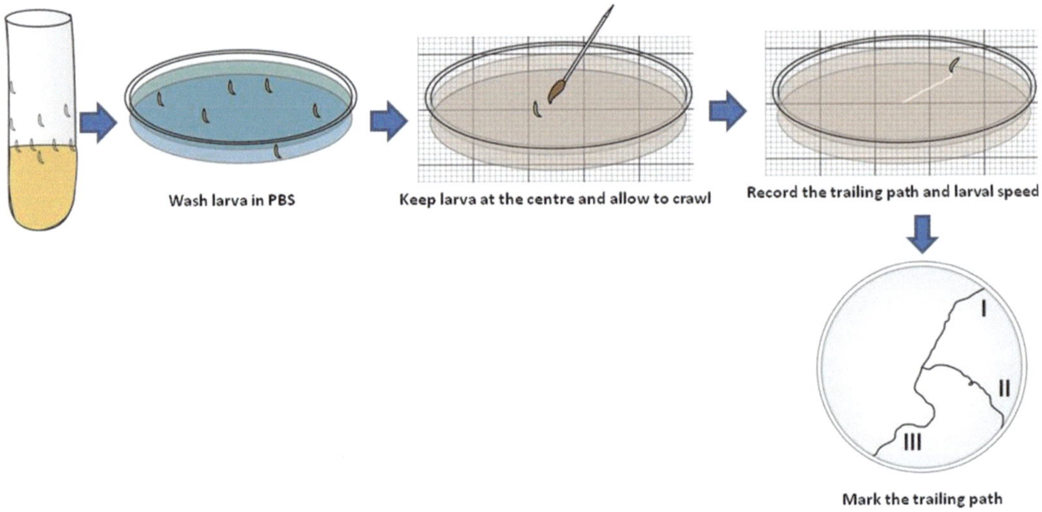

Fig. 1 Schematic diagram of larva crawling assay

3.2 Crawling Assay

1. Take an agar plate (*see* **Note 1**) and keep a graph paper under it before starting the experiment.
2. Collect six third instar larvae in a petri plate (*see* **Note 3**) and wash them with 1X PBS to remove the food particles from their body (*see* **Note 2**) (Fig. 1).
3. Transfer one larva at a time at the centre of the agar plate with a paint brush (Fig. 1).
4. Close the plate and let the larva to crawl freely for 1 min without disturbance.
5. As the larva crawls on agar, it makes a trail mark.
6. Make a record of the trailing path on the petri plate lid by seeing the trail mark (*see* **Note 4**) (Fig. 1).
7. Count the number of grid lines (0.1 cm^2 each) in a graph, the larva has crossed in 1 min from the trailing path.
8. Make a comparison of the trailing path as well as the speed of the control and treated larva.

3.3 Touch Sensitivity Test

1. Take a petri plate containing 2% of agar medium (*see* **Note 1**).
2. Take out (4–5) third instar larva from the rearing media (*see* **Note 3**) and wash it with 1X PBS (*see* **Note 2**).
3. Put the larvae on agar plates and record the activity.
4. Take a toothpick and an eyelash. Stick the thick part of eyelash with the toothpick with adhesive glue so that the fine point of the eye lash is towards the apical part (*see* **Note 7**).
5. Gently touch the thoracic portion of larva with the eyelash tip (Fig. 2) (see **Note 5**).

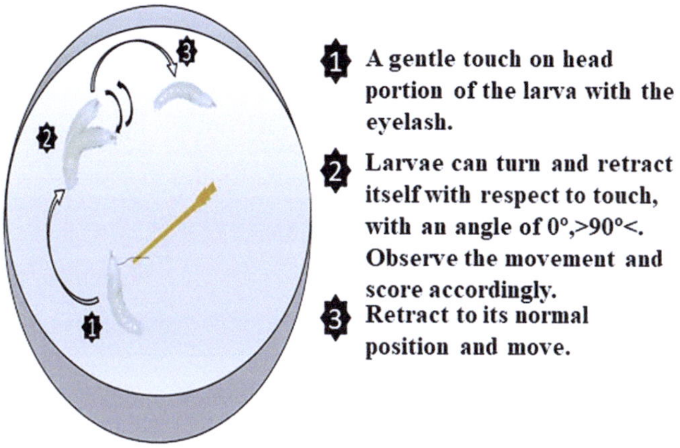

Fig. 2 Touch sensitivity assay of *Drosophila* third instar larvae

Table 1
Touch response score

Sl. No	Response of larva	Score
1	Not respond/cease its movement	0
2	Hesitate to move	1
3	Retract and then move	2
4	Retract and turn 90°	3
5	Retract and turn >90°	4

6. Mark the movement of that larva.
7. Score according to the stimulus shown by the larva [8] (Table 1).

3.4 Self-Right Test

1. Take freshly hatched first instar larva and wash it with 1X PBS.
2. Prepare a 2% agar plate and place the larva on it to acclimatize for 1 min.
3. Gently roll the larva in an inverted position with the help of a paint brush (ventral side up) (*see* **Note 6**).
4. The wild-type larva will exhibit a typical SR response. When inverted, the larva twists its head first and then grabs the substrate with the help of mouth hooks.
5. Then it rolls its body onto the ventral surface, and the dorsal side is up now (Fig. 3).
6. Record the time lapsed between the ventral to the dorsal side in a camera (*see* **Note 7**).

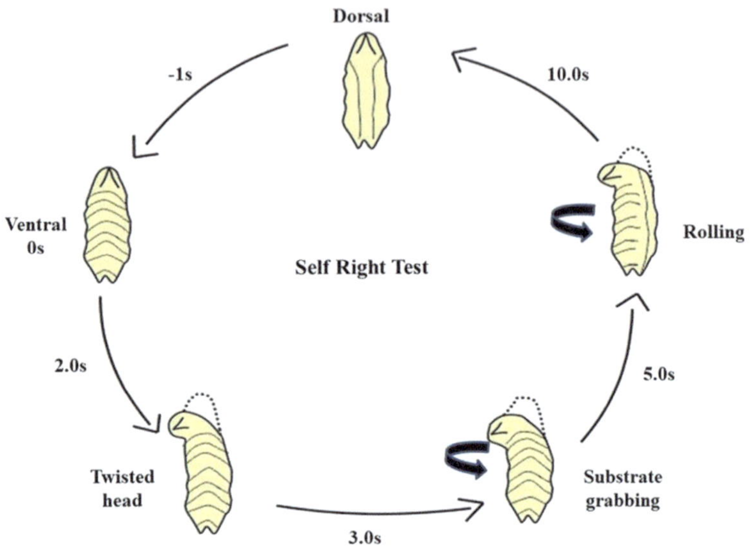

Fig. 3 The self-right test shown by the wild-type *Drosophila* larva

7. Usually, a wild-type larva takes around 10 s to complete the self-right test.
8. On the other hand, a mutant larva with a defective nervous system will exhibit a delayed SR response by being sluggish and shows backward peristaltic waves.
9. Compare the time taken by the larva for completing the self-righting behaviour between the control and the mutant larva (*see* **Note 8**).

3.5 Temperature Sensitivity Assay

3.5.1 Thermal Test

1. Take 1L of tap water in a glass beaker.
2. Warm it on a hot plate at 35 °C.
3. Measure the temperature with a thermometer.
4. Take the 2% agar containing petri plate (for agar plate preparation, see the agar petri plate preparation section).
5. Take out 10–15 third instar larvae and wash with 1X PBS.
6. Transfer the larvae to the agar petri plate.
7. Seal the agar plate with the parafilm, to avoid the entry of water vapour into the agar plate (*see* **Note 9**).
8. Put the petri plate in a 35 °C water bath (*see* **Note 10**).
9. Record the activity of larvae for about 5 min.
10. After 5 min take out the petri plate and count the number of larvae on the agar media and on the lid of the petri plate (Fig. 4a).

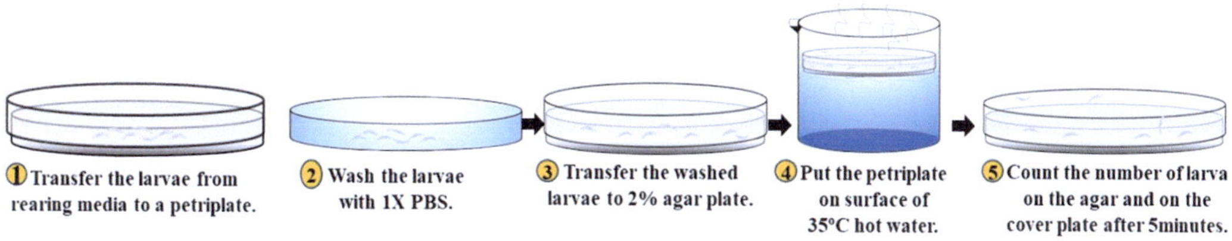

Fig. 4a Thermal sensitivity test of *Drosophila* third instar larvae

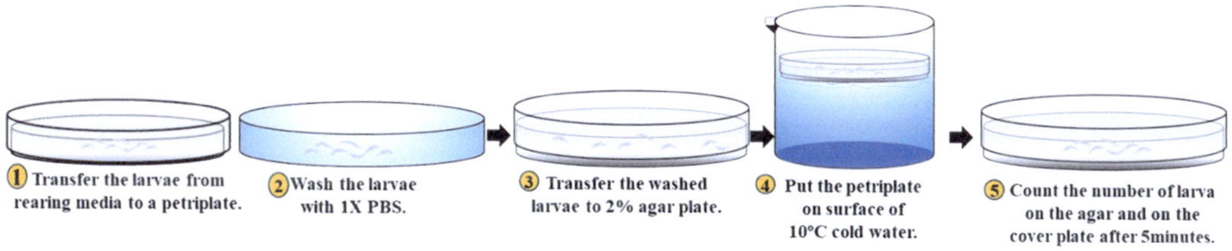

Fig. 4b Cold sensitivity test of *Drosophila* third instar larvae

3.5.2 Cold Sensitivity Test

1. Take 1L of ice-cold water in a beaker.
2. Measure the temperature with a thermometer.
3. Take the 2% agar containing petri plate (for agar plate preparation, see the agar petri plate preparation section).
4. Take out 10–15 third instar larvae and wash with 1X PBS.
5. Put the agar petri plate on the cold water for 5 min.
6. Then, transfer 10–15 numbers of larvae to the agar petri plate.
7. Record the responses made by the larvae for about 5 min.
8. After 5 min take out the petri plate and count the number of larvae on the agar media and on the petri plate lid (Fig. 4b).

3.6 Larva Choice Assay

1. Rear the control as well as the treated fly of interest in a standard yeast-cornmeal media at 25 °C with 12 h light and dark cycle.
2. Five to seven days after culturing the flies, collect 10–15 fresh wondering third instar larva from the vial (*see* **Note 22**).
3. Transfer them to a petri plate containing 1X PBS.
4. Wash them with a zero-sized brush to get rid of food particle if there is any (*see* **Note 2**).
5. Prepare 100 ml of 1% agar media, and melt it in a beaker.
6. In another beaker, liquefy 100 ml of 1% agarose, and to it, add 195.735 mg of quinine to make the concentration of quinine 5 mM.
7. Pour the normal agar and the tastant mixed agar in the opposite chamber of the petri plate. Allow it to cool down for 30 min.

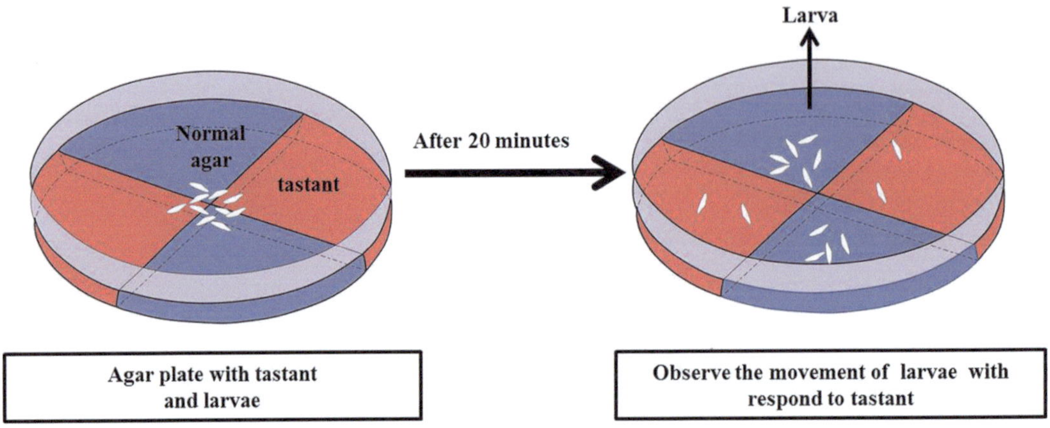

Fig. 5 A four-chambered petri plate with normal agar as well as the testant-agar mixture has been taken for the experiment. The larva is placed at the centre of the petri plate in each chamber (left). After incubating for 20 min, the migration of larva towards respective chambers occurs (right)

8. Place equal numbers of larvae in both the chambers. Cover it and keep them for 20 min at room temperature (*see* **Note 23**).
9. After 20 min, open the petri plate cover and count the number of larva on each side of the plate (Fig. 5).
10. The food preferences are calculated by counting the number of larvae present on either side of the petri plate by following the formula mentioned below (*see* **Note 24**).

$$\mathrm{PI} = \frac{\mathrm{PREFquinine} - \mathrm{PREFagar}}{\mathrm{Total\ larva\ taken}}$$

11. Usually, quinine acts as a repellent in most of the studies.
12. The negative result shows avoidance behaviour towards quinine, while positive indicates attraction towards it (*see* **Note 25**).

3.7 Larval Light Preference Assay

1. Take a petri plate and mark its lid into four quadrants of the same size.
2. Colour the alternate quadrants of the petri plate and its lid with a black permanent marker pen or paint it with black paint or cover it with aluminium foil (Fig. 6).
3. Prepare a 2% agar petri plate.
4. Keep the third instar larvae in dark for 6 h before starting the experiment.
5. Perform the experiment in a dark chamber where no other light source available except the experimental light source (*see* **Note 31**).

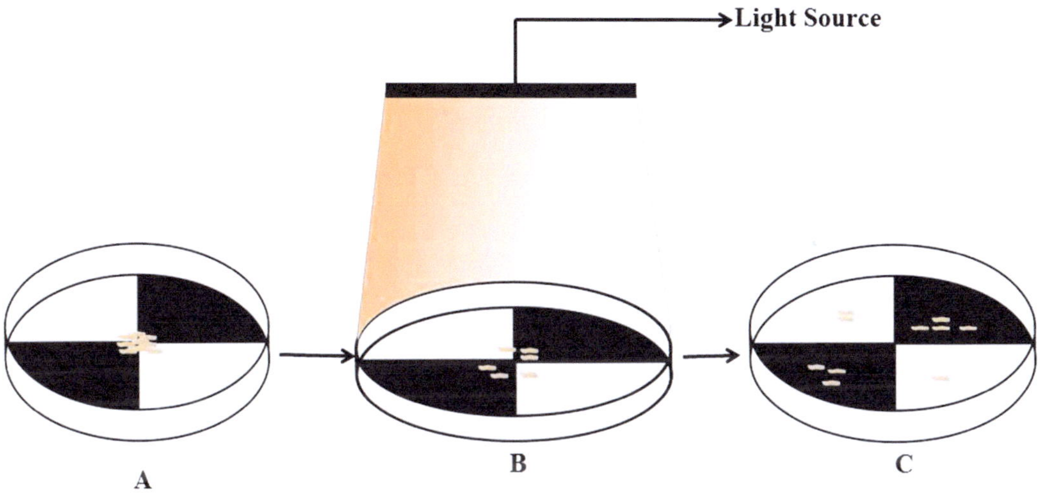

Fig. 6 Light preference test. (**a**) Larva placed at the centre of the petri plate. (**b**) Movement of larva in presence of light. (**c**) Position of the larva at different quadrants after 5 min based on their light preference

6. Keep the petri plate below the homogeneous light source.
7. Transfer around 15 third instar larvae to the centre of the petri plate with the help of a brush.
8. Cover the petri plate with its lid and turn on the light source. Black and transparent quadrants should be arranged in both halves of the petri plate. Permit the larvae to move for 5 min.
9. Remove the lid and immediately count the larvae number in the transparent and black quadrants of the agar plate.
10. Repeat the process of at least five times.
11. Late wild-type third instar larvae go towards the light, while the early third instar larvae prefer to remain in the dark (*see* **Note 32**).

3.8 Chemotaxis Assay

3.8.1 Chemotaxis in Third Instar Larvae

1. Take a 2% agar petri plate.
2. On the external surface of the plate, mark the plate so that it can be divided into two equal halves.
3. Place two microcentrifuge caps at the two opposite ends of the petri plate.
4. Now fill one of the caps with water and another cap with quinine/sucrose solution using a micropipette.
5. Place ten larvae on the centre of the plate and allow them to move freely in the plate for 5 min (*see* **Note 11**) (Fig. 7).
6. After 5 min, count the number of larvae in each shared half of the petri plate.
7. Then calculate the olfactory preference index (OPI) by using the formula as follows

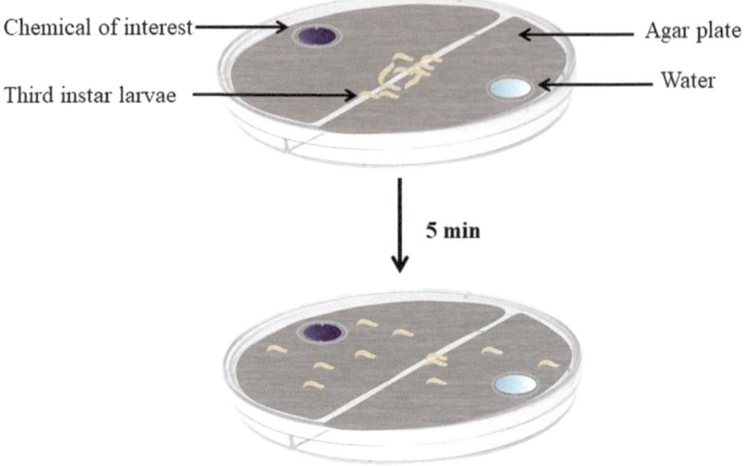

Fig. 7 Materials and steps involved to measure chemotaxis in larvae

$$\frac{\text{Larval number in chemical part} - \text{Larval number in water part}}{\text{whole larvae number used for the experiment}} \times 100$$

8. OPI is calculated for wild type as well as for treated one to mark the difference.

3.8.2 Chemotaxis in Adult Fly

1. For this assay, honey is taken as an attractant, while benzaldehyde is taken as a repellent.
2. Mix the compounds (honey and benzaldehyde) separately with 1.5% agar solution.
3. Now pour the two different solutions into two different caps named as A and B and allow the agar to cool down and solidify in the cap (*see* **Note 13**).
4. Then keep the caps (A, B) on either end of the chemotaxis equipment (Y maze) (Fig. 8).
5. Take 20 adult flies at the end of the maze (end C) and leave the flies there for 5 min (*see* **Note 12**).
6. Allow the flies to choose between the two chemicals and observe the olfactory responses of the flies.
7. At the end of 5 min, calculate the number of flies in both the end of chemicals (end A and end B).
8. Repeat the experiment six times for control and treated to determine the mean result and to examine the anomaly.
9. The wild-type flies are usually attracted towards honey, whereas they show repulsion towards benzaldehyde.

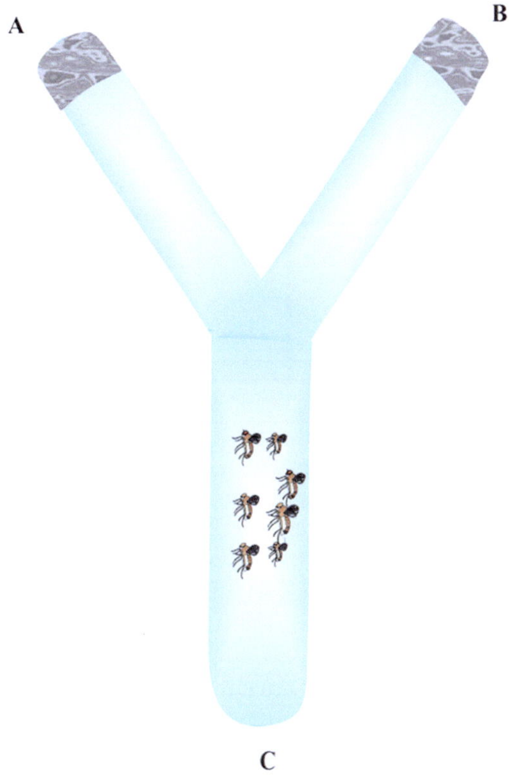

Fig. 8 Y-maze device to measure the chemotaxis behaviour of adult flies

3.9 Climbing Assay

1. Make a marking on a measuring cylinder at 80 ml (measured as 16 cm) (*see* **Note 14**).
2. Transfer 25–30 adult flies into the measuring cylinder without anesthetizing, and close the mouth with a cotton plug to avoid the escape of the flies (*see* **Note 16**) (Fig. 9).
3. Keep the flies for 10 min without disturbance to allow them to acclimatize to the new environment.
4. During this time, place a focused digital camera in front of the measuring cylinder. Set a timer of 10 s and do the recording (*see* **Note 15**).
5. Tap the cylinder sharply three times to knock down the flies to the bottom.
6. With the completion of the third tap, start the timer and observe the climbing for 10 s (Fig. 9).
7. At 10 s, take a picture.
8. Repeat the above steps (5–7) for six times.
9. Take the images to a computer and use the best images to note the number of flies that crossed the 16 cm mark in 10 s.

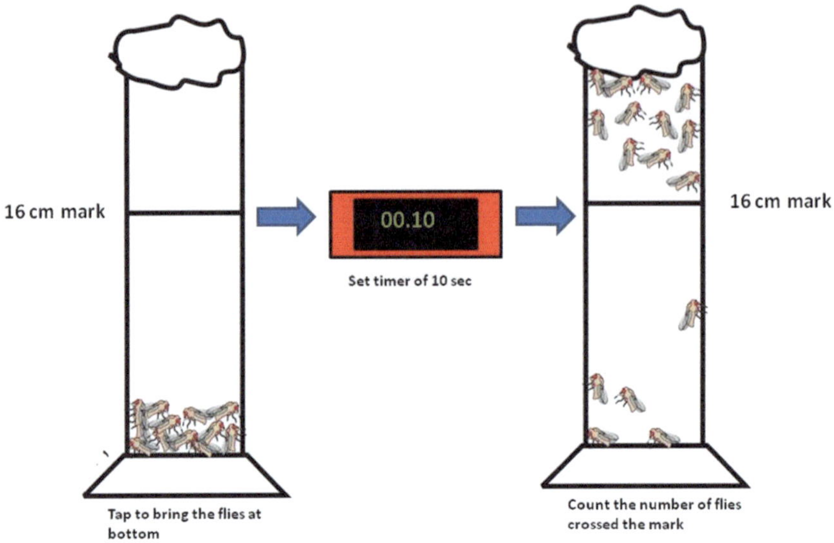

Fig. 9 Schematic diagram showing climbing assay in adult flies

10. Determine the percentage of flies that showed the negative geotaxis behaviour by the following equation:

$$\frac{\text{Number of flies above 16 cm mark}}{\text{Total number of flies}} \times 100$$

11. Generally, wild-type flies show negative geotropism and tend to move upwards. Compare your results with the wild-type flies to notice any change.

3.9.1 Thermal Avoidance Assay

1. Perform the heat assay in a dark room (*see* **Note 17**).
2. Make an apparatus with two vials that are connected by a connector.
3. Place the apparatus 0.5 cm from a hot plate set at about 100 °C (*see* **Note 19**). A stable heat gradient is generally established in 20–30 min (Fig. 10a).
4. Place a parallel tube containing the probe of a thermometer adjacent to the apparatus to measure the temperature of the air at various places in the parallel tube.
5. Put the flies to be tested in a vial plugged with cotton and keep the flies for 30 min at dark (*see* **Note 20**).
6. Insert the flies in the apparatus by separating the vials and put the flies at the warmer half.
7. Pound the flies to the warmer end and then reconnect the vials.
8. Keep the apparatus under red light (*see* **Note 18**).

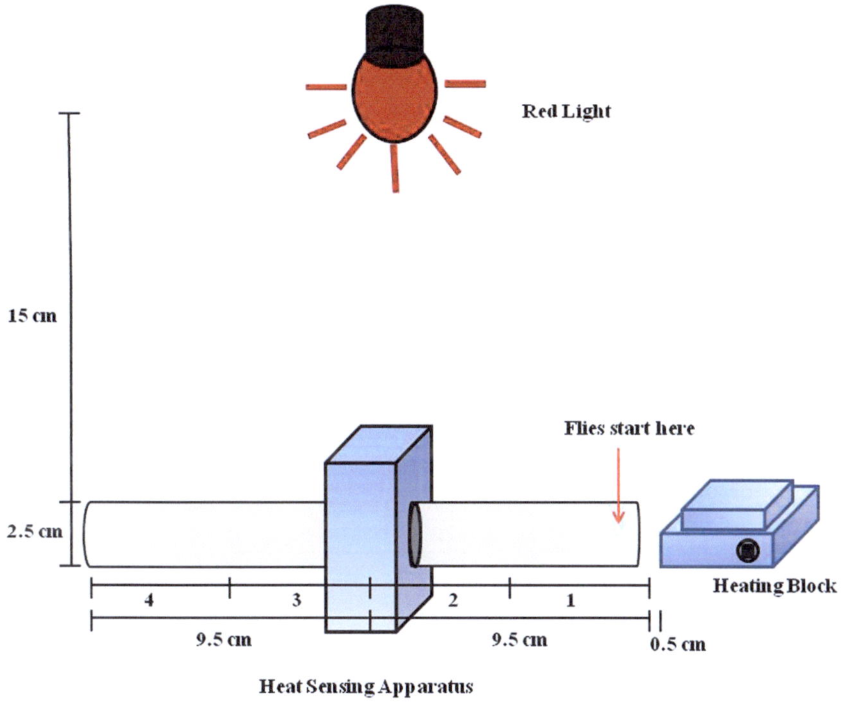

Fig. 10a Thermal avoidance assay

9. Count the number of flies in each compartment in every 5 min while briefly engaging the red light each time.
10. Repeat the experiment for at least six times.

3.9.2 Cold Avoidance Assay

1. Perform the cold assay as like the heat assay in a dark room (*see* **Note 17**).
2. For this, take a beaker full of ice cubes and place it at one end of a tube of the apparatus (*see* **Note 19**) (Fig. 10b).
3. Place a parallel tube containing the probe of a thermometer adjacent to the apparatus to measure the temperature of the air at various places in the parallel tube.
4. Keep the flies to be tested in a vial at dark for 30 min before starting the experiment.
5. Put the flies in the apparatus by removing the vials from the apparatus, away from the cold. Reconnect the vials to the apparatus.
6. Keep the apparatus under red light to provide visibility (*see* **Note 18**).
7. Count the number of flies in every 5 min in each compartment while briefly engaging the red light each time (*see* **Note 21**).
8. Repeat the assay for at least six times.

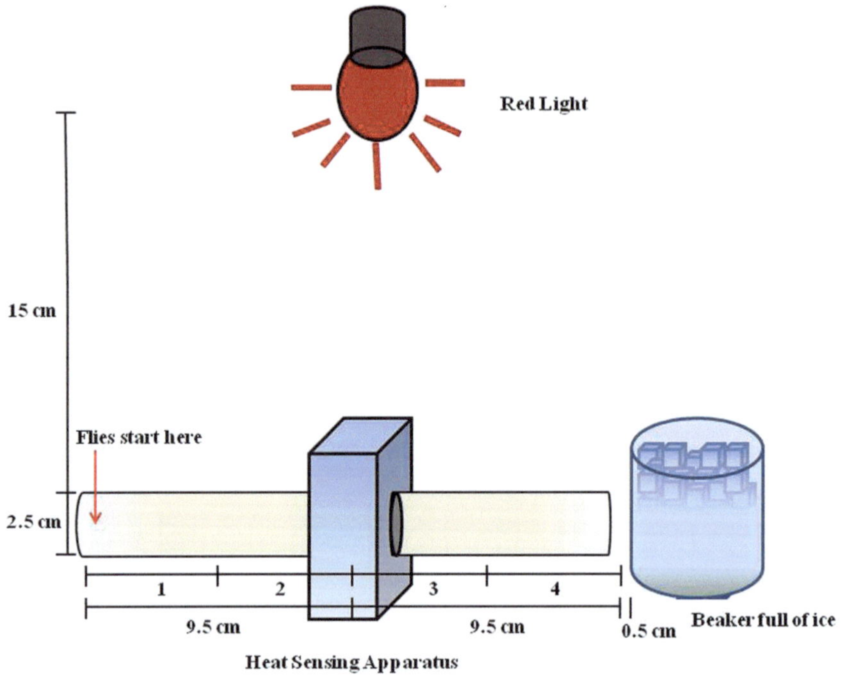

Fig. 10b Cold sensitivity assay

3.10 Adult Choice Assay

1. Culture the fly as above-mentioned conditions. Inoculate 100 μl of bacteria of interest in LB broth and keep it in 37 °C in the shaker (120 rpm) overnight to grow (*see* **Notes 26 and 27**).
2. Two days before the start of the experiment, make 4% LB agar and autoclave it. Pour it in one chamber of the three-chambered petri plate. The petri plate should fill up to more than half.
3. Once agar gets solidified, pour 50 μl of a freshly grown bacterial suspension of your interest at the centre of the plate. Cover it and keep the plate in an incubator at 37 °C for 24 h so that bacteria can grow.
4. Now take out the plate and place it under the laminar hood for further processing.
5. Melt 1% agar and pour it in remaining two chambers. After solidifying, add yeast paste to the centre of the one chamber.
6. Transfer ten flies (five male: five females) to each chamber (*see* **Notes 28 and 29**) (Fig. 11).
7. Keep it for at least 1 day. Next day remove all the files from the disc.
8. Under microscope, count the number of eggs laid by the female in each chamber.
9. Repeat each experiment five times to get statistically significant data (*see* **Note 30**).

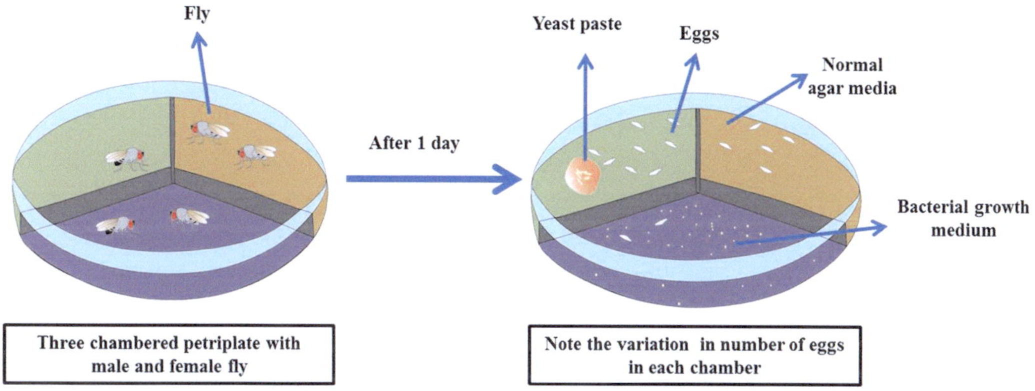

Fig. 11 A three-chambered petri plate with normal agar, yeast paste and bacterial suspension in three respective chambers is taken for this assay. To each chamber, male and female flies are transferred (left). After keeping the same for 24 h, the number of eggs laid by the female is to be counted (right)

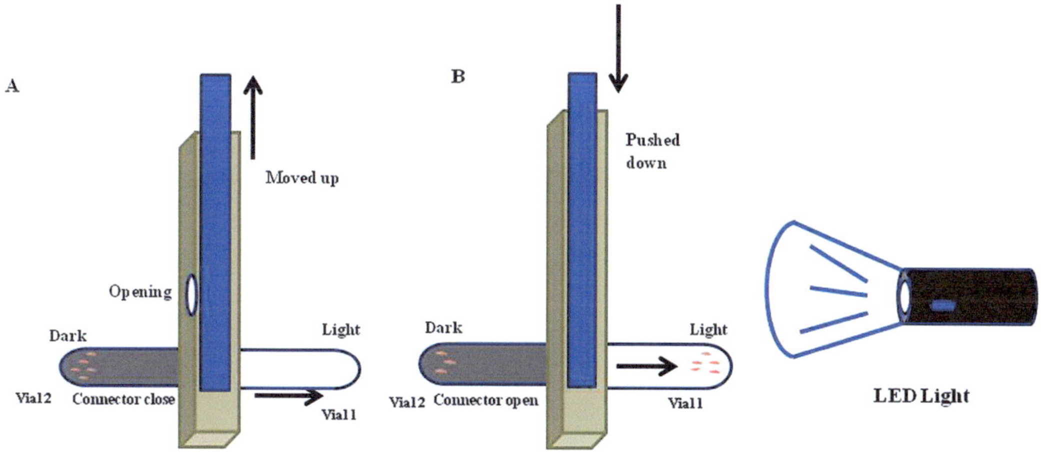

Fig. 12 Diagram representing the phototaxis assay by a T-maze device. (**a**) Flies are in dark chamber vial (**b**) Flies move towards the light as the connector is open

3.11 Phototaxis Behaviour

1. Collect 2–5-day-old *Oregon* flies around 30 in number in an empty glass vial (*see* **Note 12**).
2. Keep them in dark for 6 h prior to the start of the experiment.
3. Conduct the assay in completely dark conditions (*see* **Note 31**).
4. Take a T-maze possessing two arms containing glass vials which are separated by a connector.
5. Place the vials perpendicular to the axis of the device as shown (Fig. 12).
6. Connect an empty glass vial to one arm of the T-maze device which is the lighted chamber.

7. Take another glass vial and cover it with an aluminium foil. Put this vial to the darkened side of the chamber.
8. Transfer the flies into this covered vial and attach it to the other arm of the T-maze device.
9. Adjust the connector by pressing it such that there is the free passage of flies between the two glass vials (*see* **Note 33**).
10. Keep a LED light source near to the uncovered glass vial around 15 cm away and parallel to the device set-up.
11. The wild-type flies will move towards the light source being positively phototactic. Count the number of flies in the light chamber and also the flies remaining in the dark chamber.
12. The percentage of flies which shows positive response towards light is calculated as

$$\frac{\text{No.of flies in the lighted chamber vial}}{\text{Total no.of flies}} \times 100$$

13. Carry out the experiment for wild-type flies as well as any treated or mutant flies and compare the differences.
14. Repeat the experiment for six times for accuracy.

3.12 Sound Avoidance Assay

1. Take a spoonful food out from the vial and spread it on a petri plate (Fig. 13).
2. Collect 30 third instar larvae from the food vial and keep it on the petri plate.
3. Cover the petri plate with a lid.
4. Allow the larvae to move freely in the petri plate on the food.
5. Keep the petri plate over the speaker.
6. Then record the overall surface of the food.
7. Count the number of larvae which are present on the surface of the food.

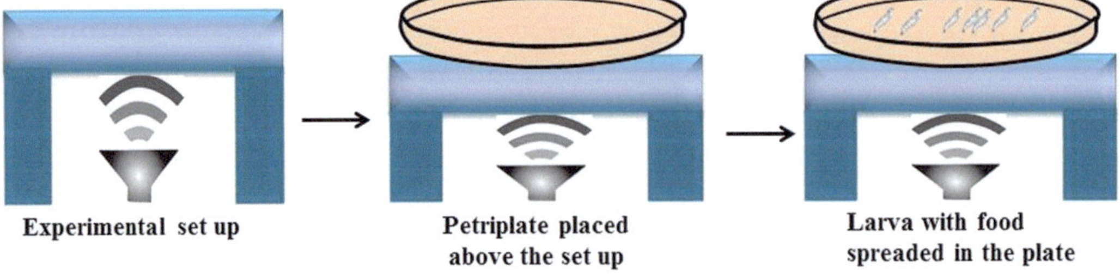

Fig. 13 Sound avoidance assay

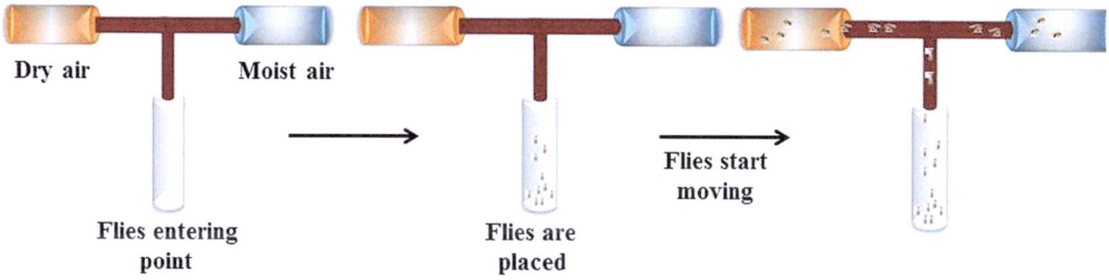

Fig. 14 Hygroscopic assay

8. Compare the behaviour during the time of sound exposure and after the termination of the sound.

3.13 Hygroscopic Assay

1. Take a T-maze apparatus having three tubes of equal size (Fig. 14).
2. Introduce the moist air (~99% relative humidity) into tube 1.
3. Fill tube 2 with dry air (~3% relative humidity). Insert tube 3 in between them.
4. Introduce the flies to the third tube and allow them to choose between the two different tubes having different humidity.
5. After 5 min, count the number of flies in each tube.
6. Calculate the preference index (PI) by using the formula:

$$\mathrm{PI} = \frac{\text{Number of flies in moist air} - \text{Number of flies in dry air}}{\text{Total number of flies for the test}}$$

4 Notes

4.1 Larva Crawling and Touch Sensitivity

1. Use the agar plates only after solidification.
2. Wash the larva properly to remove the attached food particles to the body wall. It may reduce the larval crawling speed.
3. Use larva of the same age to perform the assay.
4. Mark the trailing path clearly with the marker.
5. Use of proper stimulation with eyelash is a prerequisite for touch sensitivity experiment.

4.2 Self-Right Test

6. The larva should be handled carefully while inverting on the agar plate.
7. The recording should be done precisely as soon as the larva exhibits movement. The video can be analysed using open-source software 1.2 V code.
8. The process should be repeated at least ten times for accurate results.

4.3 Temperature Sensitivity Test

9. Avoid steam accumulation on thermal sensitivity test. Cover the plate with proper parafilm.
10. The petri plate must be in a floating condition. Care should be taken so that it will not sink.

4.4 Chemotaxis Assay

11. The larvae should be allowed to move on the agar plate before the experiment to acclimatize to the plate.
12. Same age adults should be taken for the experiment.
13. Avoid spilling of chemicals or water over the petri plate. With spilling, both water and chemicals will be mixed, and larvae cannot differentiate water from the chemicals.

4.5 Climbing Assay

14. Use a clean and dry glass measuring cylinder. Any trace of water may hinder the fly movement.
15. Keep the camera at a platform without disturbance. The distance should not be kept more than 1 m.
16. Be careful while transferring the flies to the measuring cylinder, as mechanical damage to the fly may alter the behaviour.

4.6 Thermal Avoidance Assay

17. Perform the assay in a dark room to avoid the interference of light to the movement of the flies.
18. Red light is poorly visible by the flies. The number of flies can be counted by brief exposure to the red light.
19. Keep the hot plate (or ice beaker) at a distance of 0.5 cm from the apparatus.
20. Keep an equal number of flies in each assay.
21. Control flies will avoid going to the hot and cold regions. Compare the results of the treated or mutant flies with control to get a significant information of the thermal behaviour.

4.7 Choice Assay

(Larva choice assay)

22. Early third instar larva is mostly recommended to avoid pupa formation during the experimental procedure.
23. Larva should pass from one chamber to another to get an equal chance of choosing between different foods.
24. During counting, discard the larvae which climb up to the lid.
25. Each experiment must be repeated at least six times to get a significant result.

(Adult choice assay)

26. Use only autoclaved materials for bacterial-related study under aseptic conditions.
27. A proper environmental condition should be maintained for each fly.

28. The lid of the petri plate must be placed in such a way that flies can move from one chamber to other.
29. Virgin flies must be taken for each experiment to have a proper counting of egg numbers.
30. Each experiment must be repeated at least five times to get a significant result.

4.8 Larval Light Preference and Adult Phototaxis Assay

31. Maintain the dark acclimatization properly as light adaptation is quicker as compared to the dark adaptation.
32. Choose the perfect stage of larvae. Since early and late instar larvae response differently towards the light.
33. Keep the connector firmly placed in the correct position or else there will be improper movement or escape of the flies.

Acknowledgements

SS, GD and NN are grateful to DST Inspire fellowship. SM is thankful to MHRD and JB to DBT for funding. RR is thankful to DST-WOS (A). MM lab is supported by Grant No. BT/PR21857/NNT/28/1238/2017, EMR/2017/003054, Odisha DBT 3325/ST(BIO)-02/2017.

References

1. Jakubowski BR, Longoria RA, Shubeita GT (2012) A high throughput and sensitive method correlates neuronal disorder genotypes to Drosophila larvae crawling phenotypes. Fly 6(4):303–308
2. Devaud J-M (2003) Experimental studies of adult Drosophila chemosensory behaviour. Behav Process 64(2):177–196
3. Nichols CD, Becnel J, Pandey UB (2012) Methods to assay Drosophila behavior. J Vis Exp (61)
4. Heckscher ES, Lockery SR, Doe CQ (2012) Characterization of Drosophila larval crawling at the level of organism, segment, and somatic body wall musculature. J Neurosci 32 (36):12460–12471
5. Green C, Burnet B, Connolly K (1983) Organization and patterns of inter-and intraspecific variation in the behaviour of Drosophila larvae. Anim Behav 31(1):282–291
6. Wang JW, Sylwester AW, Reed D, Wu D-AJ, Soll DR, Wu C-F (1997) Morphometric description of the wandering behavior in Drosophila larvae: aberrant locomotion in Na+ and K+ channel mutants revealed by computer-assisted motion analysis. J Neurogenet 11(3–4):231–254
7. Marder E, Calabrese RL (1996) Principles of rhythmic motor pattern generation. Physiol Rev 76(3):687–717
8. Mishra M, Barik BK (2018) Behavioral teratogenesis in *Drosophila melanogaster*. In: Teratogenicity testing. Springer, New York, pp 277–298
9. Caldwell JC, Miller MM, Wing S, Soll DR, Eberl DF (2003) Dynamic analysis of larval locomotion in Drosophila chordotonal organ mutants. Proc Natl Acad Sci 100 (26):16053–16058
10. Suster ML, Bate M (2002) Embryonic assembly of a central pattern generator without sensory input. Nature 416(6877):174
11. Marder E (2002) Developmental biology: senseless motion. Nature 416(6877):131
12. Sokolowski M, Hansell K (1992) The foraging locus: behavioral tests for normal muscle movement in rover and sitter *Drosophila melanogaster* larvae. Genetica 85 (3):205–209S

13. Gomez-Marin A, Louis M (2014) Multilevel control of run orientation in Drosophila larval chemotaxis. Front Behav Neurosci 8:38
14. Gepner R, Skanata MM, Bernat NM, Kaplow M, Gershow M (2015) Computations underlying Drosophila photo-taxis, odor-taxis, and multi-sensory integration. Elife 4:e06229
15. Fushiki A, Zwart MF, Kohsaka H, Fetter RD, Cardona A, Nose A (2016) A circuit mechanism for the propagation of waves of muscle contraction in Drosophila. Elife 5:e13253
16. Tastekin I, Riedl J, Schilling-Kurz V, Gomez-Marin A, Truman JW, Louis M (2015) Role of the subesophageal zone in sensorimotor control of orientation in Drosophila larva. Curr Biol 25(11):1448–1460
17. Loveless J, Webb B (2018) A neuromechanical model of larval chemotaxis. Integr Comp Biol 58(5):906–914
18. Bridges CB, Morgan TH (1923) Third-chromosome group of mutant characters of *Drosophila melanogaster*. Carnegie Institution of Washington, Washington, DC
19. Sánchez-Herrero E, Vernós I, Marco R, Morata G (1985) Genetic organization of Drosophila bithorax complex. Nature 313(5998):108
20. Picao-Osorio J, Johnston J, Landgraf M, Berni J, Alonso CR (2015) MicroRNA-encoded behavior in Drosophila. Science 350(6262):815–820
21. Picao-Osorio J, Lago-Baldaia I, Patraquim P, Alonso CR (2017) Pervasive behavioural effects of microRNA regulation in Drosophila. Genetics 206:1535–1548
22. Dillon ME, Wang G, Garrity PA, Huey RB (2009) Thermal preference in Drosophila. J Therm Biol 34(3):109–119
23. Barbagallo B, Garrity PA (2015) Temperature sensation in Drosophila. Curr Opin Neurobiol 34:8–13
24. Neely GG, Keene AC, Duchek P, Chang EC, Wang Q-P, Aksoy YA, Rosenzweig M, Costigan M, Woolf CJ, Garrity PA (2011) TrpA1 regulates thermal nociception in Drosophila. PLoS One 6(8):e24343
25. Lee Y, Lee Y, Lee J, Bang S, Hyun S, Kang J, Hong S-T, Bae E, Kaang B-K, Kim J (2005) Pyrexia is a new thermal transient receptor potential channel endowing tolerance to high temperatures in *Drosophila melanogaster*. Nat Genet 37(3):305
26. Li K, Gong Z (2017) Feeling hot and cold: thermal sensation in Drosophila. Neurosci Bull 33(3):317–322
27. Tracey WD, Wilson RI, Laurent G, Benzer S (2003) Painless, a Drosophila gene essential for nociception. Cell 113(2):261–273
28. Xu S, Cang C, Liu X, Peng Y, Ye Y, Zhao Z, Guo A (2006) Thermal nociception in adult Drosophila: behavioral characterization and the role of the painless gene. Genes Brain Behav 5(8):602–613
29. Sayeed O, Benzer S (1996) Behavioral genetics of thermosensation and hygrosensation in Drosophila. Proc Natl Acad Sci 93 (12):6079–6084
30. Hamada FN, Rosenzweig M, Kang K, Pulver SR, Ghezzi A, Jegla TJ, Garrity PA (2008) An internal thermal sensor controlling temperature preference in Drosophila. Nature 454 (7201):217
31. Gallio M, Ofstad TA, Macpherson LJ, Wang JW, Zuker CS (2011) The coding of temperature in the Drosophila brain. Cell 144 (4):614–624
32. Rosenzweig M, Kang K, Garrity PA (2008) Distinct TRP channels are required for warm and cool avoidance in *Drosophila melanogaster*. Proc Natl Acad Sci 105 (38):14668–14673
33. Light P, Dawson WR, Shoemaker VH, Main AR (1966) Observations on the thermal relations of western Australian lizards. Copeia 1966:97–110
34. Barbour M, Racine CH (1967) Construction and performance of a temperature-gradient bar and chamber. Ecology 48(5):861–863
35. McDaniel R, Hostert EE, Seager RD (1995) Acclimation and adaptive behavior of *Drosophila robusta* and *D. tripunctata* adults in response to combined temperature and desiccation stress. Am Midl Nat 133:52–59
36. Yamamoto AH (1994) Diallel analysis of temperature preference in *Drosophila immigrans*. Jpn J Genet 69(1):77–86
37. Yamada Y, Ohshima Y (2003) Distribution and movement of *Caenorhabditis elegans* on a thermal gradient. J Exp Biol 206 (15):2581–2593
38. Prince G, Parsons P (1977) Adaptive behaviour of Drosophila adults in relation to temperature and humidity. Aust J Zool 25 (2):285–290
39. Ueno T, Tomita J, Kume S, Kume K (2012) Dopamine modulates metabolic rate and temperature sensitivity in *Drosophila melanogaster*. PLoS One 7(2):e31513
40. Semmelhack JL, Wang JW (2009) Select Drosophila glomeruli mediate innate olfactory attraction and aversion. Nature 459 (7244):218
41. Wang Y, Chiang A-S, Xia S, Kitamoto T, Tully T, Zhong Y (2003) Blockade of neurotransmission in Drosophila mushroom bodies

impairs odor attraction, but not repulsion. Curr Biol 13(21):1900–1904
42. Kreher SA, Mathew D, Kim J, Carlson JR (2008) Translation of sensory input into behavioral output via an olfactory system. Neuron 59(1):110–124
43. Asahina K, Louis M, Piccinotti S, Vosshall LB (2009) A circuit supporting concentration-invariant odor perception in Drosophila. J Biol 8(1):9
44. Suh GS, Wong AM, Hergarden AC, Wang JW, Simon AF, Benzer S, Axel R, Anderson DJ (2004) A single population of olfactory sensory neurons mediates an innate avoidance behaviour in Drosophila. Nature 431(7010):854
45. Stensmyr MC, Dweck HK, Farhan A, Ibba I, Strutz A, Mukunda L, Linz J, Grabe V, Steck K, Lavista-Llanos S (2012) A conserved dedicated olfactory circuit for detecting harmful microbes in Drosophila. Cell 151 (6):1345–1357
46. Fraenkel GS, Gunn DL (1961) The orientation of animals. Dover Publications, New York
47. Wilson RI, Mainen ZF (2006) Early events in olfactory processing. Annu Rev Neurosci 29:163–201
48. Maimon G, Straw AD, Dickinson MH (2008) A simple vision-based algorithm for decision making in flying Drosophila. Curr Biol 18 (6):464–470
49. Ohashi S, Morimoto T, Suzuki Y, Miyakawa H, Aonishi T (2014) A novel behavioral strategy, continuous biased running, during chemotaxis in Drosophila larvae. Neurosci Lett 570:10–15
50. Aceves-Piña EO, Quinn WG (1979) Learning in normal and mutant Drosophila larvae. Science 206(4414):93–96
51. Miyakawa Y (1982) Behavioural evidence for the existence of sugar, salt and amino acid taste receptor cells and some of their properties in Drosophila larvae. J Insect Physiol 28 (5):405–410
52. Ayyub C, Paranjape J, Rodrigues V, Siddiqi O (1990) Genetics of olfactory behavior in *Drosophila melanogaster*. J Neurogenet 6 (4):243–262
53. Jenkins JB, Tompkins L (1990) Effects of amiloride on taste responses of *Drosophila melanogaster* adults and larvae. J Insect Physiol 36(9):613–618
54. Cobb M, Bruneau S, Jallon J-M (1992) Genetic and developmental factors in the olfactory response of *Drosophila melanogaster* larvae to alcohols. Proc R Soc Lond Ser B Biol Sci 248(1322):103–109
55. Singh RN (1997) Neurobiology of the gustatory systems of Drosophila and some terrestrial insects. Microsc Res Tech 39(6):547–563
56. Rodrigues V, Siddiqi O (1978) Genetic analysis of chemosensory pathway. In: Proceedings of the Indian Academy of Sciences-Section B, vol 7. Springer, pp 147–160
57. Monte P, Woodard C, Ayer R, Lilly M, Sun H, Carlson J (1989) Characterization of the larval olfactory response in Drosophila and its genetic basis. Behav Genet 19 (2):267–283
58. Carlson JR (1996) Olfaction in Drosophila: from odor to behavior. Trends Genet 12 (5):175–180
59. Park Y, Caldwell MC, Datta S (1997) Mutation of the central nervous system neuroblast proliferation repressor ana leads to defects in larval olfactory behavior. J Neurobiol 33 (2):199–211
60. Kim H, Choi MS, Kang K, Kwon JY (2015) Behavioral analysis of bitter taste perception in Drosophila larvae. Chem Senses 41 (1):85–94
61. Madabattula ST, Strautman JC, Bysice AM, O'Sullivan JA, Androschuk A, Rosenfelt C, Doucet K, Rouleau G, Bolduc F (2015) Quantitative analysis of climbing defects in a Drosophila model of neurodegenerative disorders. J Vis Exp JoVE (100)
62. Pathak T, Agrawal T, Richhariya S, Sadaf S, Hasan G (2015) Store-operated calcium entry through Orai is required for transcriptional maturation of the flight circuit in Drosophila. J Neurosci 35(40):13784–13799
63. Manjila SB, Hasan G (2018) Flight and climbing assay for assessing motor functions in Drosophila. Bio-protocol 8(5):e2742. https://doi.org/10.21769/BioProtoc.2742
64. Chakraborty R, Vepuri V, Mhatre SD, Paddock BE, Miller S, Michelson SJ, Delvadia R, Desai A, Vinokur M, Melicharek DJ (2011) Characterization of a Drosophila Alzheimer's disease model: pharmacological rescue of cognitive defects. PLoS One 6(6):e20799
65. Orso G, Martinuzzi A, Rossetto MG, Sartori E, Feany M, Daga A (2005) Disease-related phenotypes in a Drosophila model of hereditary spastic paraplegia are ameliorated by treatment with vinblastine. J Clin Invest 115(11):3026–3034
66. Thorne N, Chromey C, Bray S, Amrein H (2004) Taste perception and coding in Drosophila. Curr Biol 14(12):1065–1079
67. Siddiqi O, Joshi S, Arora K, Rodrigues V (1989) Genetic investigation of salt reception

in *Drosophila melanogaster.* Genome 31 (2):646–651
68. Ganter PF (2006) Yeast and invertebrate associations. In: Biodiversity and ecophysiology of yeasts. Springer, Berlin, pp 303–370
69. Barker J (1992) Genetic variation in cactophilic Drosophila for oviposition on natural yeast substrates. Evolution 46(4):1070–1083
70. Lam K, Geisreiter C, Gries G (2009) Ovipositing female house flies provision offspring larvae with bacterial food. Entomol Exp Appl 133(3):292–295
71. Panteleev DY, Goryacheva I, Andrianov B, Reznik N, Lazebny O, Kulikov A (2007) The endosymbiotic bacterium Wolbachia enhances the nonspecific resistance to insect pathogens and alters behavior of *Drosophila melanogaster.* Russ J Genet 43(9):1066–1069
72. Leitão-Gonçalves R, Carvalho-Santos Z, Francisco AP, Fioreze GT, Anjos M, Baltazar C, Elias AP, Itskov PM, Piper MD, Ribeiro C (2017) Commensal bacteria and essential amino acids control food choice behavior and reproduction. PLoS Biol 15(4): e2000862
73. Liu W, Zhang K, Li Y, Su W, Hu K, Jin S (2017) Enterococci mediate the oviposition preference of *Drosophila melanogaster* through sucrose catabolism. Sci Rep 7(1):13420
74. Lin C-C, Prokop-Prigge KA, Preti G, Potter CJ (2015) Food odors trigger Drosophila males to deposit a pheromone that guides aggregation and female oviposition decisions. Elife 4:e08688
75. Becher PG, Flick G, Rozpędowska E, Schmidt A, Hagman A, Lebreton S, Larsson MC, Hansson BS, Piškur J, Witzgall P (2012) Yeast, not fruit volatiles mediate *Drosophila melanogaster* attraction, oviposition and development. Funct Ecol 26(4):822–828
76. Sprecher SG, Pichaud F, Desplan C (2007) Adult and larval photoreceptors use different mechanisms to specify the same Rhodopsin fates. Genes Dev 21(17):2182–2195
77. Sprecher SG, Desplan C (2008) Switch of rhodopsin expression in terminally differentiated Drosophila sensory neurons. Nature 454(7203):533
78. Xiang Y, Yuan Q, Vogt N, Looger LL, Jan LY, Jan YN (2010) Light-avoidance-mediating photoreceptors tile the Drosophila larval body wall. Nature 468(7326):921
79. Diaz NN, Sprecher SG (2011) Photoreceptors: unconventional ways of seeing. Curr Biol 21(1):R25–R27
80. Emery P, Stanewsky R, Helfrich-Förster C, Emery-Le M, Hall JC, Rosbash M (2000) Drosophila CRY is a deep brain circadian photoreceptor. Neuron 26(2):493–504
81. Luna AJF, von Essen AM, Widmer YF, Sprecher SG (2013) Light preference assay to study innate and circadian regulated photobehavior in Drosophila larvae. J Vis Exp (74): e50237
82. Keene AC, Mazzoni EO, Zhen J, Younger MA, Yamaguchi S, Blau J, Desplan C, Sprecher SG (2011) Distinct visual pathways mediate Drosophila larval light avoidance and circadian clock entrainment. J Neurosci 31(17):6527–6534
83. Le Bourg E, Badia J (1995) Decline in photopositive tendencies with age in *Drosophila melanogaster* (Diptera: Drosophilidae). J Insect Behav 8(6):835–845
84. Seugnet L, Suzuki Y, Stidd R, Shaw P (2009) Aversive phototaxic suppression: evaluation of a short-term memory assay in *Drosophila melanogaster.* Genes Brain Behav 8(4):377–389
85. Bertholf LM (1932) The extent of the spectrum for Drosophila and the distribution of stimulative efficiency in it. Z Vgl Physiol 18 (1):32–64
86. Miller GV, Hansen KN, Stark WS (1981) Phototaxis in Drosophila: R1–6 input and interaction among ocellar and compound eye receptors. J Insect Physiol 27(11):813–819
87. Yamaguchi S, Heisenberg M (2011) Photoreceptors and neural circuitry underlying phototaxis in insects. Fly 5(4):333–336
88. Gao S, Takemura S-Y, Ting C-Y, Huang S, Lu Z, Luan H, Rister J, Thum AS, Yang M, Hong S-T (2008) The neural substrate of spectral preference in Drosophila. Neuron 60 (2):328–342
89. Bosch DS, van Swinderen B, Millard SS (2015) Dscam2 affects visual perception in *Drosophila melanogaster.* Front Behav Neurosci 9:149
90. Kain JS, Stokes C, de Bivort BL (2012) Phototactic personality in fruit flies and its suppression by serotonin and white. Proc Natl Acad Sci 109(48):19834–19839
91. Gorostiza EA, Colomb J, Brembs B (2016) A decision underlies phototaxis in an insect. Open Biol 6(12):160229
92. Zhu EY, Guntur AR, He R, Stern U, Yang C-H (2014) Egg-laying demand induces aversion of UV light in Drosophila females. Curr Biol 24(23):2797–2804
93. Benzer S (1967) Behavioral mutants of Drosophila isolated by countercurrent distribution. Proc Natl Acad Sci U S A 58(3):1112
94. Carbone MA, Yamamoto A, Huang W, Lyman RA, Meadors TB, Yamamoto R,

Anholt RR, Mackay TF (2016) Genetic architecture of natural variation in visual senescence in Drosophila. Proc Natl Acad Sci 113 (43):E6620–E6629

95. Konopka RJ, Pittendrigh C, Orr D (1989) Reciprocal behaviour associated with altered homeostasis and photosensitivity of Drosophila clock mutants. J Neurogenet 6(1):1–10
96. Benna C, Bonaccorsi S, Wülbeck C, Helfrich-Förster C, Gatti M, Kyriacou CP, Costa R, Sandrelli F (2010) Drosophila timeless2 is required for chromosome stability and circadian photoreception. Curr Biol 20(4):346–352
97. Emery P, So WV, Kaneko M, Hall JC, Rosbash M (1998) CRY, a Drosophila clock and light-regulated cryptochrome, is a major contributor to circadian rhythm resetting and photosensitivity. Cell 95(5):669–679
98. Zimmerman WF, Goldsmith TH (1971) Photosensitivity of the circadian rhythm and of visual receptors in carotenoid-depleted Drosophila. Science 171 (3976):1167–1169
99. Markow TA, Smith WL (1977) Genetic analysis of phototactic behavior in *Drosophila simulans*. Genetics 85(2):273–278
100. van Swinderen B (2011) The aversive phototaxic suppression assay for individual adult Drosophila. Cold Spring Harb Protoc 2011 (10):1203–1205

Chapter 19

Methods to Detect the Complex Behaviours in *Drosophila*

Swetapadma Sahu, Gyanaseni Dhar, and Monalisa Mishra

Abstract

The behaviour of an organism reflects physiological fitness and its response to the environment. Consequently, the behaviour redirects the functionality of the gene. *Drosophila* with its fully sequenced genome served as a model to screen the functionality of various genes using behavioural assay. Some of the behaviours reflect the functionality of a single gene or organ, whereas others reflect the coordinated action of multiple organs. Such behaviour which needs multimodal signalling is referred to as complex behaviour. The most common complex behaviour includes courtship and mating behaviour, grooming behaviour and aggressive behaviour. The courtship and mating behaviour is essential for the propagation of the species. Grooming helps to clean the surface of sensory organs from foreign particles. Aggressive behaviour helps in finding food and partner and protecting its territory. The current chapter summarizes the courtship-mating, grooming and aggressive behaviour in *Drosophila*.

Key words *Drosophila*, Courtship behaviour, Aggression behaviour, Grooming behaviour

Abbreviations

CI	Courtship index
CPG	Central song pattern generator
Dsx	Doublesex
FRU	Fruitless
GR	Gustatory receptor
GRN	Gustatory receptor neuron
LPS	Lipopolysaccharide
OD	Optical density
ORN	Olfactory receptor neuron
SEM	Scanning electron microscope
SP	Sex peptide
WEI	Wing extension index
WVI	Wing vibration index

Monalisa Mishra (ed.), *Fundamental Approaches to Screen Abnormalities in Drosophila*, Springer Protocols Handbooks, https://doi.org/10.1007/978-1-4939-9756-5_19, © Springer Science+Business Media, LLC, part of Springer Nature 2020

1 Introduction

The behaviour of an organism reflects morphology, life history and environmental factors that are associated with it [1]. Behaviour is regulated by an epistatic factor, pleiotropic effects, environment and the interaction of various genes [2–4]. With the fully sequenced genome, gene associated behavioural defects can be scored easily using *Drosophila* [5]. Some of the behaviours are associated with a single gene or organ, whereas few behaviours are linked with multiple factors and often referred to as complex behaviour. The complex behaviour includes the courtship-mating, grooming and aggressive behaviour [4].

1.1 Mating Behaviour

Mating is an innate behaviour which resulted in successive gene transfer between species through a mechanism known as courtship behaviour. In *Drosophila*, both male and female contribute differently to courtship. The males attract females to trigger their sexual response. Females also choose a fit and specific male for being fertilized and choose the area where to lay the eggs. If the pheromone he emits along with a song is suitable for her, then she allows him to mate. She spreads her wing and allows the male to mount on her abdomen. If she is interested, she started sending some cues like grooming of abdomen most of the time and walks less [6]. Male uses several cues like chemosensory, olfactory to assess female receptivity. A pheromone cis-vaccenyl acetate triggers the neuronal activity, which leads to female receptivity [7]. Once copulated, the female flies remain non-receptive up to 8–10 days until the live sperms get degenerated inside the female body [8]. This can be achieved by sex peptides [9]. A female mates six times in her lifetime. Each successful mating resulted in laying of approximately 80 eggs.

The mating behaviour is regulated by two genes: *fruitless* (*fru*) and *doublesex* (*dsx*). FruM is a fru protein transcribed through the *fru* gene expressed only in males to generate masculinity in males. Male-specific neuron in its foreleg i.e. P1 express both these proteins, so that the connection of male to female activates these neurons to respond to the olfactory cues and facilitate the patterning of songs [10]. Gustatory receptor neurons present in males identify specific information provided by females during mate selection [11]. Three types of gustatory neurons are so far discovered. The pheromone hydrocarbons of females are selected by *Gr39a* [12], while male-specific pheromones are selected by *Gr32a* and *Gr33a* [13]. CPG is present in the central nervous system in flies that induces song generation as well as wing extension [14]. Male plays a vital role in courtship behaviour. The female part of the response is a selection of a proper partner for her and exhibits the rejection after mating for production of eggs. During insemination, the male injects a pheromone sex peptide (SP) in a female's body so that for a certain time period, female will become non-receptive for mating [15]. Female contains an SP protein receptor (SPR) which is a G protein-coupled receptor to identify the SP. The fly having no SPR may behave like a virgin and remain receptive [16].

Sturtevant in 1915 first described the cues responsible for this behaviour in *Drosophila*. In the case of *Drosophila*, a desire for mating with species-specific female triggers the males to orient towards female. Orientation is the initial stage of courtship behaviour. Once the males and females approach towards each other, a number of males started following a female, but only one can able to tap at its back by its forelegs. The cuticle of female abdomen releases specific pheromones, which are being recognized by males. Tapping is required for recognizing the species and sex of the fly. Tapping triggers the activation of P1 neurons in foreleg of males which further initiates several activities of courtship behaviour [17]. Following this step, male vibrates its wing at right angles to its body to attract the female [18]. Male vibrates its one side wing which creates a courtship song which can act as a cue for mating female [19]. Then male licks female ovipositor by extending his proboscis and grabs its abdomen with its foreleg [20]. Then it bends its abdominal tip which is the initial attempt for copulation. This process repeats until successful mounting. If the female is receptive, then they will undergo successful mating [11]. By undergoing some behaviours like extruding ovipositor, kicking and decamping, a virgin female can escape from mating [21, 22]. Once it wants to accept the male, it stops the rejection behaviour and exposes its vaginal plate [23, 24]. To enhance the egg-laying mechanism, fly itself becomes unreceptive to males for a certain period of time [25].

Courtship is a combined action of various sensory organs like photoreceptor, taste receptor, gustatory receptor and hearing organ [26]. *Drosophila* uses its eye to detect the female. The stimulation of sensory organs triggers the sensation for mating [27]. Thus vision is the first step to initiate the mating behaviour [28]. Vision detects the species specificity and sex of the fly which is crucial during mating [29]. Vison differentiates the specific from non-specific function. Specific includes the identification of phenotypes of species and its postures, while non-specific includes the observation of female response while touching it whether it moves away or not, its locomotion and orientation during mating [30]. In the absence of visual signal, recognition occurs through olfaction [31]. *Drosophila* males possess olfactory neuron receptors that can identify the pheromones released by females. Thus, the gustatory neuronal circuit gets activated and triggers several mechanosensory activities [32]. Auditory stimuli have a great impact on courtship behaviour [33]. The females or males having defective auditory organs show reduced mating behaviour. The female secretes some pheromones which can be recognized by male gustatory organs. The male taste receptors are present on their foreleg sensilla which are used to taste the female pheromones [34]. Before initiation of mating, with their foreleg tarsi, males test the cuticular hydrocarbons that are 7, 11-HD (7, 11-heptacosadiene) of females [11]. These hydrocarbons are nothing but the volatile pheromone molecules expressed by both males as well as females to carry out courtship behaviour. In some cases, *Drosophila* species also show male courtship behaviour which can be suppressed by releasing pheromones like 7-T (7-tricosene) by male itself [35]. *Drosophila* produces species-specific courtship song to identify the partner. Based on the courtship song, the female decides whether to mate or not. Song types vary according to the distance from the female and speed [36]. In *Drosophila* two types of courtship songs are observed: "Sine song" and "Pulse song" [37]. The defective wing in a male is the reason for unsuccessful mating [38]. To attract a female, the male usually vibrates its wing that creates a song which makes female ready for mating. The quality of a song is necessary to make female receptive for copulation. Song listening by other males is also responsible to maintain their activeness and successful mating [23].

1.2 Aggressive Behaviour

Aggression is a complex behaviour seen against environmental, genetic and hormonal factors [39]. It is important for survival and reproduction [40]. It is seen within the individuals while competing for territory, finding partners and food [41]. Sturtevant first reported the male-male aggression, female-male rejection and fighting behaviour in fruit fly [42]. Aggression behaviour is regulated by various genetic factors. Furthermore, various neurotransmitters like monoamine oxidase A, nitric oxide, dopamine, octopamine, GABA, and serotonin regulate fly behaviour [39, 43–45]. Serotonin is the sole neurotransmitter responsible for aggressive behaviour [46]. In *Drosophila*, serotonin receptors are found in the antennal lobe, mushroom bodies and optic lobes. The *Drosophila* genome contains serotonin receptors like *5-HT7*, *5-HT1A*, *5-HT1B*, 5*-HT2A* and *5-HT2B*. Out of which *5-HT1A* and *5-HT2A* are responsible for aggression behaviour [47, 48]. High aggression level is combined with mutations that occur in the signalling of oestrogen and androgen in vertebrates and *Drosophila* sex determination. Many studies have reported the impact of various types of sensory modalities in the aggressive behaviour regulation in the fruit fly. The sensory modalities include gustatory, olfactory, visual system and hearing. Aggressive behaviour includes chasing, kicking, boxing, wing threats and head butts [49, 50].

1.3 Grooming Behaviour

Grooming is one of the complex defence mechanism observed from arthropods to vertebrates [51]. It is the type of innate immune behaviour which can be carried out by several motor systems [52] to keep the animal clean [53–55]. In the case of cockroach, the chemoreceptors are cleaned by grooming behaviour [56].The insect uses their mouthparts, leg and wings for grooming activity. Grooming helps to remove pathogenic, chemical or any foreign particles deposition from the body [57]. Some insects have a device in their legs to clean antennae by rubbing the two legs above the antennae [58]. *Drosophila* continuously rubs its leg parts to clean the surface of various sensory organs [59]. By the active defence mechanism, insects inhibit the spreading of microbe (bacterial as well as fungal growth) over the surface of their cuticle [60, 61]. Similarly, during social contact between insects, grooming reduces contact-induced disease transmission [62]. Female flies undergo more grooming after egg-laying to avoid contamination by pathogens [63]. Lipopolysaccharide (LPS) of gram-negative bacteria activates the neural circuit for grooming behaviour once it comes in contact with the fly body [64]. Besides LPS various chemicals like quinine, dyes and salt can also induce grooming behaviour by activating various chemoreceptors present in different body parts of *Drosophila* [65]. A faulty grooming behaviour resulted in defective cleaning of eye, antennae and wings which is required for proper locomotion, the formation of courtship songs, mating, flight, vision and hearing [66].

The current protocol describes the courtship-mating behaviour, aggressive behaviour and grooming behaviour in *Drosophila*.

2 Materials and Chemicals Required

2.1 Food Preparation

1. Agar-agar type 1 (HiMedia-RM666)
2. Sucrose (Fischer scientific-qualigens-28106)
3. Yeast powder (Subala Yeast, SYD-003/10)
4. Corn meal (from local market)
5. Propionic acid (99%)
6. Nipagine (99%)
7. Distilled water
8. Glass bottle (200 ml)
9. Vial (25 m × 100 mm)
10. Cotton

2.2 Courtship Assay

1. Virgin female and male *Drosophila* (*Oregon R*)
2. Glass vials (25 × 100 mm)
3. Diethyl ether (HiMedia)
4. Stopwatch
5. Dissection microscope/stereomicroscope

2.3 Aggressive Behavioural Assay

1. Adult flies
2. Food vial (25 × 100 mm)
3. Test chamber
4. Camera

2.4 Grooming Behavioural Assay

1. Drosophila stock
2. Ethanol (HiMedia; CHN01)
3. Turmeric (natural product)
4. Diethyl ether (SRL; 64665(052943))
5. Forceps (KOOCU ST-11)
6. Paintbrush (zero sizes)
7. Vial (Rivera; 25 × 100)
8. Microtiter well plate (TARSON; 980040)
9. Vortex (TARSON; SPINIX™ vortex shaker)
10. ELISA-reader (BIOBASE-EL10A)

3 Methods

3.1 Food Preparation

1. In order to make food up to 100 mL, weigh 0.8 g of agar powder, 4 g sucrose, 2.5 g yeast powder and 5 g corn meal, and add 100 mL of distilled water.
2. Shake all the above ingredients and autoclave the mixture.
3. Allow the mixture to cool.
4. Add 300 μl of propionic acid and 500 μl of nipagine to it.
5. Slowly pour it into the vial, cover the vial with cotton and allow it to cool to solidify properly.
6. Transfer the flies to the vial for the culture.

3.2 Courtship Assay

1. After eclosion, collect 20 male and female virgin flies by spotting the meconium deposition at the abdominal region. Flies can be collected during early morning as chances of getting virgin fly are more at that time (*see* **Note 1**).
2. Separate them and keep them in different vials for 3–4 days in 25 °C with 12 h light/dark cycle. Feed them yeast paste along with fly food to make them healthy (*see* **Note 2**).
3. After 4 days, transfer them to a chamber of six-well plates; transfer a single male and a female fly by anaesthetizing them. Cover the plate (*see* **Notes 3 and 4**).
4. When the flies are active, observe their behaviour under the stereomicroscope or record it (*see* **Note 5**).
5. Visualize the following steps (Fig. 1) under the stereomicroscope:

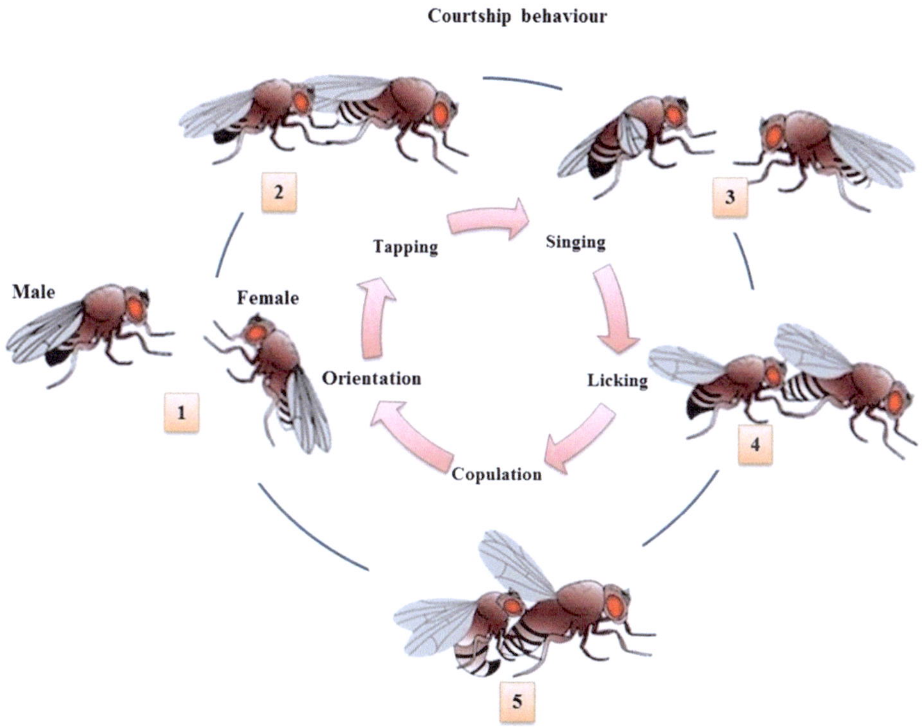

Fig. 1 Depicting the various steps of courtship behaviour like (1) orientation, (2) tapping, (3) singing, (4) licking, (5) copulation

I. **Orientation:** The male angles just before the female and follow it

II. **Tapping:** The male pats the female

III. **Wing song:** Shaking of a single wing of male and its extension towards female

IV. **Licking:** Licking of the female genitalia by a male

V. **Curling:** Bending of the abdominal portion of a male under itself

VI. **Copulation attempt:** An attempt to mount the female through curling

6. To measure the courtship behaviour, several parameters can be used. The whole performance and energy of males can be determined by calculating the courtship index (CI).

7. In this, the time spent by the male for each step up to successful copulation is calculated and the total time taken is designated as CI. The total time span at which wing vibrates is calculated as wing extension index (WEI) or wing vibration index (WVI) [67]. The number of time the wing vibrates determines the wing vibration latency before a successful mating.

8. The time between pairing and successful mounting determines the copulation latency. A number of pairs that undergo successful copulation in 30 min can be denoted as a mating success. These two parameters can suggest the receptivity of female and performance of male [68]. Space, as well as light condition, affects the latency. In low light and bigger space, more time spent by a male to find a female so latency is more.
9. Monitor the process in each vial for about 15 min to calculate the number of successful copulation, courtship index, frequency and latency of copulation. Courtship index is determined by the duration of time at which the fly engaged for copulation. Frequency is the number of pairs performs successfully.
10. $\text{Courtship index(CI)} = \frac{\text{Duration of courtship}}{\text{Total time period up to successful mating}}$

    CI in wild-type fly varies between 0.6 and 0.8.

3.3 Aggressive Behavioural Assay

1. Maintain the flies in normal food, and maintain 25 °C and 65% humidity with normal day/light cycle.
2. Collect 3–8-day-old males from the mixed group of population (Fig. 2). Do not anaesthetize the flies before 1 day of the experiment.
3. Do this assay between 10 and 11.30 AM.
4. Place the males in the vial without food for 2 h.
5. Afterwards, transfer the flies to the test chamber that contain a food droplet, and allow it to acclimatize for some time (*see* **Note 6**).
6. After acclimatisation, observe the flies for 5 min.
7. Observe and count the aggressive activity of the flies like chasing, kicking, boxing, wing threats and head butts. These steps are called as aggressive encounters.
8. Aggressive encounters can be calculated by using the following formula:

$$\text{Percentage of aggressive encounters} = \frac{\text{Aggressive encounters in the treated flies}}{\text{Aggressive encounters in the control flies}} \times 100$$

3.4 Grooming Behavioural Assay

1. Rear the flies in 25 °C with 60% humidity condition (*see* **Note 7**).
2. Take 15 anaesthetized adult flies in an empty vial (*see* **Note 8**) (Fig. 3).
3. Dust the flies using 10 mg of turmeric powder (you can use any non-toxic dye instead) (*see* **Note 9**).
4. Follow the same procedure for mutant or treated fly also in separate vials. Keep some dusted flies for SEM imaging.

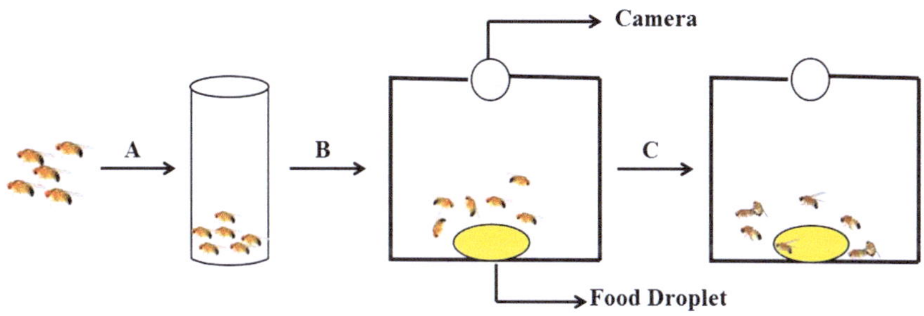

Fig. 2 Steps involved in aggressive behaviour. (**a**) Males are kept in the empty vial without food for 2 h. (**b**) Flies inside the test chamber. (**c**) Flies showing various aggressive activity like chasing, kicking, boxing, wing threats and head butts

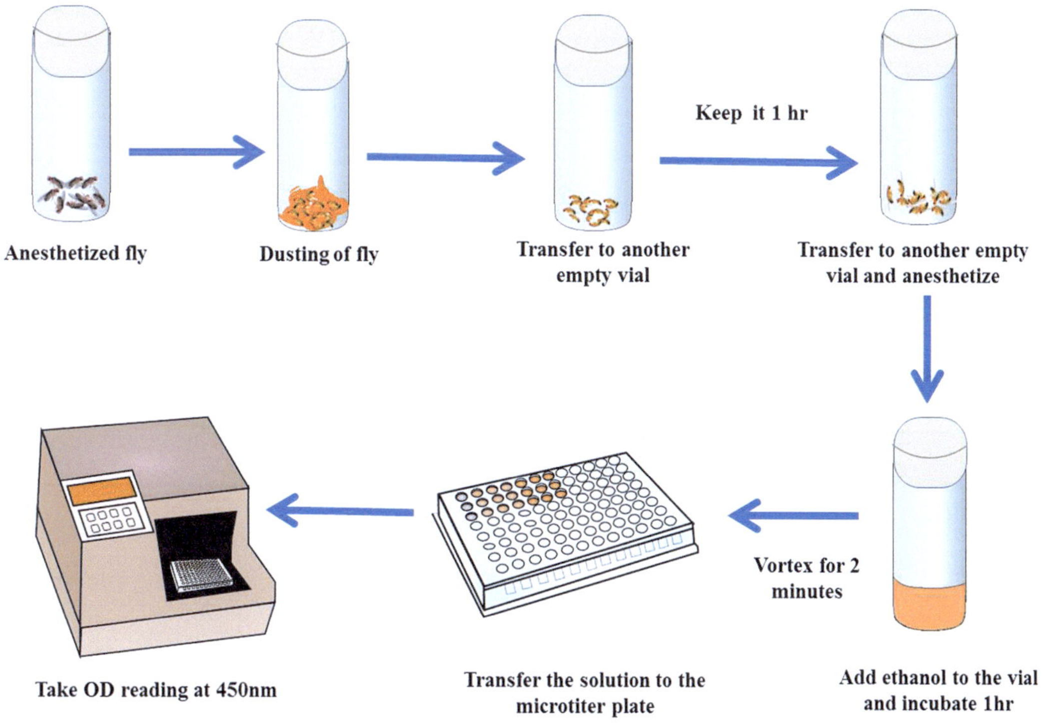

Fig. 3 Representing the steps to analyse the grooming behaviour in *Drosophila*

5. Now carefully transfer all the flies to another empty vial by forceps through their leg (*see* **Note 10**).
6. Allow the flies to be active and groom for 1 h.
7. Now allow the groomed flies to move to this new vial. Keep them separately for SEM analysis.
8. Take the anaesthetised flies in an empty food vial and add 3 ml of ethanol to it. Keep at this step for 1 h by tightly closing it with cotton plug (*see* **Note 11**).

Fig. 4 Grooming behaviour. (**a**) Depicting the turmeric powder used as a foreign substance, (**b**) digital camera images of fly body before grooming, (**c**) after grooming, (**d**) SEM images of fly wing before grooming, (**e**) after grooming

9. Next, vortex the vials containing fly and ethanol for 2 min to remove the remaining dust from their body.
10. From the above step, transfer 200 μl of the solution to the well (4nos) of 96-well plates, and take the OD at 450 nm (*see* **Note 12**).
11. Take the SEM images of the wing before and after grooming, and note the phenotypic changes (*see* **Note 13**) (Fig. 4).
12. Repeat for five times to get statistically significant data.

4 Notes

1. Virgin fly should be properly selected by observing the time of eclosion and the presence of meconium in the fly.
2. The fly should be acclimatized in particular environmental conditions like temperature and light exposure to enhance the chances of mating.
3. During anaesthetizing the fly, use ear bud dipped with ether. Dip the ear bud inside the ether, and squeeze it by pressing the bud at the tip of ether tube. The excess dose of ether may lead to the death of fly, so once they are anaesthetized, we need to keep them 20–30 min to be active for the next dose (if required).
4. The well plates must be cleaned and dried prior to experiment; otherwise the wings of fly may stick to moisture, and they can't carry out mating properly.
5. The experiment has to be repeated 5–10 times to acquire statistically significant data.

6. Carefully keep the flies in the test chamber.
7. All the flies should be grown in similar environmental conditions.
8. The empty vials need to be cleaned properly before using.
9. Shake well slowly so that all the flies get uniformly dusted all over their body.
10. Transferring of flies should be done carefully so that no dust will be removed during the time of transferring.
11. You can take samples in triplicate or more to get accuracy.
12. Avoid evaporation of ethanol by keeping the vials blocked carefully using cotton and aluminium foil.
13. Individual body portions of the fly should be removed carefully during preparation for SEM.

Acknowledgements

We thank Mr. Sumit Mukherjee for his help with Fig. 3. GD and SS are thankful to DST inspire for financial support. MM lab is supported by Grant No. BT/PR21857/NNT/28/1238/2017, EMR/2017/003054, Odisha DBT 3325/ST(BIO)-02/2017.

References

1. Higgins LA, Jones KM, Wayne ML (2005) Quantitative genetics of natural variation of behavior in *Drosophila melanogaster*: the possible role of the social environment on creating persistent patterns of group activity. Evolution 59(7):1529–1539
2. Mackay TF (2001) Quantitative trait loci in Drosophila. Nat Rev Genet 2(1):11
3. Mackay TF (2001) The genetic architecture of quantitative traits. Annu Rev Genet 35 (1):303–339
4. Anholt RR, Mackay TF (2004) Quantitative genetic analyses of complex behaviours in Drosophila. Nat Rev Genet 5(11):838
5. Adams MD, Celniker SE, Holt RA, Evans CA, Gocayne JD, Amanatides PG, Scherer SE, Li PW, Hoskins RA, Galle RF (2000) The genome sequence of *Drosophila melanogaster*. Science 287(5461):2185–2195
6. Markow TA, Hanson SJ (1981) Multivariate analysis of Drosophila courtship. Proc Natl Acad Sci 78(1):430–434
7. Kohl J, Ostrovsky AD, Frechter S, Jefferis GS (2013) A bidirectional circuit switch reroutes pheromone signals in male and female brains. Cell 155(7):1610–1623
8. Manning A (1967) The control of sexual receptivity in female Drosophila. Anim Behav 15(2–3):239–250
9. Liu H, Kubli E (2003) Sex-peptide is the molecular basis of the sperm effect in *Drosophila melanogaster*. Proc Natl Acad Sci 100 (17):9929–9933
10. Yamamoto D, Koganezawa M (2013) Genes and circuits of courtship behaviour in Drosophila males. Nat Rev Neurosci 14(10):681
11. Spieth HT (1974) Courtship behavior in Drosophila. Annu Rev Entomol 19(1):385–405
12. Watanabe K, Toba G, Koganezawa M, Yamamoto D (2011) Gr39a, a highly diversified gustatory receptor in Drosophila, has a role in sexual behavior. Behav Genet 41(5):746–753
13. Miyamoto T, Amrein H (2008) Suppression of male courtship by a Drosophila pheromone receptor. Nat Neurosci 11(8):874
14. Clyne JD, Miesenböck G (2008) Sex-specific control and tuning of the pattern generator for courtship song in Drosophila. Cell 133 (2):354–363
15. Aranha MM, Vasconcelos ML (2018) Deciphering Drosophila female innate behaviors. Curr Opin Neurobiol 52:139–148

16. Yapici N, Kim Y-J, Ribeiro C, Dickson BJ (2008) A receptor that mediates the post-mating switch in Drosophila reproductive behaviour. Nature 451(7174):33
17. Kohatsu S, Yamamoto D (2015) Visually induced initiation of Drosophila innate courtship-like following pursuit is mediated by central excitatory state. Nat Commun 6:6457
18. Shorey H, Bartell R (1970) Role of a volatile female sex pheromone in stimulating male courtship behaviour in *Drosophila melanogaster*. Anim Behav 18:159–164
19. Ewing AW (1964) The influence of wing area on the courtship behaviour of *Drosophila melanogaster*. Anim Behav 12(2–3):316–320
20. Demir E, Dickson BJ (2005) Fruitless splicing specifies male courtship behavior in Drosophila. Cell 121(5):785–794
21. Connolly K, Cook R (1973) Rejection responses by female *Drosophila melanogaster*: their ontogeny, causality and effects upon the behaviour of the courting male. Behaviour 44:142–166
22. Ejima A, Nakayama S, Aigaki T (2001) Phenotypic association of spontaneous ovulation and sexual receptivity in virgin females of *Drosophila melanogaster* mutants. Behav Genet 31 (5):437–444
23. Villella A, Hall JC (2008) Neurogenetics of courtship and mating in Drosophila. Adv Genet 62:67–184
24. Lasbleiz C, Ferveur J-F, Everaerts C (2006) Courtship behaviour of *Drosophila melanogaster* revisited. Anim Behav 72(5):1001–1012
25. Kubli E (2003) Sex-peptides: seminal peptides of the Drosophila male. Cell Mol Life Sci CMLS 60(8):1689–1704
26. Ejima A, Griffith LC (2007) Measurement of courtship behavior in *Drosophila melanogaster*. Cold Spring Harb Protoc 2007(10):pdb. prot4847
27. Connolly K, Burnet B, Sewell D (1969) Selective mating and eye pigmentation: an analysis of the visual component in the courtship behavior of *Drosophila melanogaster*. Evolution 23(4):548–559
28. Maimon G, Straw AD, Dickinson MH (2008) A simple vision-based algorithm for decision making in flying Drosophila. Curr Biol 18 (6):464–470
29. Agrawal S, Safarik S, Dickinson MH (2014) The relative roles of vision and chemosensation in mate recognition of Drosophila. J Exp Biol 217:2796–2805. jeb. 105817
30. Ewing AW (1983) Functional aspects of Drosophila courtship. Biol Rev 58(2):275–292
31. Heimbeck G, Bugnon V, Gendre N, Keller A, Stocker RF (2001) A central neural circuit for experience-independent olfactory and courtship behavior in *Drosophila melanogaster*. Proc Natl Acad Sci 98(26):15336–15341
32. Manoli DS, Foss M, Villella A, Taylor BJ, Hall JC, Baker BS (2005) Male-specific fruitless specifies the neural substrates of Drosophila courtship behaviour. Nature 436(7049):395
33. Liimatainen J, Hoikkala A, Aspi J, Welbergen P (1992) Courtship in *Drosophila montana*: the effects of male auditory signals on the behaviour of flies. Anim Behav 43(1):35–48
34. Bray S, Amrein H (2003) A putative Drosophila pheromone receptor expressed in male-specific taste neurons is required for efficient courtship. Neuron 39(6):1019–1029
35. Ferveur J-F, Cobb M (2010) Behavioral and evolutionary roles of cuticular hydrocarbons in Diptera. Insect Hydrocarb Biol Biochem Chem Ecol 325–343
36. Coen P, Clemens J, Weinstein AJ, Pacheco DA, Deng Y, Murthy M (2014) Dynamic sensory cues shape song structure in Drosophila. Nature 507(7491):233
37. Clemens J, Coen P, Roemschied FA, Pereira TD, Mazumder D, Aldarondo DE, Pacheco DA, Murthy M (2018) Discovery of a new song mode in Drosophila reveals hidden structure in the sensory and neural drivers of behavior. Curr Biol 28(15):2400–2412. e2406
38. Bussell JJ, Yapici N, Zhang SX, Dickson BJ, Vosshall LB (2014) Abdominal-B neurons control Drosophila virgin female receptivity. Curr Biol 24(14):1584–1595
39. Zwarts L, Magwire MM, Carbone MA, Versteven M, Herteleer L, Anholt RR, Callaerts P, Mackay TF (2011) Complex genetic architecture of Drosophila aggressive behavior. Proc Natl Acad Sci 108 (41):17070–17075
40. Dukas R (2008) Evolutionary biology of insect learning. Annu Rev Entomol 53:145–160
41. Benelli G, Desneux N, Romano D, Conte G, Messing RH, Canale A (2015) Contest experience enhances aggressive behaviour in a fly: when losers learn to win. Sci Rep 5:9347
42. Sturtevant A (1915) Experiments on sex recognition and the problem of sexual selection in Drosoophilia. J Anim Behav 5(5):351
43. Zhou C, Rao Y, Rao Y (2008) A subset of octopaminergic neurons are important for Drosophila aggression. Nat Neurosci 11 (9):1059
44. Miczek KA, Fish EW, Joseph F, De Almeida RM (2002) Social and neural determinants of aggressive behavior: pharmacotherapeutic targets at serotonin, dopamine and γ-aminobutyric acid systems. Psychopharmacology 163(3–4):434–458

45. Baier A, Wittek B, Brembs B (2002) Drosophila as a new model organism for the neurobiology of aggression? J Exp Biol 205(9):1233–1240
46. Narvaes R, Martins de Almeida RM (2014) Aggressive behavior and three neurotransmitters: dopamine, GABA, and serotonin—a review of the last 10 years. Psychol Neurosci 7 (4):601
47. Huser A, Eschment M, Güllü N, Collins KA, Böpple K, Pankevych L, Rolsing E, Thum AS (2017) Anatomy and behavioral function of serotonin receptors in *Drosophila melanogaster* larvae. PLoS One 12(8):e0181865
48. Saudou F, Boschert U, Amlaiky N, Plassat J-L, Hen R (1992) A family of *Drosophila* serotonin receptors with distinct intracellular signalling properties and expression patterns. EMBO J 11(1):7–17
49. Hoyer SC, Eckart A, Herrel A, Zars T, Fischer SA, Hardie SL, Heisenberg M (2008) Octopamine in male aggression of Drosophila. Curr Biol 18(3):159–167
50. Versteven M, Broeck LV, Geurten B, Zwarts L, Decraecker L, Beelen M, Göpfert MC, Heinrich R, Callaerts P (2017) Hearing regulates Drosophila aggression. Proc Natl Acad Sci 114(8):1958–1963
51. Zhukovskaya M, Yanagawa A, Forschler B (2013) Grooming behavior as a mechanism of insect disease defense. Insects 4(4):609–630
52. Szebenyi AL (1969) Cleaning behaviour in *Drosophila melanogaster*. Anim Behav 17 (4):641–651
53. Anselme P (2008) Abnormal patterns of displacement activities: a review and reinterpretation. Behav Process 79(1):48–58
54. Zhukovskaya MI (2012) Modulation by octopamine of olfactory responses to nonpheromone odorants in the cockroach, *Periplaneta americana* L. Chem Senses 37(5):421–429
55. Root-Bernstein M (2010) Displacement activities during the honeybee transition from waggle dance to foraging. Anim Behav 79 (4):935–938
56. Böröczky K, Wada-Katsumata A, Batchelor D, Zhukovskaya M, Schal C (2013) Insects groom their antennae to enhance olfactory acuity. Proc Natl Acad Sci 110(9):3615–3620
57. Rath W (1999) Co-adaptation of *Apis cerana* Fabr. and Varroa jacobsoni Oud. Apidologie 30 (2–3):97–110
58. Basibuyuk HH, Quicke DL (1999) Grooming behaviours in the Hymenoptera (Insecta): potential phylogenetic significance. Zool J Linnean Soc 125(3):349–382
59. Phillis RW, Bramlage AT, Wotus C, Whittaker A, Gramates LS, Seppala D, Farahanchi F, Caruccio P, Murphey R (1993) Isolation of mutations affecting neural circuitry required for grooming behavior in *Drosophila melanogaster*. Genetics 133(3):581–592
60. Baracchi D, Mazza G, Turillazzi S (2012) From individual to collective immunity: the role of the venom as antimicrobial agent in the Stenogastrinae wasp societies. J Insect Physiol 58(1):188–193
61. Lusebrink I, Dettner K, Seifert K (2008) Stenusine, an antimicrobial agent in the rove beetle genus Stenus (Coleoptera, Staphylinidae). Naturwissenschaften 95(8):751–755
62. Hughes WO, Eilenberg J, Boomsma JJ (2002) Trade-offs in group living: transmission and disease resistance in leaf-cutting ants. Proc R Soc Lond Ser B Biol Sci 269(1502):1811–1819
63. Yang C-H, Belawat P, Hafen E, Jan LY, Jan Y-N (2008) Drosophila egg-laying site selection as a system to study simple decision-making processes. Science 319(5870):1679–1683
64. Hampel S, Franconville R, Simpson JH, Seeds AM (2015) A neural command circuit for grooming movement control. Elife 4:e08758
65. Yanagawa A, Guigue A, Marion-Poll F (2014) Hygienic grooming is induced by contact chemicals in *Drosophila melanogaster*. Front Behav Neurosci 8:254
66. Sibley D, Elphick C, Dunning JB (2001) The Sibley guide to bird life & behavior, vol Sirsi. National Audubon Society. i9780679451235
67. Villella A, Ferri SL, Krystal JD, Hall JC (2005) Functional analysis of fruitless gene expression by transgenic manipulations of Drosophila courtship. Proc Natl Acad Sci 102 (46):16550–16557
68. Eberl DF, Duyk GM, Perrimon N (1997) A genetic screen for mutations that disrupt an auditory response in *Drosophila melanogaster*. Proc Natl Acad Sci 94(26):14837–14842

MIX
Papier aus verantwortungsvollen Quellen
Paper from responsible sources
FSC® C105338
FSC
www.fsc.org

If you have any concerns about our products,
you can contact us on
ProductSafety@springernature.com

In case Publisher is established outside the EU,
the EU authorized representative is:
Springer Nature Customer Service Center GmbH
Europaplatz 3, 69115 Heidelberg, Germany

Printed by Libri Plureos GmbH
in Hamburg, Germany